THE
DAGUERREOTYPE

Stereo daguerreotype of an
early darkroom with da-
guerreotype and calotype
equipment with paper neg-
atives tacked to the shelf
for drying. Private collection; courtesy of William
L. Schaeffer.

THE
DAGUERREOTYPE

NINETEENTH-CENTURY
TECHNOLOGY AND
MODERN SCIENCE

M. Susan Barger and William B. White

The Johns Hopkins University Press
Baltimore and London

© 1991 by the Smithsonian Institution
Printed in the United States of America on
   acid-free paper

Originally published in hardcover by
Smithsonian Institution Press,
Washington, D.C., 1991

Johns Hopkins Paperbacks edition, 2000
9 8 7 6 5 4 3 2 1

The Johns Hopkins University Press
2715 North Charles Street
Baltimore, Maryland 21218-4363
www.press.jhu.edu

For permission to reproduce illustrations
appearing in this book, please correspond
directly with the owners of the works, as
listed in the individual captions. All the micrographs in chapters 8 through 12 were made by
M. Susan Barger. Unless otherwise noted, the
micrographs are secondary electron images.
The daguerreotypes shown in these figures are
from the Materials Research Laboratory Study
collection unless otherwise noted.

Library of Congress Cataloging-in-Publication
Data

Barger, M. Susan, 1949–
    The daguerreotype : nineteenth-century
technology and modern science /
M. Susan Barger and William B. White.
       p.   cm.
    Reprint. Originally published: Washington,
D.C. : Smithsonian Institution Press, 1991.
    Includes bibliographical references and
index.
    ISBN 0-8018-6458-5 (pbk. : alk. paper)
    1. Daguerreotype—History. I. White,
William B. (William Blaine), 1934–
II. Title.

TR365 .B37 2000
772´.12—dc21

                                99-086934

A catalog record for this book is available from
the British Library.

List of Figures and Tables **vii**
Preface to the Johns Hopkins Edition **xi**
Preface to the First Edition **xvii**
Acknowledgments **xix**

Contents

1
Beginnings 1

2
Image Making before Photography 4

3
Toward Points de Vue *and the Daguerreotype* 11

4
The Technological Practice of Daguerreotypy 28

5
Scientific Interest in the Daguerreotype during the Daguerreian Era 55

6
The Daguerreotype as a Scientific Tool 72

7
Scientific Interest in Daguerreotypy after the Daguerreian Era 98

8
The Daguerreotype Image Structure 117

9
Image Formation 135

10
Image Deterioration 160

v

11

Corrosion Removal   183

12

*Daguerreotype Preservation
and Display*   201

13

*Daguerreotypy: A Model
for the Interpretation of
Early Technology and
Works of Art*   216

Notes   219
References   239
Name Index   243
Subject Index   246

Figures

List of Figures and Tables

Frontispiece: Daguerreotype darkroom.   ii
Acknowledgments: Daguerreotype of Materials Research Laboratory staff.   xiv

1.1. Engraving from Figuier of August 19, 1839, meeting.   xviii

2.1. Thomas Allason's "perspectograph."   5
2.2. Architectural camera obscura.   6
2.3. Portable camera obscura with a Wollaston landscape lens.   7
2.4. Ray diagrams for a camera lucida.   7
2.5. Camera lucida drawing by Hershel.   8

3.1. Portrait of Niépce.   17
3.2. The first photograph.   19
3.3. Portrait of Daguerre.   20
3.4. Portrait of Arago.   23
3.5. Daguerreotype by Daguerre.   24
3.6. Portrait of Talbot.   25

4.1. Voigtländer camera.   31
4.2. Portrait of Saxton.   32
4.3. View of the Central High School, Philadelphia.   33
4.4. Portrait of Cornelius.   33
4.5. Portrait of Paul Beck Goddard.   34
4.6. Portrait of Fizeau.   36
4.7. Daguerreotype coloring box.   39
4.8. Portrait of Niépce de St. Victor.   40
4.9. Copy daguerreotype.   42
4.10. Reversing prism.   43
4.11. Daguerreian outfit.   44
4.12. Treadle buffing machine.   44
4.13. A Smee's galvanic battery.   45
4.14. A Coad's patented galvanic battery.   45
4.15. Contents of a halogen coating box.   46
4.16. Halogen coating box.   46
4.17. Daguerreotype camera.   47
4.18. Engraving of a mercury bath.   47
4.19. Photograph of a mercury bath.   47
4.20. Portrait of Alter.   53

5.1. Portrait of Draper.   57
5.2. Portrait of Becquerel.   59
5.3. Illustrations made from daguerreotype micrographs by Donné and Foucault.   60
5.4. Illustration made from tithonographs and spectra taken by Draper.   63

5.5. Illustration from a daguerreotype spectrum taken by Draper.   64

5.6. Illustration from daguerreotype spectra taken by Becquerel.   67

5.7. Illustration of crystallization drawings by Waller.   69

6.1. Photogenic drawing by Lea.   73

6.2. Lithograph from a daguerreotype by Eliphalet Brown taken during the Perry expeditions.   76

6.3. Illustration from a Catherwood daguerreotype and camera lucida drawings taken in the Yucatan.   77

6.4. Engraving from a phrenological daguerreotype by Brady.   79

6.5. Portrait of Laura Bridgeman.   79

6.6. Zealy daguerreotype made for Agassiz.   80

6.7. Daguerreotype of a laboratory, the Academy of Natural Sciences of Philadelphia.   81

6.8. Daguerreotype illustrating reconstruction surgery.   82

6.9. Daguerreotype of operation performed with anesthesia.   82

6.10. Daguerreotype of a man with a birth defect.   83

6.11. Daguerreotype of a leper.   83

6.12. Postmortem portrait.   83

6.13. Daguerreotype of a dentist.   84

6.14. Portrait of Foucault.   85

6.15. Illustration made from Foucault-Fizeau daguerreotype of the sun.   86

6.16. Daguerreotype of the moon by Humphrey.   87

6.17. Portrait of Whipple.   87

6.18. Whipple-Bond daguerreotype of the moon.   89

6.19. Daguerreotype of a solar eclipse by Whipple.   90

6.20. Langenheim brothers' daguerreotype series of a solar eclipse.   91

6.21. Daguerreotype apparatus for recording the Transit of Venus.   92

6.22. Daguerreotype of the Transit of Venus.   93

6.23. Magnetograph.   94

6.24. Claudet's photographometer.   95

6.25. Alter's emission spectra.   96

6.26. Cornelius daguerreotype of chemical apparatus.   97

6.27. Engraving from Cornelius daguerreotype.   97

7.1. Portrait of Lea.   101

7.2. Portrait plaque of Wiener.   102

7.3. Interference bands (plate 1 from H. Scholl).   104

7.4. Interference bands (plate 2 from H. Scholl).   105

7.5. Portrait of Poboravsky.   108

7.6. Poboravsky's spectra.   109

7.7. Scanning electron micrograph images of daguerreotype plate.   115

7.8. Swan image particle cross section.   116

8.1. Daguerreotype image structure.   119

8.2. Daguerreotype image structure.   120

8.3. Energy-dispersive x-ray spectra for daguerreotypes.   121

8.4. CESEMI image particle counting.   121

8.5. Theoretical "white" material.   123

8.6. Scattering curves for daguerreotype microstructure.   125

8.7. Image structure for conventional photographic materials.   125

8.8. Image structure of daguerreotypes.   126

8.9. Reflectance spectra of daguerreotypes.   127

8.10. Characteristic curve of a daguerreotype.   128

8.11. Reflectance spectra of daguerreotype color casts.   129

8.12. Goniophotometer.   130

8.13. Ray diagram of goniophotometer.   131

8.14. Ideal reflectance curves of daguerreotypes.   131

8.15. Reflectance data for an unsealed nineteenth-century daguerreotype.   132

8.16. Reflectance data for a scratched nineteenth-century daguerreotype.   133

8.17. Reflectance data for an overcleaned nineteenth-century daguerreotype.   133

9.1. Schematic of daguerreotype processing.   136

9.2. Schematic of development mechanisms.   137

9.3. Microstructure of daguerreotype treated with an acid-iron developer.   138

9.4. Microstructure of daguerreotype treated with a mercury intensifier and pyro developer.   139

9.5. Microstructure of a daguerreotype with a Metz low-potential developer.   139

9.6. Microstructure of a daguerreotype treated with a Metz high-potential developer.   140

9.7. Microstructure of initial stage of image formation.   141

9.8. Microstructure of mercury initiation.   141

9.9. Microstructure of platelets.   142

9.10. Microstructure of freshly made image particles.   142

9.11. Highlight, midtone, and shadow areas of a new image.   143

9.12. Schematic of image particle growth.   143

9.13. Silver-mercury phase diagram.   144

9.14. Week-old image particles.   145

9.15. Shadow particle agglomerates.   145

9.16. Microstructure of singly sensitized daguerreotype.   147

9.17. Microstructure of cold mercury treatment.   148

9.18. Microstructure of Becquerel-developed daguerreotype.   149

9.19. Energy-dispersive x-ray spectra of Becquerel-developed daguerreotype.   149

9.20. X-ray diffraction spectra of silver halide layers.   151

9.21. Micrograph of silver halide layer.   155

9.22. Schematic of sensitization and image particle formation.   155

9.23. Platelets.   156

9.24. Portrait of Du Ponceau.   158

9.25. Highlight area of Du Ponceau portrait.   158

9.26. Shadow particle agglomerate on Du Ponceau portrait.   159

10.1. Daguerreotype corrosion fronts.   162

10.2. Blank plate corrosion experiment (1.5 years).   163

10.3. Blank plate corrosion experiment (1 month).   163

10.4. Reflectance spectra of silver sulfide.   164

10.5. Infrared spectra of daguerreotype tarnish.   168

10.6. Raman spectra of daguerreotype plate.   168

10.7. Etch patterns on daguerreotype plate.   171

10.8. Portrait of Paul Beck Goddard by Cornelius.   172

10.9. Corrosion from a mat of buffered board.   172

10.10. Daguerreotype step tablet with corrosion caused by a mat of buffered board.   173

10.11. Malachite-like corrosion on a daguerreotype plate.   173

10.12. Compositions of daguerreotype cover glasses.   175

10.13. Micrograph of a daguerreotype cover glass.   176

10.14. Energy-dispersive x-ray spectra of daguerreotype cover glass.   177

10.15. Micrograph of the interior surface of a daguerreotype cover glass.   177

10.16. Micrograph of sodium sulfate blisters.   178

10.17. Daguerreotype with corrosion debris from the cover glass.   178

10.18. Micrograph of copper-bearing crystals on a daguerreotype cover glass.   179

10.19. Composite micrograph of silicate growths on daguerreotype plate surface.   179

10.20. Daguerreotype with dendritic corrosion.   180

10.21. Micrograph of daguerreotype corrosion debris.   180

10.22. Micrograph of silicified bugs on daguerreotype cover glasses.   181

11.1. Micrograph of cyanide etching.   187

11.2. Micrograph of daguerreotype surface before and after sputter cleaning.   190

11.3. Silver blank before and after sputtering.   191

11.4. Daguerreotype half cleaned by sputtering.   191

11.5. Pourbaix diagram for silver-oxygen-hydrogen system.   193

11.6. Corrosion diagram for silver.   193

11.7. Phase diagram for the silver oxide-ammonia-water system.   193

11.8. Electrical diagram for the daguerreotype electrocleaning circuit.   194

11.9. Plate holder for electrocleaning daguerreotypes.   194

11.10. Micrograph of pit on daguerreotype surface.   195

11.11. Daguerreotype with peeling and delamination.   195

11.12. Micrographs of surface before and after electrocleaning.   197

11.13. Micrograph of cyanide-etched plate surface after electrocleaning.   197

11.14. Daguerreotype half cleaned using electrocleaning.   198

11.15. Daguerreotype before and after electrocleaning.   199

11.16. Daguerreotype with filiform delamination.   200

12.1. Diagram of daguerreotype package.   202

12.2. Daguerreotype case.   202

12.3. Lockets with daguerrreotypes.   203

12.4. Diagram of sputtering apparatus.   206

12.5. Reflectance spectra for daguerreotype with silica coating.   206

12.6. Micrographs of coated and uncoated daguerreotypes.   207

12.7. Morphology of silica coating on daguerreotypes.   208

12.8. Micrographs of coated and uncoated daguerreotypes.   209

12.9. Reflectance spectra of daguerreotypes with a boron nitride coating.   209

12.10. Parylene film morphologies.   209

12.11. Diagram of arrangement for photographing daguerreotypes.   211

12.12. Image enhancement of a surface-damaged daguerreotype.   214

13.1. Daguerreotype of the moon by the Lunar Daguerreian Society.   217

9.1. Structure-Growth Relationships for Condensed Species Produced by Chemical Vapor Deposition (CVD)   153

10.1. Analysis of Corrosion on Daguerreotypes: Elemental Data   166

10.2. Analysis of Corrosion on Daguerreotypes: Compound Data   167

11.1. Sputter Cleaning Conditions   189

12.1. Optical Properties of Coating Materials   205

Tables

8.1. Average Image Particle Dimensions of Gilded Daguerreotypes   122

8.2. Average Image Particle Dimensions of Ungilded Daguerreotypes   123

8.3. Average Image Particle Dimensions of Becquerel-Developed Daguerreotypes   123

THE MANUSCRIPT FOR this book was originally completed in 1990, during the 150th anniversary of the introduction of photography, with this paperback edition being brought out a decade later. Since 1990, sparked by activities marking that photographic milestone, daguerreotypes have been increasingly sought after as objects of desire. Beginning with the many exhibitions set in 1989–90 to mark the daguerreotype's sesquicentennial year, there have been several blockbuster shows of daguerreotypes. Along with this, books on various aspects of daguerreian history and new scholarly work addressing the historical and artistic impact of the daguerreotype have appeared in the last decade. The Daguerreian Society was founded in the United States in 1989, and its membership has grown every year since. The French Société Daguerre is a few years older and it, too, has grown in size. These and similar societies have promoted and cultivated new daguerreotype collectors and scholars.

There is much to attract this new group of collectors. When we began collecting daguerreotypes to use as samples for our original study in 1979, they could be had for as little as $5 to $25 apiece. Rare daguerreotypes sold for several hundred dollars. Several dealers and collectors gave us daguerreotypes for our study with a generosity that probably would not be possible in today's market. Since 1979, however, the prices for daguerreotypes, even for garden variety images, have soared. And, on April 27, 1999, Sotheby's Auction House in New York broke all records at the auction of the David Feigenbaum Collection of Southworth & Hawes daguerreotypes. The highest amount paid for a single daguerreotype in that sale was $387,500 (with the buyer's premium)—a price that broke all previous records. Various photographic and art journals claimed that the Feigenbaum sale was the photographic event of the decade and that

Preface to the Johns Hopkins Edition

daguerreotype prices would only climb as result. This prediction seems accurate because the number of daguerreotype collectors is increasing while the number of extant daguerreotypes remains finite; there is no end to the rise of prices in sight.

Along with the increased interest in daguerreotype collecting, there has also been a parallel increase in the number of people making daguerreotypes. In 1989, there were probably two dozen people who could claim that they had ever made daguerreotypes, a few people who exhibited contemporary daguerreotypes as a portion of their oeuvre, and one person who claimed to be a full-time professional daguerreotypist. Now there is a growing group of contemporary daguerreotypists and even an occasional newsletter addressed to the problems of modern daguerreotypy.[1] Both the Daguerreian Society's *Annual* and *Newsletter* have published articles on daguerreotype production that range from descriptions of specific applications of daguerreotypy, to recommendations on how to set-up and make daguerreotypes using the Becquerel process, to development using cold mercury. Workshops on making daguerreotypes have been offered at the International Museum of Photography at George Eastman House and elsewhere. Custom production of some of the specialized equipment necessary to make daguerreotypes has begun, and there are now several platers in the United States who can manufacture daguerreotype plates of sufficient quality to meet a maker's requirements. The work of these contemporary daguerreotypists spans the gamut from very traditional images, which would not have been out of place during the daguerreian era, to extremely modern and abstract images, which could never be mistaken for nineteenth-century daguerreotypes.

Since the first edition of this book, there has been very little additional scientific work on daguerreotypes. We want to take note of the published work that is specifically relevant to the topics covered in this book with this preface to the paperback edition.

With the increasing number of practitioners of this art, there have been some publications and works on the methods for producing daguerreotypes, most of which are similar to those written in the nineteenth century in that they describe the specific working methods of individuals. The same problems confronting contemporary makers plagued their nineteenth-century forebears, as well: what is the best way to use iodine or bromine?[2] what is the best timing for halogen sensitization? how do you handle mercury? are there ways to avoid using mercury? where do you get chemicals and plates?[3] and finally, a more modern concern, what are the safety issues when using silver, mercury, gold, bromine, iodine, and chlorine?[4]

One way to lessen the chance of exposure to mercury vapor during daguerreotype processing is to use cold, rather than hot, mercury for development. A few nineteenth-century reports noted experiments using cold mercury, and we show micrographs of the image structure of daguerreotypes processed in this way (fig. 9.17). We discuss how image formation takes place in the mercury vapor and how the mercury acts as a solvent for the formation of the silver crystals that make up the image (see chap. 9). This requires that there be enough vapor-phase mercury available at the plate surface for silver crystals to grow to the proper size, thus producing a visible image in a reasonable time. Heating mercury, of course, makes more mercury vapor available for image formation. Another method for making more mercury available for development while not heating it is to reduce the air pressure above the mercury in the development box—essentially processing a daguerreotype in vacuum. A recent study has shown vacuum processing to be feasible.[5] In the article describing this method, there are scanning electron micrographs of the resulting image structures, and, interestingly, one of these micrographs shows mercury platelet formation in a transition phase to well-formed image particles. The notations in the article concerning the range of color and warmth of tone possible in daguerreotypes processed over cold mercury in a vacuum demonstrate that this method can be successfully used to achieve good silver crystal growth and, thus, good-looking daguerreotype images.

Daguerreotype images can be processed either conventionally, using hot mercury vapor, or by placing a plate that has received an image-forming exposure in a camera in red light for a prolonged period until a visible image appears. This last method is called "Becquerel development." It was not much used during the nineteenth century (see chap. 5), however, there are contemporary daguerreotypists who employ this type of development in order to avoid the use of mercury.[6] We describe and show scanning electron micrographs of the type of image structure that results from daguerreotypes that have been processed in this way (fig. 9.18). Now a group from Appalachian State University has made an interesting set of time-lapse micrographs of image particles growing over a two-hour period during Becquerel development using an atomic force microscope.[7] Scanning probe microscopy, of which the atomic force microscope is part, has been developed since the late 1980s. Because the physical characteristics that define a daguerreotype are surface phenomena, the scanning probe inquiry is well suited to the daguerreotype. Further, since the Becquerel process doesn't require mercury vapor, these micrographs could be made without damage to the probe instrument.

Despite years of warning about inappropriate cleaning methods, collectors still use thiourea-mineral acid solutions to treat daguerreotypes. We cannot say often enough that thiourea cleaners cause irreparable damage to daguerreotypes. Unfortunately, it is visually rewarding to clean daguerreotypes by these methods, so asking collectors to refrain from using something that is cheap, available, rapid, and seemingly so simple and straightforward to apply, rather than taking daguerreotypes needing treatment to a qualified conservator, is difficult. However, as the value of daguerreotypes increases, it is even more important that they be treated in a way that does not do them unalterable harm. In our original study, we examined the deterioration mechanisms of daguerreotypes, the historical methods for cleaning, and investigated new cleaning methods that we based on the properties of daguerreotypes. We found that daguerreotypes treated in thiourea cleaners will eventually cloud over and the image will be obscured. This phenomenon has been observed by many. In 1991, we had the opportunity to re-examine a set of eight daguerreotypes that had been cleaned using thiourea in 1977. These daguerreotypes had begun to cloud over as expected. We used Fourier transform infrared spectroscopy to obtain an infrared spectra from all of the previously treated daguerreotypes. For the first time we were able to identify the four-step thiourea-breakdown cycle.[8] Thiourea (I) forms a tightly bonded monolayer on the surface of the silver daguerreotype plate. This thiourea layer breaks down in the presence of oxygen and light to form cyanamid (II) and sulfuric acid. Cyanamid is unstable and dimerizes to dicyanamid (III). Finally, in the presence of sulfuric acid, the dicyanamid undergoes hydrolysis to dicyandiamidinsulfate (IV), which causes the clouding of the daguerreotype surface. The set of reactions that lead to the formation of dicyandiamidinsulfate will continue until the entire daguerreotype surface is covered. When a thiourea-cleaned daguerreotype is "recleaned" in thiourea, the breakdown products already on the surface of the daguerreotype act as catalysts to accelerate the thiourea-corrosion cycle. Many factors affect the rate of thiourea corrosion, including storage conditions, the effectiveness of the daguerreian seal, and whether or not a daguerreotype is displayed. Further, we found that thiourea-cleaned daguerreotypes also had silver phosphate, silver oxides, and sulfuric acid on their surfaces.

In the same set of experiments described above, we tested the effectiveness of electrocleaning to remove the thiourea-corrosion products and found that electrocleaning (see chap. 11) did successfully remove most of these materials and that any remaining material was removed by using a hot-water (90°C) rinse in the final cleaning step. The thiourea-corrosion products that are formed in II, III, and IV are more or less soluble in hot water; however, we found that they cannot be removed by treatment in hot water alone. The ammonium hydroxide used in the electrocleaning process also causes the thiourea-silver complex to break down, and the combination of ammonium hydroxide and the hot-water rinse seems to be necessary to effect the total removal of the thiourea-corrosion products.

In our original work, we tested and reported on a variety of cleaning methods that might be applicable to the removal of corrosion from daguerreotype surfaces (see chap. 11). Our cleaning experiments were done between 1982 and 1984. The daguerreotypes in that sample have been stored together in rather haphazard conditions, much like those that might be found in a private collection. In the intervening time, some daguerreotypes that were sputter-cleaned show slight recorrosion, especially in areas that had been more heavily corroded before treatment. This may be an indication that sputtering leaves the daguerreotype surface more active after treatment. Daguerreotypes from our original study that were electrocleaned show no visible signs of recorrosion. In addition, many people have used the simpler process of electrolytic cleaning successfully.

Still, we know that cleaning in a solution may not be suitable for colored daguerreotypes or for daguerreotypes that are not gilded. Two groups have investigated the use of lasers for nonaqueous cleaning of daguerreotypes.[9,10] In our work with lasers to measure the optical properties of daguerreotypes, we observed that it was possible to remove corrosion products from the surface of daguerreotypes using lasers. However, at that time (1981), lasers would not have been a practical or easily controlled technology to apply to cleaning daguerreotypes. Laser technology has greatly advanced over the last two decades, and lasers are now used for the routine cleaning of some works of art, especially for stone and building conservation. Two groups have reported on the use of lasers for cleaning daguerreotypes. Lasers produce coherent, highly directional radiation of a wavelength that depends upon the material used to produce the lasing effect. This means that it is theoretically possible to choose radiation of a wavelength that will remove specific corrosion products from a daguerreian surface, but will leave other products, such as pigments, alone. The two groups have demonstrated laser cleaning for daguerreotypes using two different eximer lasers[8] and a Nd:YAG laser.[9] Laser cleaning is not yet practical for routine cleaning because these specialized lasers are expensive and require a skilled technician to perform the cleaning operation.

Finally, we want to review some of the issues involved in daguerreotype care. As we pointed out years ago, the major cause of deterioration in daguerreotypes is related to the failure of their packaging and, in particular, to the aging of their cover glasses. These observations have led to many suggestions for better packaging methods. All of these methods are more or less sufficient for the safe housing of daguerreotypes, but some make it possible to maintain the traditional casing while others are more contemporary and may only be suitable for specific types of display. Some of these package designs go to great lengths to produce impermeable seals on daguerreotypes, which we feel are unnecessary. It is clear that impermeable seals break down and that in the long run, they provide no long-term advantage in preservation. Of the tapes that have been suggested for sealing daguerreotypes, we prefer self-adhesive tapes over tapes used with a wet-adhesive solution because self-adhesive tapes will not introduce unnecessary moisture into the daguerreian package. Any archival tape is suitable for daguerreotype seals, and it might be best to chose sealing tapes for their ease of removal. Mylar® tapes seem to be the best in terms of easy removal after some period of aging.

It has been suggested that specialty glasses be used to glaze daguerreotypes, for example, glasses with antireflection coatings or a special ultraviolet-filtering coating. In our experience, these specialty glasses are not necessary—in order to be viewed properly, daguerreotypes have to be displayed so that there are no extraneous reflections from their surface and they are unaffected by ultraviolet radiation. We have observed that daguerreotypes packaged with thinner mats and, therefore, smaller spaces within the daguerreian package, show signs of glass corrosion fairly quickly regardless of storage conditions. This is in keeping with what we previously described: the amount and rate of glass corrosion is in opposite proportion to the size of the interior space of the glass container. We have seen good results with polycarbonate glazing materials that have an antiscratch coating (e.g., Lexan MR-10®).

The illumination of displayed daguerreotypes is a critical factor in being able to see and enjoy their beauty. It has become almost de rigueur to exhibit daguerreotypes using fiber optic bundles. This type of lighting can be used very effectively for daguerreotype display, however it tends to be fairly expensive and difficult to arrange so that the full surface of the daguerreotype is illuminated. Daguerreotypes are unaffected by UV radiation and can be effectively displayed using fluorescent lights or tungsten spot lights, but daguerreotype case materials can be affected by UV illumination. Cases for displaying daguerreotypes that use fluorescent lamps are commercially available.

The original motivation for our work was to devise better ways to preserve and care for daguerreotypes. As materials scientists, we knew that we needed to understand exactly what a daguerreotype is and how it is formed before we could determine how best to care for these images. Thus, we investigated the daguerreotype process, the daguerreian image structure, and the mechanisms of image formation and deterioration. We used those empirical findings to describe what a daguerreotype is from a materials point of view, and using that information, to define what is meant by daguerreotype preservation. We were able to evaluate previous daguerreotype preservation methods and to devise new methods for their care. Our work also gave us the opportunity to take a new look at the scientific and technological literature on the daguerreotype and to re-evaluate its technical history. We are pleased that this book is being brought back into print by the Johns Hopkins University Press in this paperback edition. The text is unchanged except for minor corrections.

M. S. B.

Notes

1. For more on this phenomenon and to see issues of the *Daguerreotypist Newsletter*, check the Daguerreian Society's website: http://www.daguerre.org/resource/dagtypist/dagtypist.html.

2. Charlie Schreiner. "Iodine and Bromine: Part I," *Daguerreian Annual* (1998): 165-174.

3. Irving Pobboravsky. "The Daguerreotype Plate as Seen by a Contemporary Daguerreian Artist," *Daguerreian Annual* (1991): 114-122.

4. Kenneth E. Nelson. "Mercury and the Daguerreotypist: A Modern Assessment," *Daguerreian Annual* (1994): 118-143.

5. John R. Hurlock. "Warming Up to Cold Mercury," *Daguerreian Annual* (1998): 230-240.

6. Gerard Meegan. "Becquerel-Development 'In a New Light,'" *Daguerreian Annual* (1991): 161-168.

7. Joe Pollock and David Mahesh. "The Growth of Daguerreotype Image Particles, *The Movie*," *Daguerreian Annual* (1997): 226-234.

8. Thomas M. Edmondson and M. Susan Barger. "The Examination, Surface Analysis, and Retreatment of Eight Daguerreotypes which Were Thiourea Cleaned in 1977," *Topics in Photographic Preservation* 5 (1993): 14-26.

9. I. Turovets, A. Lewis, and M. Maggen. "Cleaning of Daguerreotypes with an Excimer Laser," *Studies in Conservation* 43 (1998): 89-100.

10. DaNel L. Hogan et al. "Laser Ablation Mass Spectroscopy of Nineteenth-Century Daguerreotypes," *Applied Spectroscopy* 53 (1999): 1661-1668.

THE DAGUERREOTYPE PROCESS was the first practical method of producing photographs. Since its introduction in 1839, the unequalled beauty of daguerreotypes has held a special place and fascination among those who love photographs. Even though daguerreotypy was superseded by other photographic processes in the mid-1850s, daguerreotypists have continued to carry on the tradition.

During the last two decades there has been a revival of interest in the daguerreotype, as well as other forms of photography. This has led to a new understanding of the importance of the daguerreotype both as a historic recording medium and as an art form. Many recent books and exhibitions attest to these facts. This new awareness of the importance of daguerreotypes has also led to concerns for their care and preservation. However, many recommendations made for the care of daguerreotypes have been based on traditionally held assumptions about the behavior and chemistry of the daguerreotype and not on sound scientific analysis. Indeed, little attention was paid to the scientific characterization of the daguerreian image either during the nineteenth century or in more recent times, perhaps because the daguerreotype process produced no derivations that have survived as modern photographic methods. In spite of this, traditionally held assumptions about the daguerreotype process are based on the idea that the daguerreotype system fits into the conventional photographic model.

This book concentrates on the more scientific aspects of the daguerreotype process. It is primarily a description of the findings of more than six years of analytical, scientific work on the daguerreian system. It is also a history of previous scientific work on the daguerreotype process and the use of daguerreotypes as scientific tools. These seemingly disparate topics go hand in hand and are each necessary to understand how

Preface to the First Edition

we went about conducting our investigations. In our case science is informed by history, and historical work brings a depth and grounding to the scientific work.

At this point one may wonder why a group of materials scientists would choose for study an obsolete and outmoded photographic process. From a practical point of view, we were interested in devising better ways of preserving these images based on a thorough understanding of their material properties. From a more romantic point of view we, like other scientists before us, were seduced by the incredible beauty of these images and the scant information about what they are and how they work. We were interested in understanding the scientific gleanings about daguerreotypes and using that earlier work to construct our own investigations about these illusive images. It is not surprising that many of the same questions about the daguerreian system have been investigated again and again.

The book contains three main parts. The first seven chapters deal mainly with the history of the daguerreotype and its technology as it was known and understood through the closing years of the nineteenth century. The state of scientific understanding changed little from the late nineteenth century until the work of a few mid-twentieth century researchers immediately prior to the beginning of our own investigations.

Chapters eight and nine address the scientific explanation for the daguerreotype and summarize the results of contemporary laboratory investigations. The daguerreotype image is shown to arise from a characteristic microstructure consisting of small silver spheres dispersed over a polished silver surface. The optical properties created by this structure explain the daguerreotype's visible appearance. The detailed chemistry of image formation during plate preparation, exposure, development, and fixing has been worked out, and we now know how nineteenth-century technology worked. The interplay of chemistry, microstructure, and optical physics is quite intricate and illustrates the clever good fortune (or dogged persistence) of the nineteenth-century daguerreotypists, who got on quite nicely without benefit of modern scientific knowledge.

The problems of conservation and preservation of daguerreotypes are the subject of the final chapters. These address the mechanism of image deterioration, review some old methods and describe some new methods for cleaning tarnish from the daguerreotype surface, and discuss some entirely new techniques for the protection, packaging, and display of daguerreotypes.

The book is addressed to historians of technology and science, to the curious general reader, and, most important, to those charged with the care and preservation of photographs, such as curators and conservators. We have attempted to construct a hybrid of historical summary, scientific investigation fresh from our own laboratory, and a practical application of the information at the nitty-gritty level of the working conservator. As a result the three main sections of the book can be to some extent read independently. However, the book as a whole is intended to be an integrated account of the entire daguerreotype story.

Acknowledgments

THE WORK THAT led to the production of this book began in 1978, and since then we have enjoyed the help and assistance of many individuals and institutions. Over the years the National Museum Act and the Andrew W. Mellon Foundation provided the major funding for this project. In addition, in 1982 the State College Branch of the American Association for University Women awarded Barger the Lucretia V. T. Simmons Project Renew Grant, which was used to purchase some of the nineteenth-century daguerreotypes that were used in our study collection. The following year Barger also received a Research Fellowship from the American Philosophical Society, which furnished funding for many of the illustrations found in this book. Support from the Barra Foundation made possible the analytical work done in conjunction with the exhibition "Robert Cornelius: Portraits from the Dawn of Photography," held at the National Portrait Gallery, Smithsonian Institution, Washington, D.C., from October 20, 1983, through January 22, 1984.

The setting for most of this work, the Materials Research Laboratory of The Pennsylvania State University, has always maintained an active interest in the relationship between science and art. This attitude facilitated our work because the entire laboratory has been as open to this study as it is to other areas of more conventional scientific interest. While we have always enjoyed the interest and encouragement of our friends and colleagues in the laboratory, we benefitted in other ways because the seductive beauty of our daguerreotypes drew many colleagues into joining us in various aspects of the experimental work. We particularly wish to thank Russell Messier, Deane K. Smith, S. V. Krishnaswamy, and Ajay Giri, who contributed to our work as coauthors on some of our papers. In addition, Mary Bliss, Craig Bohren, Kate Chess, Phelps Freeborn, Patrick

Daguerreotype, made November 1979 by Kenneth E. Nelson, of some members of the Materials Research Laboratory of The Pennsylvania State University. Source: Collection of Rustum Roy.

McMarr, Norman Macmillan, Eric Plesko, Randy Ross, Ron Roy, Rustum Roy, Karl Spear, and Don Strickler have all contributed through useful discussions and arguments and by providing a helping hand with the work itself. "Toozer" Wilson and Sonny Gross constructed many of the gadgets and the equipment we needed to make and handle daguerreotypes. Brad Wilt administered the dispersion of funds used in this project, and Lois Annichini Moore typed many of the early manuscripts. Kathye Kissick produced untold numbers of photographic copies of daguerreotypes and micrographs.

Friends outside the university community have assisted us in many ways. From the beginning one of our strongest champions has been William F. Stapp of the Smithsonian Institution's National Portrait Gallery. Will not only provided unfailing support early on when this work was regarded with some askance; he also made it possible for many of our experimental observations about the daguerreian system to be verified using some of the earliest and most important of American daguerreotypes. The analyses made as part of the preparations for the exhibition "Robert Cornelius: Portraits from the Dawn of Photography" were really a gift. Working with these daguerreotypes was like having Joseph Saxton, Robert Cornelius, and Paul Beck Goddard join us in the laboratory to tell us that our operating theories, especially about image formation, were on the right track. For many years the National Portrait Gallery has been Barger's second home in Washington.

The modern daguerreotypists Irving Pobboravsky and Kenneth E. Nelson assisted from the very beginning with technical advice on the practice of daguerreotypy. They both helped Barger in numerous ways as she was learning to make daguerreotypes in the laboratory. They have also been good sounding boards as we worked to understand our experimental findings and relate them to the practical problems of making daguerreotypes. Ken came to the laboratory and made daguerreotypes before Barger learned to make them on her own. While there, he also made the daguerreotype of the Materials Research Laboratory included in these acknowledgments.

The conservator Thomas M. Edmondson once stopped to visit the project while returning home from a cross-country trip. He saw our early work with the electrocleaning process for daguerreotypes and got roped into being our field tester. We gave him daguerreotypes, and he returned to his own conservation studio in Torrington, Connecticut, to run trials. Then, he came back to the Materials Research Lab and worked with us side by side until the process had been fully analyzed and *he* was satisfied that the process was successful. Tom was just the sort of tester we needed. He is brutally honest, he followed our instructions, but he made significant contributions because of his skill in treating objects and his own curiosity and creativity.

William Ginnel and Frank Preusser of the Getty Conservation Institute were involved in the work on protective coating for daguerreotypes.

At various times friends have served as sounding boards for working out ideas for interpreting data and merging the scientific information from our laboratory work with the historical portions of the study. Barger wants to thank especially Doug Corbin, Eastman Kodak Company, Rochester, New York; Jan Kenneth Herman, the Old Naval Observatory, Washington, D.C.; Matthew R. Isenberg, Hadlyme, Connecticut; Reese V. Jenkins, Thomas A. Edison Papers, New Brunswick, New Jersey; Mary Panzer, University of Kansas, Lawrence, Kansas; Françoise Reynaud, Musée Carnavalet, Paris; Willman Spawn, Bryn Mawr College, Philadelphia; Roger Taylor, National Museum of Photography, Film, and Television, Bradford, England; and John Wood, McNeese State University, Lake Charles, Louisiana. Various photographic scientists have been interested in our work and have led us to pertinent work in related areas. These men include Dr. Burt H. Carroll, Grant Haist, Tom Hill, Paul Gilman, Dr. John Mitchell, and Hanoch Shallit.

Many of the historical daguerreotypes that illustrate the text are used by the kind permission of various institutions and private collections. We would like to thank the following people and institutions for allowing us to use images from their collections or for their assistance in locating and identifying many of these scientific daguerreotypes: the American Philosophical Society, The Athenaeum (Boston), James D. Barger, Laurie Baty, Janet Buerger, Marian S. Carson, Brenda Corbin, Cambridge Observatory, Kenneth Finkel, Roy Flukinger, The Fogg Art Museum (Harvard University), The Franklin Institute, Gilman Paper Collection, Martha Hazen, The Harry Ransom Humanities Research Center (University of Texas at Austin), Jan Kenneth Herman, Matthew R. Isenberg, Stephen Joseph, Carol Johnson, Paul Katz, Dorothy Kilgo, Charlie Mann, Musée Carnavalet (Paris), l'Observatoire (Paris), The Graham Nash Collection, Kenneth E. Nelson, The Peabody Museum (Harvard University), The Philadelphia Academy of Natural Sciences, Sally Pierce, Alex Ross, Richard Rudisill, William L. Schaeffer, Sandy Stelts, and John Wood.

The following people either contributed daguerreotypes to the study collection or loaned daguerreotypes for the electrocleaning trials: Laurie Baty, William Becker, Joe Buberger, the Canadian Conservation Institute, Henry (Dan) Deeks, Betty Fiske, Huntingdon (Pa.) County Historical Society, Michael Kamins, David Kolody, Lou Kontos, Debbie Hess Norris, Irving Pobboravsky, Edwin E. Pontius, Grant B. Romer, William L. Schaeffer, Ric Smaus, Willman Spawn, and William F. Stapp.

Don Clark, Franz Huber, and Charlotte Schupert assisted at various times with translations of German papers. Barger wishes to thank her friends and students, especially Mary Becker, Marilyn Karr, Eilene Perkins, and Claire Rutiser, who helped with preparing the manuscript by picking up photographs, checking references, reading drafts, xeroxing, and entertaining the cats. Barger also wants to thank Mary Bliss, Else Brevel and Carsten Pederson, Nancy Brown and Chuck Fergus, Elizabeth and Jon Wilson, and Celia Wyckoff, who allowed her to be a bad guest in their homes after she moved from State College to Washington and Baltimore. We both thank our students and friends who were neglected or who let us work during our sometimes frequent meetings to struggle with this manuscript. Amy Pastan, our acquisitions editor at the Smithsonian Institution Press, has always been patient and encouraging.

We both wish to thank our families. They have always been supportive of, if not genuinely interested in, the work. Robert Shlaer read the manuscript and made many helpful comments and additions from his pragmatic viewpoint as a practicing daguerreotypist. We owe a special debt of gratitude to Elizabeth L. White. Bette has made invaluable contributions to keeping us on track and seeing this project through to the end.

THE
DAGUERREOTYPE

Fig. 1.1. The joint meeting of the Académie des Sciences and the Académie des Arts on August 19, 1839, at which the daguerreotype was introduced to the world. Standing at the dias is François Arago; Louis Jacques Mandé Daguerre is seated at the center with his partner, Isidore Niépce, at his side. Source: Figuier, *Les Merveilles de la Science*, fig. 3. Courtesy of the Gernsheim Collection, Harry Ransom Humanities Research Center, The University of Texas at Austin.

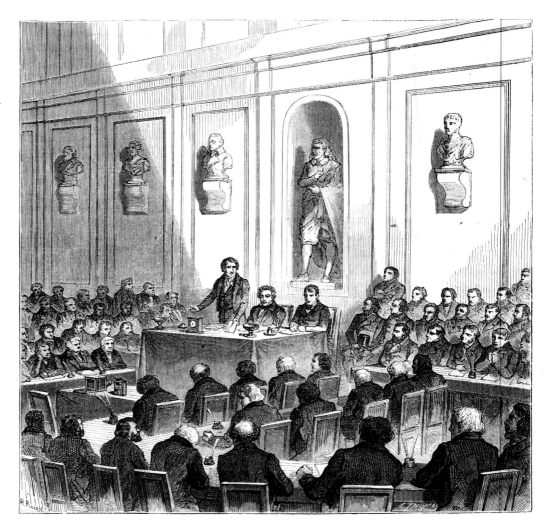

FROM THE EARLY MORNING of August 19, 1839, people had been arriving at l'Institute in Paris to hear the public disclosure of Louis Jacques Mandé Daguerre's method for capturing images using a camera. By 3:00 in the afternoon every seat in the hall was occupied, the adjoining courtyards were filled, and a large crowd had gathered in the streets outside. Seated at the front of the room were François Arago, head of the Académie des Sciences, Daguerre, and Isidore Niépce, son of Daguerre's deceased partner. Daguerre declined to speak, in spite of Arago's considerable urging, so Arago described the daguerreotype process and delineated the scientific details and applications of the process (fig. 1.1). There was a great deal of excitement, and the crowd's reaction to the announcement was intense and immediate: within hours every optician in town was besieged with people trying to obtain cameras in order to share in the wonder of the new art-science.

In an age when photographic and electronic images are so commonplace, it is hard to remember that the ubiquitous photograph is produced by a relatively new technology. It has been just 150 years since photography was first introduced, and the promulgation of that discovery has changed the way people view the world. Daguerreotypes—jewellike images produced on polished plates of silver-plated copper—were the first practical method of producing photographs. After the daguerreotype process was introduced in August 1839, and for most of the next twenty years, the production of daguerreotypes accounted for the majority of photographic images made throughout the world.

Daguerre's process, as described by Arago, had five steps:

1. A piece of silver plate was cleaned to a mirror finish using a slurry of pumice in oil followed by various washings in nitric acid and water to remove the oil residue.

2. The prepared plate was sensitized by fuming—that is, exposing it to iodine

1

Beginnings

Two centuries ago the daguerreotype art would have been looked upon as the work of witchcraft, but in our age of improvement we are accustomed to extraordinary discoveries; we are capable of admiring and appreciating these huge efforts of genius, and nothing surprises us.

Antoine Claudet, "The Progress and Present State of the Daguerreotype Art"[1]

vapor until the plate had taken on a bright golden color. The color indicated that a layer of silver iodide of the right thickness had formed on the plate. Arago stressed that this was a critical step and that the silver iodide layer needed to be perfectly uniform.

3. The sensitized plate was placed in a camera and exposed to light. Exposure times varied according to the time of day, the season of the year, and the weather. Recommended times were within the range of three to thirty minutes.

4. The exposed plate was placed over hot mercury vapor until an image appeared.

5. The silver iodide was desensitized by placing the plate in a hot solution of common salt or removed using a solution of sodium thiosulfate. This last treatment was followed by a series of rinses in water and drying of the plate.

The image formed in this way was extremely fragile and could be removed from the silver plate with the slightest touch. Daguerre recommended that these images be protected either by enclosing them in a small frame or by gluing them under glass. He also tried several common varnishes—amber, copal, caoutchouc, wax, and other resins—as protective coatings for the image. He found that the varnished specimens were not as satisfactory as those enclosed in a frame. The varnishes caused the white portions of the image to become dull, and eventually the images disappeared as the varnishes aged.

As the announcement of the process was being made, a manual prepared by Daguerre was being published. It gave the history of Nicéphore Niépce's work, the joint work of the two partners, and the agreements leading to the public announcement as well as the details of the daguerreotype process, with plates and descriptions of the special apparatus needed to accomplish the art. It took a few weeks for the manual to appear, and in the meantime newspaper accounts of Arago's presentation appeared. Some of the reports were harsh, claiming

that if one were to take up daguerreotypy one must have a complete grasp of chemistry and physics, a reflection of Arago's emphasis on the scientific underpinnings of the daguerreotype process. A private demonstration of the process made by Daguerre to some newspaper editors was reported in a more sympathetic manner on September 1. Daguerre also began giving public demonstrations of his process on September 3. As time went on, more and more people had success with the process with or without lessons by the master.

The process as it was introduced had three drawbacks: (1) the exposures were very long, precluding the use of the new art for practical portraiture; (2) the image was mechanically fragile after it was made; and (3) the image was not colored but appeared as a gradation of tones. The challenge to overcome these obstacles was immediately taken up by those experimenting with the process, and their efforts were met with varying degrees of success. The solutions to the first two problems were the only two actual changes made to the daguerreotype process. The third problem remained difficult and the answer elusive.

For most of the next twenty years the daguerreotype process was the most practical method of producing photographs. The practice of daguerreotypy spread as fast as the news of its announcement. Daguerreotypes were made in the South Sea Islands, in all parts of North America, in Africa and the Mideast, and in China and Japan, as well as in Europe. During the daguerreian era, from 1839 to 1860, as many as 30 million daguerreotypes were made in the United States alone, and one estimate puts the yearly American trade solely for the direct purchase of daguerreotypes between 8 and 12 million dollars a year by 1850.[2] Political cartoons in the newspaper mocked "Daguerreotypomania." Most of the important historical documents recorded by the new medium of photography during the 1840s and 1850s were made on daguerreotypes.

After the process was improved so that portraits could be made using the new process, everyone—including those recently dead—was subjected to the daguerreotypist's art. Daguerreotypes have a historical importance that belies their jewellike beauty, and recognition of this phenomenon motivates this account of the process and how it works.

Before beginning this excursion through the daguerreian process, it is necessary to review the cultural, scientific, and technological framework out of which photography and, in particular, daguerreotypy emerged. A great deal has already been written especially in the last twenty years about the daguerreotype's artistic and sociological impact, which need not be reviewed. This book does deal, in part, with the history of the science and technology of the daguerreotype, both topics that have not previously been discussed at length. The introduction of photography in 1839 had a powerful effect on all areas of human endeavor, particularly in scientific and technological pursuits.

2

Image Making before Photography

I am to blame because I can't draw; and second
because I am resolved not to let anybody know the
method I use for this purpose [i.e., using a
microscope] and so I just make only rough and simple
sketches with lines, mostly in order to assist my
memory, so that when I see them I get a general idea
of the shapes: besides, some of the forms I see are so
fine and small, that I don't know how even a good
draughtsman could trace them, unless he made them
bigger.

Anthony van Leeuwenhoek to Henry Oldenburg, March 26, 1675[1]

THIS PASSAGE IS PART of a letter sent by Anthony van Leeuwenhoek, the Dutch microscopist, to Henry Oldenburg, the secretary of the Royal Society, in response to a complaint by the Royal Society that Leeuwenhoek's illustrations were not always clear and usable for generating reproductions. The ability to make illustrative drawings of scientific observations became an increasingly important skill beginning about Leeuwenhoek's time, primarily because of the rise during the seventeenth and eighteenth centuries of scientific fields based on visual observation. These fields included systematics (or the taxonomy of biological organisms), topology, and geology. The Dutch school of realistic and descriptive art of the seventeenth century in many ways helped raise popular interest in the types of science that could be described pictorially.

The ability to make realistic, accurate drawings was also important in the mechanical arts and engineering. The rapid industrialization brought about by the Industrial Revolution in Britain and its influence elsewhere during the late eighteenth century increased the need for draftsmen to produce drawings to aid in the production and use of the ever-expanding number of machines needed for the new industries.

These factors increased the demand for people who were able to record images directly as seen in nature without interpretation by an artist or draftsman. Indeed, the ability to draw was of such significance that this talent could provide upward social mobility. This exact style of drawing is derived from art conventions of perspective drawing set forth in the early Renaissance, which were in turn derived from Euclid's geometry. The rules of perspective aimed to remove the artifice from art through the codification of a system of drawing that produced what was understood to be scientifically accurate reflections of the world. These Renaissance rules were brought to their apex by the Dutch artists of the seventeenth century. In addition to rules of perspective, artists and draftsmen were aided by a variety of mechanical drawing devices or drawing machines.

Drawing Machines

The drawing machine not only was an aid to the artist but also helped convey the impression of scientific objectivity in drawing. Essentially, drawing machines were any of a number of tools used to "facilitate the work of artist and contribute . . . to that accuracy and precision of execution by which the present race of artists are so remarkably distinguished even from their immediate predecessors."[2] Improvements and innovations in these devices were actively sought by both private concerns and government bodies, and prizes or premiums were offered as incentives for those who worked to develop these instruments.[3] Simplicity in use, economy, and portability were the qualities actively sought in these machines and encouraged in the descriptions found in the yearly solicitations by the premium organizations, such as the Society of Arts[4] in Great Britain and similar organizations on the Continent and in the United States.

Diverse drawing machines were available during the early decades of the nineteenth century, but since none was entirely satisfactory, refinements were continually attempted. In his letter of application to the Society of Arts, for example, Thomas Allason described the lack of satisfactory drawing machines:

The want of an Instrument of this kind is universally felt and acknowledged; many attempts have been made, and contrivances of great merit have resulted from the attention given this important subject. In fact, we have numberless optical and mathematical instruments for this purpose, but the whole of them I conceive liable to many objections; of the former, the most in use are the common camera obscura, Wollaston's camera lucid, with its application to the telescope by Varley;—of the

latter, there is a more considerable number, such as Maltrus, Ferguson's, and Turrell, &c., with others of great ingenuity, in the possession of private individuals.

To the first I principally object either from want of portability, simplicity, &c., and what perhaps may not constitute the least object, their great expense; to the second I attach nearly the same objections, in addition to which their only ascertaining one point at an operation entirely precludes their being used, except by persons of considerable leisure.[5]

Allason sought to remedy the situation with his "perspectograph" (fig. 2.1), which was awarded a premium in 1817. He described the operation of his instrument in the following way:

It principally consists of a metal frame, 9 inches by 6, having two graduated scales moving upon its surface, the one vertical to the other, for the purpose of finding both the horizontal distances, with the height and pro-

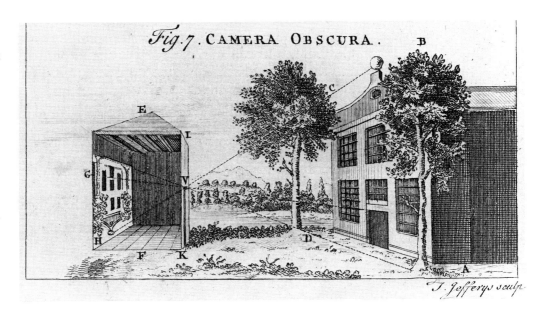

Fig.7. CAMERA OBSCURA.

T. Jefferys sculp.

Fig. 2.2. "The camera obscura, or darkened room, is made after two methods; one the camera obscura, properly so called, that is, any large room made as dark as possible, so as to exclude all light, but that which is to pass through the hole and lens in a ball, fixed in the window of said room.

"The other is made in various forms, as that of a box whose sides fold out, etc. for the conveniency of carrying it from place to place.

"For the construction of a camera obscura, 1. Darken the room EF leaving only one little aperture open, in the window, at V, on the side IK, facing the prospect ABCD. 2. In this aperture fix a lens, either plane convex or convex on both sides. 3. At a due distance, to be determined by experience spread a paper or white cloth, unless there be a white wall for the purpose: then on this, GH, the desired objects, ABCD, will be delineated invertedly. 4. If you would have them appear exact, place a concave lens between the center and the focus of the first lens; or receive the image on a plane speculum, inclined to the horizon, under the angle of 45°; or by means of two lenses included in a draw-tube, instead of one. If the aperture does not exceed the largeness of a pea, the objects will be represented without any lens at all." Source: Engraving by Thomas Jeffreys, *A New and complete dictionary of arts and sciences*, pl. 35.

portions of the different objects to be drawn to the lower surface; and at right angles with the frame is attached a small beam with a sight point at the other extremity, which can be lengthened or shortened agreeably to the quantity of objects desirable to be introduced in the drawing, and the whole supported by a triangle, forming a walking stick.[6]

Other drawing instruments also offered for premiums during the same year included a machine for drawing ellipses, a parallel rule, a drawing board with a T-square, and an angulometer, a device for marking out and measuring angles. These instruments would have all been used for recording landscapes ("views"), natural history subjects, and the like and for copying, enlarging, and reducing drawings.

Another type of mechanical drawing aid was devised and developed to assist in the production of portraits. By the end of the eighteenth century, the rising middle class clamored for portraits, stimulating the development of new methods of producing quick and inexpensive portraits. In particular, silhouettes and profiles came into vogue.[7] This type of portraiture was embraced by all social classes, and artists at all levels of skill responded to the demand. Interest in the emerging science of phrenology, the study of character based on the shape of the head, also helped promote this type of portraiture.

Many drawing machines were used for "taking" these types of silhouettes and profiles. The most well known was the physiognotrace, a profile machine invented in 1786 by Gilles Louis Chrétien. To use this machine, the artist placed the sitter in a frame so that a part of the machine actually touched the sitter's face and guided the artist as a direct life-size tracing of the profile was made. At the same time, an attached pantograph simultaneously made reductions of the silhouette. The physiognotrace was only one of many types of profile machines, which ranged from complicated machines like the physiognotrace to very modest devices that simply allowed the artist to trace the sitter's shade without interference from extraneous shadows.

Cameras

An additional class of drawing machines was the camera. The most common camera was the camera obscura, which, although referred to in writings dating from antiquity, was first described in detail by Johann Baptista Porta in 1533. *Camera obscura* means "dark room"; thus, a camera obscura is basically a dark chamber, or box, with an opening at one end through which light passes. The light entering the camera obscura falls onto the wall opposite the opening to form an image. If the opening is small, it needs no lens; however, larger openings outfitted with a lens will produce brighter images. Whole buildings were built as camerae obscurae (fig. 2.2). Some of these buildings became popular tourist attractions, especially during the nineteenth century. Today, tourists may still visit an architectural camera obscura on top of Outlook Tower in Edinburgh, Scotland.

Camerae obscurae were also made as portable boxes used to make tracings of the scene projected on the back wall of the camera box. The portable camera obscura (fig. 2.3) shows a late innovation added to this type of camera, the Wollaston landscape lens (B), introduced in 1812 by William Hyde Wollaston. This innovation reduced the curvature of field at the edges of the image, a problem associated with earlier camerae obscurae that used biconvex lenses. The Wollaston landscape lens had a front stop and a meniscus-shaped lens positioned so that the concave portion of the lens faced the stop.[8]

The camera obscura could be used alone or in conjunction with other instruments, such as the solar microscope, for sketching and drafting. In order to use the camera obscura to trace the image formed on its back panel, the entire instrument or at least the panel needed to be darkened; otherwise, the projected image would lack sufficient contrast to be seen clearly. Thus, the instruments were constructed with housings that could accommodate this restriction. These housings made portable camerae obscurae heavy and awkward devices to carry around. These cameras also required some training and skill for their fullest potential

to be exploited. In spite of many improvements made to increase their portability and facilitate their use in the field, the camera obscura remained an inconvenient instrument to use.

Another type of camera, the *camera lucida*, was first described in 1807, also by William Hyde Wollaston. Although the exact source of this name is unknown, it means "light room" and may refer to the fact that this instrument did not require an enclosure, box, or lens as in the camera obscura. There are two basic types of camerae lucidae (fig. 2.4). The first is a device with a reflective glass mounted on a stand. This glass is positioned at a 45-degree angle to the paper so that the artist or draftsman can see the reflection of the scene on the glass and also look through the glass to the paper. The scene is thus perceived to be superimposed on the paper and can be easily copied. Images made using this type of camera lucida are laterally reversed. The second

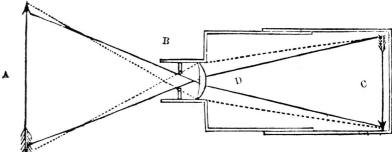

Figure 2.3. Diagram of a portable camera obscura with a Wollaston landscape lens. Light from the object at *A* is reflected through the Wollaston lens and stop, *B* and *D*, and an inverted image is formed at the camera back at *C*. Focusing is effected by moving the two boxes that form the camera body in and out. Source: Snelling, *A Dictionary of the Photographic Art*, p. 28.

Fig. 2.4. Ray diagrams for the two most common types of camerae lucidae. In *A* the image of the scene is reflected onto glass surface inclined at 45 degrees to the scene. The artist sees the reversed reflection of the image on the glass surface but also perceives the image on the drawing surface below. In *B* Wollaston's four-sided prism reflects the image of the scene twice, and the artist splits the field of view so that both the reflection from the image and the drawing surface are seen at once. An image made using this type of camera lucida is not laterally reversed. Source: Drawn after John H. Hammond and Jill Austin, *The Camera Lucida in Art and Science.*

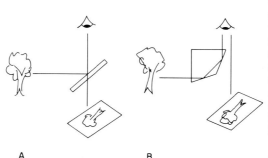

A   B

type of camera lucida uses a four-sided prism fastened to a small stand. To use this type of instrument, the artist or draftsman positions the eye so that it sees both the image of the scene reflected directly on the retina and the paper on which the scene is to be copied.

The camera lucida was more portable and convenient to use than the camera obscura because it was a very simple device with only a few parts, it did not require a darkened surround, and once its operation had been learned it was easy to use. For these reasons it quickly found wide use among draftsmen, especially for tasks such as enlarging or reducing drawings and for copy work. Its portability made the camera lucida useful to both professional artists and amateurs making topological and architectural views (fig. 2.5). This instrument had special appeal to travelers because it could be used to make "correct representations"[9] of the places visited even by those who could not draw well. The prominent traveler and British naval officer Col. Basil Hall wrote a laudatory book about the camera lucida, claiming that once the operation of

this camera was mastered the user could "rove where he pleases, possessed of the magical secret for recording the features [of a scene] freed from the triple misery of Perspective, Proportion and Form,—all responsibility respecting these being taken off his hands."[10] Both the camera obscura and the camera lucida were items of commerce and could be easily purchased from opticians and instrument makers.

Lithography

During this period there was a demand not only for unique works of art such as portraits and drawings but also, increasingly, for reproductions of works of art—that is, prints. The increased desire for art reproductions accompanied a similar rise in the demand for printed books. Thus, new printing methods and new methods of making the paper needed for mass production of cheap art works and books were developed during this time. The only new printing method discovered since the invention of movable type was introduced during this period. This process, called li-

Fig. 2.5. Camera lucida drawing made by John Herschel, June 25, 1824, in Selinunte, Sicily, at the ruins of the First or Great Temple. The circle at the lower center of the drawing marks the location of the "eye," or the camera's central viewpoint. Courtesy of the Graham Nash Collection.

thography, revolutionized printmaking because it was a faster and cheaper method of producing prints than any previous printing processes.

What is lithography? Why or how is it different from other types of printing? This process was first discovered in Germany by Alois Senefelder in 1798. Originally, it was called variously *steindruck, steindruckerey, chemische druckerey,* polyautography, engraving or printing from stone, and *imprimerie chemique.* The French were the first to call the process by its current name, *lithographie,* which translates as "writing with stone." In this process a drawing is made directly on or transferred to a polished, fine-grained limestone using a greasy crayon. The stone is then wiped with a slightly acid, dilute solution of gum arabic and set aside. Before a print is to be pulled from a prepared stone, the stone is wetted with water and inked with a greasy ink. Excess ink is wiped away, and what is left sticks only to the crayon tracings and not to the portions of the stone that are wetted. The inked stone is placed in a press, and prints are pulled from it. This is a planographic process; that is, there is no relief. The areas that are printed are on the same porous ground of the stone at the same level as the areas that are not printed. The difference in areas that are printed and those that are not is the difference in their attraction toward or repulsion from oil and water.

The success of the process depends on having a highly polished, porous support and an ink that sticks only to the design drawn on the support and that can be easily wiped away from other areas. The best lithographic stones were fine-grained limestone from quarries around Bavaria. These stones were in high demand, but their availability was restricted by political and economic events such as the continental blockade and the various military campaigns of the period. The need for these stones was so great that some people were able to maintain businesses based on renting out stones to artists. Thus, early on it became important to find either local sources of limestone or suitable substitutes for printing stones in order to free countries like England and France from their dependence upon German limestone.

The tremendous growth of lithography was largely due to the fact that it caught the attention of artists and scientists alike. Not only was it the first printing process to be introduced after the invention of movable type three hundred years earlier; it was also based on chemical, not mechanical, principles. Descriptions of the process were published widely in scientific and technical journals, as well as in the popular press. By 1815 Mulhouse, France, had become a major center for lithography, and many people came there, even from England and Germany, to learn the process.

Prominent artists were solicited to make drawings either on paper or directly onto stone to be made into lithographs that were sold by subscription. These prints attained a remarkable popularity because they could be put out more quickly and cheaply than was possible using the traditional printing methods of etching or woodcuts. Because lithography does not require cutting or etching a printing plate or block, prints produced in this way have no printing impression (except around the edges of the frame); thus, to the untrained eye lithographs appear to be more like drawings than prints. Indeed, innovations made to the process throughout the first half of the nineteenth century produced lithographs that more and more closely resembled drawings and watercolors.

The demand for more prints created a need to devise better ways to transfer designs not only to lithographic stones but also to copper engraving plates and wood blocks. Lithographic designs could be drawn directly onto the stones, a technique that was useful for producing original works quickly. When lithography or traditional printing techniques were used to reproduce drawings or prints, however, making exact duplications was not so expeditious or straightforward. Elaborate and complicated machines were devised for making exact transfers of designs to stone, copper plates, and wood blocks. Some of the growing literature about the lithographic process suggested that it might be possible to transfer designs directly onto the stone if the design could be made translucent and if the lithographic stone could be varnished with a light-sensitive substance. Light passing through the design could then be transferred to the prepared stone below.

Just as with other types of drawing machines and art work, the premium societies encouraged the development of these new areas of the graphic arts. Further, the ever-increasing demand for inexpensive prints and reproductions and the successes in these areas led to a blossoming of the arts and prepared the way for even greater developments in printing and the graphic arts during the nineteenth century. These conditions also set the stage for the discovery and introduction of photography.

GENERALLY, DISCOVERIES ARE made and new inventions promulgated in a setting that encourages the particular innovation. Discovery is the outcome of problem solving. A discovery may not answer the problem being addressed, but the process of discovery revolves around some premise or mindset that allows the discoverer to recognize what has been discovered. The availability of relevant information, previous scientific knowledge, or artistic bias does not necessarily lead to discovery. If that was the case, then many discoveries, including photography, should have been made long before they were. One vital requirement for the ultimate success of a discovery or invention is that it must be recognized and made public, because unless a new idea or object is subjected to public scrutiny and testing, it cannot take on authenticity. In effect, without some public acknowledgment a discovery is not really a discovery. Often this acknowledgment is made by some person (usually not the discoverer or inventor) who recognizes the value of an earlier discovery and promotes it.

To understand the discovery of photography, it is necessary to look at the work of those who had been conducting experiments that could lead to permanent imaging with light: Joseph Nicéphore Niépce, Louis Jacques Mandé Daguerre, and William Henry Fox Talbot. These three men, working in France and England, are the major discoverers of photography. They set out to solve different problems and came up with unique solutions. While Niépce and Daguerre eventually became partners, they were brought together by chance, and each had different reasons for working with the other. These three men were not the only discoverers of photography, but public reports of their work gave others a nudge to announce their own successes. This discussion would be incomplete without mention of the role of François Arago, without

3

Toward *Points de Vue* and the Daguerreotype

M. Daguerre, it appears, struck by some hints he had received from a friend, has steadily pursued his experiments for the last twenty years, and having at length attained his object has declared his discoveries and claimed the invention his own.

"Pictorial delineations by light," 1839[1]

whom photography would have waited even longer for its introduction. The specific events of the discoveries of photography have been described in great detail elsewhere. It is not the authors' goal to reiterate these facts; rather, we will cover these events in a general way and concentrate on the discovery of photography itself.

Studies of Light Phenomena

At the beginning of the nineteenth century, the study of light was a fundamental problem that consumed the attention of most scientists in some way. Why study light? It is an old question. Light provides the primary information about and experience of the world around us. Thus, understanding what light is and how it works should provide the key to understanding a diversity of natural phenomena. Light was the purview not only of astronomers and physicists; it also captured the attention of mathematicians, chemists, biologists, and physicians. Other related artisans such as opticians, glass and lens makers, and instrument makers pursued theories of light derived by others for specific applications and improvements to their crafts. This urge to understand what light is and how it works prompted some of the work that eventually resulted in the discovery of photography; however, that knowledge is only one of the prerequisites necessary to devise a process that uses light to make drawings.

At the outset of the nineteenth century two opposing theories of light were being investigated. These fall into two broad categories: (1) the particle (also called the corpuscular or emission) theory and (2) the wave (or undulatory) theory. The first, expounded by Sir Isaac Newton (1642–1727), states that light is made up of very small corpuscles or fiery particles that are emitted from one body and move at great speed through space and strike some receptive body. During most of the eighteenth century this was the dominant working theory

for natural philosophers concerned with questions of light.

The undulatory theory, which had first been suggested by the Dutch astronomer Christiaan Huygens (1629–95), a contemporary of Newton, was revived by Thomas Young (1773–1829), an English physician. Although he had previously published work on sound and light during the 1790s, Young's primary work on light and color was first published in 1801. He was the first to suggest the trichromatic theory of color vision. He also laid the foundation for the wave theory of light, stating that when a body becomes luminous it sends out waves of particular amplitude and wavelength. The length of these waves determines their color in much the same way that the wavelength of sound waves determines their pitch. Further, Young felt that light waves had to be carried in a special medium, the luminiferous aether, which vibrates and transports the waves to some receptive body at a distance.

Aethers of all kinds have been used through the ages to explain various natural phenomena. The aether of interest here is a fluid material substance of a very subtle nature called the luminiferous aether, which fills all apparently empty spaces and is a medium for propagation of light waves. The aether is a real substance, not merely an idea. This medium carried light, and its properties were deduced from the study of light. The luminiferous aether had dynamic properties and, consequently, allowed the transmission of physical motion from one place to another, just as movement is conveyed in a solid or liquid. In Young's theory the idea of light as waves and the luminiferous aether had to go together.

When Young first presented his ideas, the main problem facing scientists was that these two theories about light could not be resolved using available experimental knowledge. Important concepts about light were being investigated, defined, and redefined. Young's ideas were not immediately accepted; however, powerful mathematical derivations based on experimental work in France and England helped sway many experimenters to the undulatory theory. In particular, work in the areas of color, interference, polarization, refraction, diffraction, and absorption were often described better by those who espoused Young's ideas.

The experimental and theoretical work of Young and, later, the Frenchman Augustin Fresnel (1788–1827) on interference and polarized light during the first few years of the nineteenth century provided new evidence for the wave theory of light. Both Young and Fresnel observed that under certain conditions light appeared to interfere with and be cancelled out by other light that arrived at an eye or a lens at the same time. Young called this phenomenon *interference.* Fresnel was able to determine a better mathematical description of the process of interference than had Young, and Fresnel's work lent more proof to the wave theory. Further work on interference by Fresnel, Étienne Louis Malus, François Arago, David Brewster, and others using various mineral crystals such as topaz and Icelandic spar led to discoveries about polarized light.

Polarized light had been discovered by Huygens. He and others after him used all sorts of transparent, natural crystals as prisms to study how these crystals changed light that passed through them. One of these crystals—natural calcite, or Icelandic spar—has particularly intriguing properties. This crystal occurs as a rhomb. When the crystal is placed on a paper that has been marked with a small, geometric figure, two figures will appear when one looks at the figure through the rhomb. Moreover, as the crystal is rotated, one figure will appear to rotate around the other. It was recognized that the Icelandic spar broke the incoming light into two separate rays: the ordinary ray and the extraordinary ray. These two components of the light ray were likened to the opposite poles of a magnet, and

this intriguing property was called *polarization.* Not until the work of Malus and Arago in particular was polarization determined to be a property characteristic of the interaction of light with media other than Icelandic spar.

The success of the wave theory can be measured by similar work in other areas of optics, which also proceeded at a rapid pace during this period. This very success stimulated the demand for new and better lenses and optical instruments, which, in turn, fed the increasing demand for instruments to aid in the accurate depiction and recording of nature. Even though the understanding of how an image is formed by a lens was not a new concept, new mathematics derived from the study of other light phenomena led to improved lens design and, likewise, improved instruments—telescopes, microscopes, and cameras. Three examples were the camera lucida, the "Wollaston doublet" lens for microscopes, and a singlet landscape lens corrected to produce a flat field of view, all developed by William Hyde Wollaston (1766–1828). New manufacturing methods for glasses were devised, and a more empirical understanding of the relationship between glass composition and the quality of lenses was established. The Royal Society in London even formed the Optical Glass Committee to work on improvements in lens manufacture, including the chemical analysis of glass. This committee and its subcommittees were active from 1824 to 1831.[2] All these activities led to the improvement of optical lenses.

Along with the search for the elucidation and control of the physical characteristics of light, active research was being conducted on the chemical effects of light. In 1727 Johann Heinrich Schulze (1687–1744) discovered that if a bottle of silver nitrate crystals was placed in the sun and part of the bottle covered with a stencil, the area not covered would blacken. He could then shake the bottle so that the black design would disappear and repeat the process. Schulze was thus the first to observe the light sensitivity of silver nitrate, opening a new area of research now called *photochemistry*. Following Schulze other experimenters discovered various metal compounds and other silver compounds that were also light sensitive, and these were used in the production of silver nitrate ink, which was used to produce permanent designs on a variety of substrates including paper, textiles, bone and ivory, and leather. Silver nitrate was also used as a hair dye. Other metal salts, such as gold chloride and mercury chloride, were used to make colored designs on cloth.

In 1777 Karl Wilhelm Scheele (1742–86) published *Aeris atque ignis examen chemicum*, the first comprehensive book on the chemical effects of light. Scheele, considered the father of photochemistry, used his experiments to demonstrate that light was composed of or contained phlogiston.[3] He based this observation on the fact that when silver oxide, gold oxide, or mercury oxide was placed at the focus of a burning glass these salts are changed to metal. Applying the chemical system of Joseph Priestly, he deduced that this reduction to simple metal indicated that these salts had absorbed phlogiston. Scheele made an extensive investigation of silver chloride. Using paper saturated with silver chloride for some of his experiments, he discovered the solvent action of ammonium hydroxide on silver chloride, and he investigated its spectral sensitivity.

The work of Antoine Lavoisier (1743–94) established that his method of systematic investigations of compounds and elements was a powerful way to approach the study of chemistry. He used changes in the weight of reactants as a means to define elements. He also included light and heat in his table of elements even though he knew that they could not be weighed and, thus, did not fit his own definitions. However, Lavoisier and other chemists after him believed that the coloration of light-sensitive materials when irradiated was analogous to other reactions that they produced in the laboratory. This meant that light was a true reagent, which when added to some compounds could alter their basic properties or cause a reaction. Testing for reactivity in light became part of the regular chemical routine of investigating new compounds and elements. Because of this routine, almost all reports on the chemical properties of newly discovered chemicals made during this period included mention of their reactivity in light.

Another important area of interest at this time was the new science of *photometry*—the measurement of the effect of light intensity, whether transmitted or absorbed, on various materials. These studies were broad based and concerned those making astronomical observations as well as those investigating the chemical effects of light. Crude photometers were made using paper saturated with silver nitrate, silver chloride, or other light-sensitive solutions. At best, the earliest experiments in this area were crude, and the results obtained were questionable and difficult to analyze. Nonetheless, many investigators dabbled in photometry, including Young, Arago, John Herschel, and Simeon Poisson.[4]

Another area somewhat related to photometry was *spectral analysis*, the forerunner of modern spectroscopy. This is the study of the physical and chemical effects of light that has been dispersed into a spectrum by being passed first through a slit and then through a prism. Spectral light was used to study gases and flames colored by the addition of various metal salts and transparent liquids and to monitor the effects of photochemical changes. Because of his work with spectral analysis, Scheele had recognized that there is a difference between the actions of light and heat. Later, this research led to the discovery of the infrared region of the spectrum in 1800 by Friedrich Wilhelm Herschel, father of John Herschel, and the ultraviolet region in 1801 by J. W. Ritter. William Hyde Wollaston also discovered the ultraviolet region in 1802 and referred to his discovery as chemical (or refrangible) rays. Likewise, the infrared region of the spectrum was commonly referred to as caloric rays. Even these names gave weight to the recognition that light and heat behave in chemical systems in different and distinct ways.

The sun and stellar studies of Joseph von Fraunhofer (1787–1826) published in 1814–15 gave spectral analysis a real boost, and his work was known throughout the scientific world by the 1820s. Fraunhofer was an optician and physicist working outside Munich whose skill in making achromatic telescope lenses and prisms was widely acclaimed. He made his discoveries using the telescope of a theodolite set up to receive light that had been passed through a slit and a prism. A theodolite is a precision instrument equipped with a telescope sight and used to establish horizontal and sometimes vertical angles. It is used in both transit astronomy and surveying. Because Fraunhofer used this particular instrumental arrangement, he was able to construct for the first time a map of the line spectra from the sun, the moon, and the planets. He named the principal lines from these spectra with the letters *A* through *G*—the now-familiar Fraunhofer lines.

Similar investigations on the absorption of light were done concurrently in England during the 1820s by David Brewster (1781–1868), Sir John Herschel (1792–1871), and William Henry Fox Talbot. Independently they observed that flame spectra obtained by burning compounds of different compositions varied in systematic ways. For Herschel, in his capacity as a member of the Royal Society's Optical Glass Committee, part of the motivation for these investigations was to devise monochromatic light sources for microscopes and new glasses to make lenses and prisms free of unwanted absorption.

These three men all produced line spectra, and they tried to relate these spectra to the dark lines observed by Fraunhofer from his solar spectrum. Two themes emerged from this work. On the one hand, there was some agreement that the spectra they produced contained chemical information characteristic of the flames analyzed. However, they could not agree about the role of light in the production of these spectra, and their discussions of results dissolved into arguments about whether the wave theory or the corpuscular theory of light was correct. These arguments diverted them from the real issue of using line spectra for chemical analysis, since no light theory can account for line spectra. These lines are not the result of light interaction with particular compounds but rather are related to their atomic structures and, therefore, to chemistry.[5]

Talbot was interested in devising some theoretical basis for using line spectra in chemical analysis. He was more of a mathematician than a chemist, however, and did not have the chemical expertise to devise a method for chemical analysis based on using line spectra. (Talbot knew his limitations in this area and made reference to them in a letter to Herschel: "I am not much of a chemist, but sometimes amuse myself with experiments."[6]) During this period a great deal of Talbot's time was also taken up with politics in his role as a member of Parliament, and this slowed his experimental work.[7]

In the early part of the nineteenth century, then, there was a great deal of both theoretical and empirical activity concerning light and its effect on a wide variety of materials. This work was widely published and available in both technical publications and more popular publications, such as the *Literary Gazette* in England and the *Edinburgh Review* in Scotland. There were also many chemical discoveries that showed the range of both light-sensitive materials and their properties in a variety of conditions.

These discoveries included light-sensitive materials such as silver nitrate, silver chloride, silver oxalate, mercury chloride, mercury oxalate, chromate and citrate salts, iron salts (including Prussian blue pigment), gold salts, and platinum salts; compounds that react with light-sensitive materials, such as sodium thiosulfate; and elements that react with other materials to form light-sensitive compounds, such as chlorine, iodine, and bromine. This list is by no means exhaustive. The use of optical instruments was widespread among artists, artisans, and scientists. In spite of the wealth of data available, the application of this information to the possible production of images was scant and not very successful.

Fathers of Photography

The first attempts at imaging using light-sensitive materials were made by Johann Heinrich Schulze, who discovered the light sensitivity of silver nitrate in 1723. Subsequently others used light-sensitive materials as part of light-sensitive inks. Imaging on a more sophisticated level was first done sometime during the 1790s by Thomas Wedgewood (1771–1805).[8] Wedgewood, the second son of the famous potter Josiah Wedgewood, attempted to use the sun to produce images on light-sensitive paper and leather. He was successful in producing silhouettes; however, he did not have a way to stop the action of the light on his papers, and his silhouettes gradually darkened and disappeared. The primary light-sensitive material that he used was silver nitrate; he also mentioned using silver chloride, stating that it offered no advantage over silver nitrate. Further, he found that images formed in a camera obscura were too faint to be copied (i.e., recorded) using his sensitive paper.

An account of Wedgewood's work with a commentary by Humphrey Davy was published in 1802 in a paper entitled "An account of a method of copying paintings on glass and of making profiles by the agency of light upon nitrate of silver, invented by T. Wedgewood, Esq., with observations by H. Davy."[9] Davy and Wedgewood met at Dr. Thomas Beddoes's Medical Pneumatic Institution in Bristol, where Wedgewood had gone for medical help and Davy was the superintendent. This institution was devoted to the investigation of the medicinal effects of various gasses. (The most famous work done at Beddoes's Institution was on the effects of breathing nitrous oxide.) Davy repeated some of the Wedgewood experiments but did not carry them further.

This work was not continued for several reasons. Although a young man, Wedgewood was sickly, and his health was a constant obstacle to any work he undertook. He died in 1805, shortly after the publication of this paper. Davy, an even younger man than Wedgewood, was consumed by other interests. Davy had left Beddoes's in 1801 for a position at the newly established Royal Institution in London, and he carried on an active life lecturing and researching many topics. In addition, he maintained an active involvement with various scientific societies and institutions in England and abroad. Perhaps the most important obstacle to the continuance of Wedgewood's work was that it had produced images that faded, and therefore its results were not very promising. The knowledge of a substance to stop the action of light was lacking, and further work was precluded until such a substance was discovered.

The Wedgewood-Davy work did not spark any new investigations along this line. The circulation of the *Journal of the Royal Institution* was limited to members of the Royal Institution, and the journal ceased publication only a few years after it had begun. Before the announcement of the discovery of photography several people tried to use sunlight to make silhouettes like Wedgewood's without discovering the photographic process. For instance, Samuel F. B. Morse (1791–1872) reported that he had experimented using silver nitrate paper in a camera obscura as a student at Yale in 1812.[10] While the idea of capturing the images produced in a camera seemed to be an intriguing question, attempts to do so did not produce promising or encouraging results. The fact remains that many people got very close to what we now think of as photography but did not or could not push the idea to fruition.

What experimental work produced results worth pursuing? What was it about the successful experimenters that led them to chase the idea of producing permanent images with a camera when, for most, this seemed a problem with too many obstacles? Did the successful ones set out to devise the process of photography? What problems were they looking to solve, and how did these affect their results?

Nicéphore Niépce

The first father of photography was Nicéphore Niépce (1765–1833) (fig. 3.1). Born in Chalon-sur-Saône, France, to a wealthy bourgeois family, he was trained for the priesthood, but during the French Revolution religious orders were dispersed. In addition, the Niépce family had to leave Chalon for Nice to avoid possible persecution because of the father's former position as a lawyer and counselor to the king. In 1792 it was decreed that all able-bodied men of a certain age must enter the military service, and accordingly Niépce joined the army. His military service was cut short because of a bout with typhoid. In 1794 he returned to Nice, where he married and became a petit bureaucrat. In 1801 Niépce and his family returned to the family estate in Gras outside of Chalon-sur-Saône. There they were joined by his older brother, Claude.

The two brothers occupied themselves with inventions. Their aim was to rebuild their family fortune by exploiting one of their discoveries. Their outlook in taking up the profession of invention was a mod-

ern one; however, they never seemed to rise above a certain nostalgia for their family's former position. Their primary invention effort was the *pyréolophore*, a boat with an internal combustion engine that was patented in 1807. About the same time they made a public demonstration sailing the *pyréolophore* up the Saône. Their efforts to market and sell their invention lasted for more than twenty years and took Claude to Paris in 1816 and then to England in 1817. They also worked on and in 1811 received recognition from the French government for new methods for the cultivation and extraction of dye from the woad plant. Indigo had become scarce and expensive as a result of the continental blockade, while the demand for blue dye for military uniforms did not diminish. In 1817 Nicéphore also invented a machine that was a forerunner of the bicycle.

Fig. 3.1. Portrait of Joseph Nicéphore Niépce, produced about 1800 by C. Languiche. Courtesy of the Gernsheim Collection, Harry Ransom Humanities Research Center, The University of Texas at Austin.

In 1813 the two brothers became interested in lithography. The lithographic process was first patented in France in 1802 and initially was primarily used for printing music. Its spread was directly related to French contacts made in Germany during Napoleon's occupation of Bavaria. As a result of the occupation, lithography aroused more interest in France than it did in England; however, the French political situation hindered the practice of lithography until after the restoration of Louis XVII to the throne in 1815. After that time French lithography came into its own. It became so popular that by the end of 1817 the government, alarmed by the spread of the new technology and the relative ease of making broadsides using the new technique, placed restrictive regulations on the process for reasons of state security. Despite these restrictions the French were the leading practitioners of the art by 1820.

Another great encouragement for lithography in France and one incentive for the Niépce brothers was the institution of premiums by the Société d'Encouragement des Arts et Métiers for improvements to the lithographic process. Premiums were consistently offered for devising better methods for transferring designs to lithographic stones, finding a suitable substitute for the German limestone traditionally used as lithographic stones, and producing better materials to ink lithographic stones. These and similar premiums were offered yearly from the time of the introduction of lithography in France until well into the 1840s. They carried prizes ranging from 1,000 to 3,000 francs, depending on the year and the objective. The English Society of Arts also offered similar premiums throughout the 1820s and 1830s. The English premiums carried prizes ranging from 20 to 50 pounds.

Nicéphore and Claude Niépce took up the work on the lithographic process with the same inventiveness that characterized their previous work. No record exists of their joint work on the process before Claude left for Paris and England to promote the *pyréolophore*. Precisely because of Claude's departure, there is some documentation of Nicéphore's work from late 1818 through the end of 1825, for the two brothers carried on a careful correspondence. Aside from those few letters, however, there are no records about the progress of the work. Part of the motivation for their work in this area was that lithography had been included in the yearly lists of premiums offered by the Société d'Encouragement. Besides, lithography was and is a very interesting process and at the time promised to be a lucrative business for those who could master and improve the process.

During the course of his research on the lithographic process, Nicéphore Niépce wedded the two objectives of improving both lithography and the methods of transferring designs to stone. For several years he worked sporadically on various aspects of lithography. In September 1816 he sent a letter to the Société d'Encouragement with local stone samples suitable for lithography found in the region of Chalon-sur-Saône. Much to his pleasure, these stones were judged by the Société to be of sufficient quality to use for lithography.

At the same time Nicéphore successfully experimented with capturing images using a camera obscura and silver chloride papers. However, he did not find a way to stop the action of the light, nor did he overcome a second serious problem—the images he produced were negative mirror images, reversed in both tone and geometry, of the scenes captured by his camera.

Since his experiments on paper had not produced promising results, Nicéphore abandoned that avenue of investigation and turned to producing images on stone, metal plates, and glass. He tried using various resin and varnish concoctions—such as gum guaiacum, which turned color and had altered solubilities when exposed to the sun—to produce these images. He also looked for materials that would bleach rather than darken in the sun, so that the image created by the camera could be viewed as a positive. Further, if a method that was more along the lines of printing could be devised, it would not matter that the images recorded by the camera were tonally reversed because they would be reversed again in the printing process. It would also be possible to obtain multiple copies from an original.

Nicéphore had made the fundamental leap—he had come to understand that these plates could be objects in and of themselves, not just intermediaries to something else. He was successful in copying engravings onto light-sensitive plates. These engravings were oiled to make them transparent and then were laid on lithographic stones or glass plates coated with his various concoctions of light-sensitive varnishes. After sufficient exposure the plates were treated with various solvents and the unexposed portions of the varnishes washed away. These could then be etched in acid and used for printing. In 1824 Nicéphore wrote to Claude that he had been successful in obtaining direct images on stone and glass that then could be etched and used for printing. He spoke too soon about direct images, but success in this area did not elude him for long.

In 1825 two events occurred that altered the course of Nicéphore's work. First, he established a working relationship with the Parisian lithographer A. F. Lémaitre. For the next several years Lémaitre received plates from Niépce, which he would etch and print. Niépce took Lémaitre into his confidence and relied on him for technical and business advice. Second, that same year a relative of Niépce was in Paris and went to the opticians Charles and Vincent Chevalier to purchase a good camera to take back

Fig. 3.2. The first photograph from nature, a view from the courtyard taken from Niépce's workroom window at his estate, Gras, near Chalon-sur-Saône, taken in 1826 or 1827. This was made on a pewter plate using heliography. Courtesy of the Gernsheim Collection, Harry Ransom Humanities Research Center, The University of Texas at Austin.

to Gras. The relative mentioned that Nicéphore had had some success in producing permanent images using a camera and showed the Chevaliers examples of his work. They were quite impressed and mentioned this work to another customer, Louis Jacques Mandé Daguerre, who was also interested in recording images made using a camera. Almost as soon as Daguerre heard of Nicéphore's work, he resolved to make contact with him and sent a letter in January 1826 asking to hear more about it.

About that same time Nicéphore began using pewter plates for his experiments. Pewter has a lighter and more neutral color than the copper and zinc plates he had been using and allowed him to attain a greater range of tones. In 1826 or 1827 he made the first successful direct camera view of nature: an image of his garden taken from his window (fig. 3.2). He called his process *heliography* and the direct camera images *points de vue*. This image was made on a pewter plate using a light-sensitive substance called asphaltum or bitumen of Judea. Bitumen of Judea is a tarlike substance that has been used since Egyptian times both as a building material and as a black varnish material. It has the peculiar property of softening when heated but hardening when exposed to light. The Niépce brothers had discussed using this material as part of a possible fuel mixture for the *pyréolophore*. Its black color and the fact that it hardened when exposed to light made it an obvious choice for Nicéphore's experiments.

Nicéphore received the letter from Daguerre with some suspicion and wrote to his son, Isidore, that "one of these Parisians wants to pump me for information. . . ."[11] He did not reply to Daguerre's letter. More than a year later Daguerre sent a second letter to Nicéphore, who this time sent a second letter to Lemaitre asking who this Louis Jacques Mandé Daguerre was and whether he knew him personally. Poor Nicéphore seemed quite taken aback by Daguerre's second letter and described him to Lémaitre as a person who appeared to be incoherent.[12] A third letter from Daguerre followed close on the heels of the second,

and this time Daguerre enclosed an example of a design done by "his" process. Daguerre's sample was a *déssin fumé* that had been reworked considerably, so it was difficult to tell what Daguerre had done. This time Nicéphore decided to send Daguerre an example of one of his heliographs on pewter and a print pulled from the plate.

Louis Jacques Mandé Daguerre

The second father of photography is Daguerre (1789–1851), who was born in Cormeille-en-Parisis, in northwest France (fig. 3.3). His father was a royalist and a minor clerk at the royal estates in Orléans. Daguerre received his elementary education at the École Publique in Orléans, but because of the upheaval caused by the Revolution his education was spotty. He had a talent for drawing, and at the age of thirteen he was apprenticed as a draftsman to an architect in Orléans. In 1804 he went to Paris to study art and was apprenticed to and lived with Ignace Eugène Marie Dégotti, the stage designer at l'Opéra. Daguerre was a flamboyant character, and during his tenure with Dégotti he became known as a dancer and even was an extra at l'Opéra. After three years Daguerre became an assistant to the panorama painter Pierre Prévost and worked with him until 1816; he also worked as a free-lance designer for stage sets in various Paris theaters. He was asked, along with hundreds of other artists, to contribute topological drawings for "Voyages Pittoresques," the massive French lithographic project. This series of lithographs was issued yearly for several decades in part to promote French lithography. While working for Prévost Daguerre married Louise Georgina Smith, the French-born daughter of English parents, thus gaining English connections.

Panoramas—large paintings depicting both fictitious and real events—were popular entertainments during the last portion of the eighteenth century and for the first

Fig. 3.3. Daguerreotype portrait of Louis Jacques Mandé Daguerre, taken in 1844 by E. Thiesson. Courtesy of the Musée Carnavalet, Paris.

half of the nineteenth century. At Prévost's Daguerre met Charles Marie Bouton, a long-time collaborator of Prévost and a former student of the painter Jacques Louis David. Daguerre was an outstanding draftsman and was known for his skill in theatrical effects, but his skill in painting did not match his other talents. Bouton and Daguerre invented and patented a wildly popular new entertainment called the Diorama.

The Diorama was a picture show somewhat like a panorama, except that special lighting effects and real props in the foreground gave the illusions of movement and reality. Large trompe l'oeil paintings (approximately 14 meters high by 21 meters wide) were painted on both sides of a transparent screen with opaque paints. By controlling the direction of the lighting from the front or the back of the painting, these scenes appeared to change; for instance, the scene could change from night to day. These shows had names like "A Midnight Mass at St. Étienne-du-Mont," "The Interior of Trinity Chapel, Canterbury Cathedral," "The Valley of Sarnan," "Cathedral of Chartres," "The Inauguration of the Temple of Solomon," and "View of Brest Harbor." Some of these dioramas were painted by Daguerre and some by Bouton. After the success of the Paris Diorama, another was built in London and others were licensed in Berlin, Liverpool, and in other cities. Paintings from the Diorama also toured the Continent and were even shown in the United States. Thus, the two men attained a great deal of publicity.

At some point Daguerre had become interested in permanently capturing the images made in the camera obscura. He used a camera as part of the routine of his work in making the large trompe l'oeil paintings for the Diorama. Thus, he was quite excited to hear from the Chevaliers that someone else was also pursuing the same line of investigation. It is not known how much progress Daguerre had made on his own

before he heard of Nicéphore Niépce, except that he had been experimenting with silver chloride paper and phosphorus for perhaps a year.

In September 1827 Niépce made a trip to England to see Claude, who was very ill and probably dying. The trip brought grave disappointment. Claude's ill health alone must have been devastating because the two brothers were the closest of friends and confidants. Nicéphore also found that Claude had made no progress in promoting the *pyréolophore*. For more than thirty years the family fortune and the family debt had been devoted to promoting this machine, and during this trip it became clear that there was to be no return on these investments—the family fortune was lost.

While in England Nicéphore made the acquaintance of Francis Bauer, a prominent botanical illustrator at the Royal Botanical Gardens at Kew and a fellow of the Royal Society. He showed Bauer some examples of heliography, including the view taken from the window at Gras. Bauer encouraged Nicéphore to prepare a memoir about heliography to be presented at the Royal Society. This memoir was never given because Niépce did not want to divulge his process publicly without making some financial arrangements. His description of the workings of the process were intentionally vague to prevent anyone from stealing his ideas. This secretiveness, however, was counter to the Royal Society's rules of scientific disclosure, and thus his memoir could not be presented. Niépce also tried to gain the royal patronage of George IV, but this plan failed also. He then approached the Society of Arts, hoping to find some interest in his process there, but to no avail. This lack of interest was very disappointing because Niépce thought that he had a better chance of exploiting his new process in England than in France. He returned home, leaving with Bauer his memoir along with some examples of heliography, including the view of the garden, the first successful photograph.[13]

Nicéphore stopped in Paris on his way home and had several meetings with Daguerre. Claude died shortly after Nicéphore returned to France, and much of the next several months was taken up with settling his brother's affairs. Once he did return to his work, he ceased copying engravings and devoted his time to perfecting the method of taking views using a Wollaston periscopic camera obscura, an improved camera obscura he had purchased in England.

Niépce began to use polished, silver-plated copper plates both because silver was a whiter metal than pewter and thus would increase the number of tones in his heliographs and because the copper made a more suitable plate for printing. Gradually, he made improvements in the process and received encouragement from both Lémaitre and Daguerre. He discovered that he could make another kind of image on silver plate by treating a bitumen of Judea image with iodine vapor to blacken the uncovered areas of the plate. After removing the bitumen he would be left with a positive silver iodide image. Niépce also recognized the need for a camera that was adapted to his purpose and had more light-gathering power than that needed in a camera used principally to make drawings. Thus, late in 1829 he suggested to Lémaitre that an improved camera would be necessary in order to have a process that could be considered practical. He decided once again to try to publish the process, and Daguerre suggested that there might be a way to exploit the process of heliography for money before making the details of it public. That December Niépce and Daguerre formed a partnership to promote and develop Niépce's process.

Niépce had come a long way—from making improvements to the lithographic process to using the sun to make permanent images not intended to be intermediaries for some printing process. He also had turned

outside his immediate family circle to exploit the process by joining Daguerre, who was so different from him that it must have been somewhat overwhelming. Niépce must have also been disappointed that he had received so little recognition for this work and had not been able to sell it himself, but he recognized that Daguerre did have great skill as a promoter and that he could be a great help in bringing to this work the recognition it deserved. Daguerre embraced Niépce's work with all the enthusiasm and passion characteristic of his flamboyant personality.

Niépce prepared a complete description of the process for Daguerre. Using Niépce's work as a foundation, Daguerre pushed forward and developed what is now known as the daguerreotype process. He discovered that silver iodide (made by fuming the silver plates with iodine as Niépce had done) was sensitive to light and that it alone could be used to make images. He also discovered that if an exposed plate with no apparent image was exposed to mercury vapor an image appeared. He thus discovered the latent image. How he decided to use mercury is something that will probably never be known. One romantic story is that he left a plate in his chemical cabinet and returned to find that an image had appeared on the plate. He then removed the bottles one by one until nothing was left in the cabinet, yet the image still appeared on the plate. He looked more carefully, discovered spilled mercury, and correctly attributed the appearance of the visible images to that metal.

Unfortunately, Nicéphore never got to see Daguerre's results. He died in 1833, and Daguerre came upon these improvements in 1835. At Nicéphore's death his son, Isidore, became Daguerre's partner, and the contract between them was modified in 1835 to reflect Daguerre's discoveries since Nicéphore's death. At this point Daguerre still could not produce permanent images, but in 1837 he finally discovered that he could stop the action of light by putting his plates in hot, saturated salt water.

This new process clearly was not the same as Niépce's original process, and Daguerre felt that it should be given the name *daguerreotype* to reflect the difference from the process of heliography. Isidore was reluctant to go along with this last modification of the original contract, but he acquiesced when he realized that Daguerre's new process probably had more commercial application than the slower process of heliography. This last change occurred in 1837, and in 1838 the partners set out to sell the two processes. From March to August they offered the processes by public subscription to four hundred subscribers who would put up 1,000 francs each. One provision was that the two processes would not be revealed if there were fewer than one hundred subscribers. Another provision was that the processes would be sold outright to anyone who paid at least 200,000 francs. Even though Daguerre went around Paris transporting his camera, making daguerreotypes, and attracting all kinds of publicity, neither subscribers nor a buyer for the processes emerged. So Daguerre approached various scientists in hopes of getting one of them to champion his cause with the French government.

Why weren't people interested in buying the processes? Evidently, many people could not imagine what these images were good for. The two partners were fortunate to catch the attentions of François Dominique Arago (1786–1853), then not only the director of l'Observatoire in Paris and the permanent secretary of the Académie des Sciences but also a member of the Chamber of Deputies in the French Parliament (fig. 3.4).

As mentioned before, Arago was interested in light and its properties, so he had an immediate, personal interest in these processes. He felt that the daguerreotype in

Fig. 3.4. Portrait of Dominique François Jean Arago. Source: The frontispiece to Arago's *Oeuvres Complètes de François Arago, Tables.*

particular could be used as a tool to unlock some of the mysteries of light and that it also would provide an invaluable aide to scientists as a recording medium. But aside from his personal interest, Arago had the vision to see that this process would have an impact far beyond the mere fact of being able to record images. He set about to see that Daguerre and Niépce were rewarded and that this discovery brought honor and glory to France. In January 1839 a brief notice about the daguerreotype process was published in *Comptes Rendus*, the journal of the Académie des Sciences. It stated merely that a process had been devised and that details would be forthcoming. It also mentioned that a striking view of the Notre Dame cathedral had been made using the process (fig. 3.5).

This notice sparked a great deal of public interest and brought about two particular responses. First, Francis Bauer sent a letter to *The Literary Gazette* describing Nicéphore Niépce's work and offering to show the heliographs that had been given to him

Fig. 3.5. Daguerreotype view of Ile de la Cité and Notre Dame, Paris, by L. J. M. Daguerre, thought to be one of the daguerreotypes mentioned by Arago at Académie des Sciences on January 7, 1839. Courtesy of the Gernsheim Collection, Harry Ransom Humanities Research Center, The University of Texas at Austin.

eleven years before. Bauer wanted to make sure that Niépce was not slighted or his contribution forgotten. Second, William Henry Fox Talbot rushed to publish an account of his process for making images with light in order to establish that he was the first discoverer of such a process.

William Henry Fox Talbot

Talbot (1800–77), the third father of photography, was born and raised at his grandfather's estate, Melbury, in Dorsetshire, England (fig. 3.6). A gentleman and a scientist, he was educated at Harrow and at Trinity College, Cambridge, where he received prizes for his work in mathematics and Greek. In 1826 Talbot moved to the Talbot family estate, Lacock Abbey, in Wiltshire. As mentioned earlier, he immersed himself in studies of the spectrum and also mathematics. For his activities in these areas he was made a fellow of the Royal Society in 1831. He was briefly involved in politics and represented Chippenham in Parliament during the beginnings of the reform legislation passed during the 1830s.

While traveling to Lake Como in 1833 with his wife, Talbot was disappointed that his inadequate drawing skills hampered his ability to record images using his camera lucida. He resolved to attempt to produce images that would be recorded by light itself and thus would be perfect replications of what he had seen in his camera. He decided to try this using silver nitrate, but papers prepared with silver nitrate were not very sensitive to light, and he soon switched to silver chloride paper. After some experimentation he, too, discovered that he could stop the action of light on his sensitive papers by soaking them in a strong solution of salt water.[14] During the summer of 1835 Talbot succeeded in making images of his house using his sensitive paper and a camera obscura. The exposure times for these initial images were on the order of several hours. He also tried using a solar microscope in conjunction with his sensitive paper and made the first photomicrographs sometime during this period.

Talbot chose not to promulgate his success at that time, possibly because he was not satisfied with the results and wanted to work further and possibly because he was quite busy with other things. During the summer of 1836 many scientific luminaries stopped at Lacock Abbey on their way to the second meeting of the British Association for the Advancement of Science in Bath, and Talbot showed his results to David Brewster at least, if not to others. He still did nothing to make his results public. However, when he heard of the French announcement in January 1839, Talbot tried to establish his priority to the discovery. He rushed to publish an account of his process and also sent letters to Arago and J. B. Biot, another prominent French scientist, claiming credit for the discovery and saying that a public disclosure was being prepared. An exhibit of Talbot's images was mounted in the library of the Royal Institution at the end of January.

The events of 1839 have been carefully documented. To those previously indifferent to the first images made using cameras, Arago's announcement took the world by storm. It forced the hand of those who had been working on light-sensitive imaging processes and also stimulated the support and interest needed to sustain those results. Arago orchestrated the events in France with such skill that, despite the great publicity surrounding the daguerreotype process, there were no public showings of the process until August. Between January, when the first announcement was made, and August Arago and his friend and colleague Joseph Louis Gay-Lussac, a member of the Chamber of Peers, carefully guided bills through the French Parliament. Arago recognized that Daguerre and Isidore Niépce should receive some monetary reward from the government as compensation for giving the process to the world. A striking part of his plan was to have the government reward Daguerre and Niépce by granting them pensions for life rather than patents, the usual form of reward. Because France and the rest of the world were in the midst of an economic depression, he knew that he could not expect the Parliament to make this monetary grant if the processes' only promise was that they were wonderful and would bring glory to France. Wonder and glory were not enough.

Arago was a visionary: he realized that these fledgling processes had not only intellectual promise but also, and more important, economic promise. Of the two processes from Daguerre and Niépce, the daguerreotype was the most promising, and as a result Arago pushed its cause with more fervor. It must have been apparent that his case would be strengthened if he presented only one process to the government with a request for support, rather than a multiplicity. He and Gay-Lussac promoted the daguerreotype process first, on the premise that this process would provide untold benefit if made public and supported by the government. They argued that if the discovery was treated in the usual manner—that is, granted a patent and left for private development—it would most likely not grow to match its promise, nor would its inventors ever receive proper remuneration for their great genius. This process held potential not only for artists and architects but also for scientists, archaeologists, and travelers. Arago reminded his colleagues of the number of artists that Napoleon took into Egypt and the Levant to record the archaeological materials they found there during the Egyptian campaigns. He claimed that in the future it would be possible for one person with a daguerreotype apparatus to record these types of data with greater accuracy and speed and less expense than heretofore possible.

The reports were persuasive and glowing. Implicit but not explicitly stated was

Fig. 3.6. Portrait of William Henry Fox Talbot, taken about 1864 by John Moffat. Source: The International Museum of Photography, George Eastman House, Rochester, N.Y. Courtesy of the Gernsheim Collection, Harry Ransom Humanities Research Center, The University of Texas at Austin.

the idea that France would benefit economically from supporting this legislation. Clearly, the need for the supplies and materials necessary to make daguerreotypes would encourage new economic growth for France. This tack was successful, and in July bills were passed awarding pensions of 6,000 francs per year to Daguerre and 4,000 francs to Isidore Niépce and half that amount to their widows at their death.

Meanwhile, in England the publicity surrounding the events in France mounted. Talbot had to concede that the discoveries of Niépce and Daguerre did indeed have priority over his own. Speaking as one scientist to another, Biot reassured Talbot that he could be the first to allow scientists to know of these discoveries for, unlike Daguerre, Talbot was free of constraints about releasing the information about his discovery. (Because of the contract with the Niépce family, Daguerre was not free to publish the details of his process until monetary arrangements had been made.) Biot clearly felt that Talbot's case was superior because Daguerre was merely an artist and not from the higher calling of science, as was Talbot. Repeatedly, Biot urged Talbot to make his process public for the benefit of physics and science. A year later Biot wrote again to Talbot lamenting, "It is unfortunate for Science to see a man [Daguerre] with such ability always considering the results from the artistic point of view, and never at all from the higher purpose of contributing to the progress of discovery in general."[15]

Talbot did publish an account of his methods, and other people began to do their own experimentation. In Talbot's case, however, the government failed to acknowledge the importance of this discovery and did not offer him any reimbursements. Therefore, Talbot patented his process, obtaining over the years a total of twelve patents on different aspects and variations of his method. Daguerre also obtained an En-

glish patent on his process just days before the public announcement of photography in France. Patents did protect each of these inventors to some extent, but in England the burden of enforcing the patent was on the patentee, and Talbot especially grew to be hated because of the patent disputes he brought to court.

The publicity surrounding the discoveries of Niépce, Daguerre, and Talbot also brought out others who had been thinking along the same lines. In May 1839 Mungo Ponton (1802–80), a member of the Edinburgh Society of Arts, published a process based on dichromate salts that he thought was a variant of Talbot's.[16] In France on Bastille Day Hippolyte Bayard (1801–87) exhibited examples of a direct positive photographic process that he had devised. Bayard, disappointed that the French government had chosen to back only one discovery, tried to embarrass the authorities and was successful in receiving a small amount of money given in order to quiet his counterclaims to Daguerre's process.

Dignitaries from various countries visited France to preview Daguerre's work. S. F. B. Morse visited Daguerre in March and sent back a rapturous report to his brother about the splendor of Daguerre's images. A group of scientists from England and Scotland, including James Watt and Sir John Herschel, visited Daguerre in May, a visit reported in *Comptes Rendus*. The committee found the results beyond their expectations. The French mathematician Augustin Louis Cauchy, who was present at the meeting, recounted that Herschel had exclaimed to him that the English samples were but child's play compared to those of Daguerre. Herschel said that Talbot himself would think the same and that he would write to him to come to Paris as soon as possible.[17] Herschel wrote to his friend Talbot: "It is hardly saying too much to call them [the daguerreotypes] miraculous. Certainly they surpass what I could have conceived as within the bounds of reasonable expectation. . . . if you have a few days at your disposition I cannot counsel you better than to *come and see*. Excuse my exultation. . . ."[18] With testimonies like this it is no wonder that the meeting rooms at l'Institut were crammed with people on August 19 when Arago gave his talk about the daguerreotype process and heliography.

Thus, in 1839 a new era began—the world was introduced to the possibility of using several processes to capture permanent images using a camera. These discoveries were not the result of a concerted scientific research program based on the systematic application of a knowledge of light-sensitive materials to the problem of imaging. Rather, they were the result of three men solving problems—one an inventor trying to improve lithography and competing for one of the premiums offered by the Société d'Encouragement des Arts; one an artist and promoter trying to fulfill the dream of capturing the images in his camera; and one a gentleman-scientist trying to make up for his deficiencies in drawing. These men all drew on knowledge that was widely available. They were able to apply that information in novel ways and devise solutions to obstacles that had prevented others from reaching the goal before. All three were able to produce images that had relatively short exposure times (i.e., on the order of minutes), and then they were able to make these images permanent by stopping the action of light with some subsequent treatment.

None of them was very successful at promoting the discoveries on his own. While in all three cases other people outside the discoverer's immediate circle were aware of his work, the application of the work as something more than a parlor trick or entertainment was not immediately apparent. Even the concerted effort to offer the discoveries of Niépce and Daguerre by sub-scription or sale was a failure. Clearly, Nicéphore Niépce and Daguerre understood the importance of their discoveries, but even Daguerre, with his experience in promotion and cultivation of audiences, was unable to present these discoveries in a way that predicted their future impact.

It was Arago, an outsider, who immediately understood the value of these discoveries and was able to envision applications for them in his own work and in the life of France and the world. He paved the way for these explorations to come to light. Arago recognized that these discoveries were of such importance that they should be singled out and that the discoverers be given special recognition and remuneration by the government. He took a gamble and won. How long it would have taken for the birth of photography had Arago not served as the midwife speeding the process on its way is not clear. Certainly, once the possibility of permanent imaging had been demonstrated, variations and new processes literally tumbled out of other experimenters who embraced this new art-science.

4

The Technological Practice of Daguerreotypy

The daguerreotype is going to be all the go here for a time and they have to have plated metal to get the Picture on. . . . Professor Morse called on me today he had . . . ordered 38 Plated 6½ x 8½ in for a trial . . . . Mr. Morse thought the quantity wanted would be great as soon as it was got in operation.

Letter from J. M. L. Scovill to W. H. Scovill, October 15, 1839[1]

THE EXCITEMENT GENERATED by the meeting at l'Institut on August 19, 1839, radiated from Paris. During his talk at the meeting, François Arago had enumerated the steps of the daguerreotype process. Giving directions, however, is easier than actually making a daguerreotype, and even in Paris people had such difficulty making daguerreotypes that Daguerre himself had to give public demonstrations to teach the process. With or without the master the process was reported and disseminated around the world as fast as the news could be carried. Often the process was spread with only the barest of instructions on how to produce these beautiful images. Enthusiasm probably helped sustain many of the earliest practitioners of the art because daguerreotypy requires exacting and careful manipulations, and the supplies needed to produce these wonders of science and art were hard to obtain.

As noted before, the initial daguerreotype process has five steps: A piece of silver-plated copper is cleaned and polished until it has a mirrorlike finish and appears black when viewed from an oblique angle. This prepared plate is put over iodine vapor in a closed box until the silver layer takes on a yellow-rose appearance. The iodized plate is then put in a camera and exposed to light for a time that varies with the season and the ambient conditions. After the camera exposure, the plate is put over hot mercury vapor until an image appears. In the last step unexposed silver iodide is removed by placing the plate in hot, concentrated salt water or in a solution of sodium thiosulfate (also known as hypo). This image is fragile and has to be packaged in some way to be protected from mechanical damage.

During the August meeting Arago also noted the few drawbacks to the process but told the crowd that these were only minor points that Daguerre would solve in a matter of weeks. The three major drawbacks were that the exposures were overly long for application of the process to portraiture, the image was not mechanically stable and so was liable to physical damage during handling, and the images were not colored. Also there was no easy way to make multiple copies of a daguerreotype image. Daguerre did not solve these problems but left them to those who embraced the new art. Various solutions were tried; some were successful, while others were not. The resolutions for the first two difficulties were the only actual changes made to the daguerreotype process. Solving the other problem was not as straightforward or successful.

Attempts to Reduce Exposure Times

Reducing exposure times was the first priority for those working on the new process. Exposures that varied from three to fifteen minutes were clearly beyond the bounds of reasonable expectation for practical portraiture or, for that matter, most other recording from life. The problem was approached from both mechanical and chemical points of view. The mechanical approach involved altering camera and lens design and also the use of elaborate lighting schemes so that more light reached the daguerreotype plate during the camera exposure. The chemical approach involved altering the light-sensitive coating on the daguerreotype plate so that the plate was more sensitive to light.

Mechanical Improvements

Clearly, the easiest solution to long exposures is simply to devise ways in which more light can reach the daguerreotype plate when it is in the camera. Before the advent of photography cameras and lenses merely had to admit sufficient light so that the person using the camera could see the image projected on the camera's viewing screen. This approach was no longer sufficient for the practice of photography. Photography placed different requirements on lenses than had been necessary when they were used merely for viewing images through a microscope, telescope, camera obscura, or even spectacles. A lens suitable for photography has to cover a large, flat field without distortion or loss of sharpness; at the same time it must have a medium-to-large aperture. Niépce and Daguerre both were limited by the lenses available and the stringent requirements for their experiments. They used lenses produced by the optician Charles Chevalier, and Niépce also purchased a Wollaston lens and prism while he was in England. Chevalier was commissioned to produce the lenses used in the first daguerreotype cameras manufactured by Giroux. These lenses had small apertures and ultimately were not sufficient for producing photographs with reasonable exposure times.

The first breakthrough for photographic lenses was the introduction of the Petzval portrait lens designed by Joseph Max Petzval (1807–91), professor of mathematics at the University of Vienna. One of his colleagues, A. F. von Ettingshausen, had been in Paris for the announcement of the daguerreotype, and after returning to Vienna he enlisted Petzval to design a lens that could be used for portraiture. Petzval made the calculations for a lens in about six months with the aid of several soldier-calculators from the Engineering Corps. The portrait lens designed in this way cut exposure times dramatically. Petzval also designed a landscape lens. These designs were given to the optician Peter Friedrich Voigtländer to manufacture. No arrangements were made to protect Petzval's design, and it was widely pirated. Voigtländer mounted the Petzval lens on a barrel-shaped camera made for producing daguerreotypes called the Voigtländer camera (fig. 4.1). The camera itself was never very popular, but the

Petzval portrait lens was widely used, became very popular, and remained so until well into this century.

Experimenters in the United States took another approach to this problem. Alexander S. Wolcott (1804–44), an instrument maker in New York City, and his partner, John Johnson, became interested in the new process. Wolcott was well versed in optics and set about to design a camera very different from the Giroux camera used by Daguerre. The Wolcott camera used a large concave mirror placed at the back of the camera box to focus the light back onto the daguerreotype plate, which was placed on an adjustable rail at the center of the camera. This camera used no lens, and the image formed was not reversed as in more conventional cameras. Daguerreotypes produced in this camera were small; despite this fact, Wolcott and Johnson were able to open up the first American daguerreotype portrait studio in the spring of 1840. In 1841 Wolcott and Johnson patented their camera design, the first patent granted for photographic apparatus in the United States.

In addition to designing a new camera, Wolcott and Johnson also used an elaborate lighting system to help get as much light on their subjects as possible. They used a system of adjustable mirrors outside the studio to concentrate and direct light into the studio. If all the light directed into the studio were concentrated on the sitter, the heat would have been unbearable. So the sitter was shielded by a blue filter made of a plate glass trough filled with a copper sulfate (i.e., blue vitriol) solution. An iodized daguerreotype is primarily sensitive to blue light; the filter helped shorten exposures because the daguerreotype was sensitive to all the light reflected from the sitter into the camera. In his original manual on the daguerreotype process Daguerre himself had suggested the use of blue glazing for studios. There were also claims that yellow light would interfere with image formation.[2] Many studios adopted the lighting

schemes used by Wolcott and Johnson, and blue glass became a standard feature in some photographic studios.

The Société d'Encouragement in Paris offered a premium for improvements in photographic lenses. Chevalier won the platinum prize for a double lens with a variable focal length. Voigtländer entered his Petzval portrait lens and won the silver prize. The Chevalier lens never enjoyed the success of the unpatented Petzval lens, which was widely copied and sold as the "German system" lens. Other improvements to lens systems during the daguerreian era included mirror and prism systems used to reverse the image in the camera so that it would appear as seen in nature, improved diaphragms for increasing depth of field and sharpness in an image, and improvements in the achromatic lenses. "Actinically correct" lenses were finally developed to overcome the problem of differing optical and "chemical" focus or the difference in the focal points of blue light and white light. This problem was not as serious as it might seem from the modern point of view because all daguerreian operators took the time to determine the specific focusing corrections needed for their own cameras as a matter of routine, and the methods for making these corrections were published in most photographic manuals.[3]

Chemical Alterations

The chemical alteration of exposure times was more tricky and, ultimately, more important. It remains a field of intense investigation. Today, various combinations of silver halides are used in photography as a matter of course. From the modern point of view it would seem that the use of various silver halides would be an obvious first choice if one was trying to increase the light sensitivity of a photographic material. The original daguerreotype process devised by Daguerre used only silver iodide, the least sensitive of the silver halides.

It is apparent from contemporary literature that experimenters recognized almost immediately that other silver salts should be tried in the daguerreotype process, but this was easier said than done. The specific light sensitivity of the different silver halides was unknown in 1839; indeed, silver iodide and silver bromide were relatively recent discoveries at that time. Further, daguerreotype processing precluded the use of salt solutions to precipitate silver halides from solution. The daguerreotype process is based on bulk silver metal, which is not reactive in salt solutions unless the silver is dissolved. In order to react with the silver in the daguerreotype plate, any halogen treatment has to be used in a vapor state; consequently, only the elemental halogens themselves or certain of their salts that are volatile crystals or liquids at room temperature are suitable for use in the daguerreotype process. Most of these were not readily available in the early 1840s.

Niépce first introduced iodine into the process. He chose iodine as a blackening agent for his heliographs because it was known that this newly discovered element would combine with silver to form a dark compound. At the time he first used iodine, he was still using bitumen of Judea as the light-sensitive material in his process, so it would appear that he did not pick iodine with the hope of producing a light-sensitive silver salt. In the version of heliography that used iodine, Niépce would spread bitumen of Judea over a polished silver plate and expose the plate to light. After that, he dissolved the unexposed portions of the bitumen of Judea image away using oil of lavender. He then exposed the plate to iodine vapor until it was blackened, and, finally, he dissolved the remaining portions of bitumen of Judea from the plate, leaving a positive image of black silver iodide (or, more probably, silver) on a white silver-metal substrate. Daguerre discovered that the iodized silver plate could be used as the light-sensitive medium omitting the bitumen of Judea.

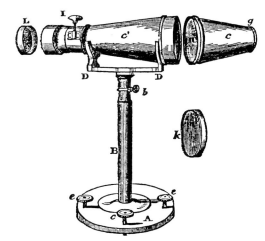

Fig. 4.1. A Voigtländer, or a German camera, showing its mounting pedestal (*A*, *B*, and *D*) with leveling screws (*e*). The lens cap (*L*) indicates the lens end of the camera body (*C*, *C'*). The screw (*I*) is used for focusing the images at the ground glass (*H*), which is also where the daguerreotype is placed during exposure. There is also a small lens (*g*). This camera produced round daguerreotypes. Fisher, *Photogenic Manipulation*, p. 33.

Although in 1839 it was known that silver salts in general were light sensitive, all the halogens were relatively recent discoveries. Chlorine, the oldest of them all, was first discovered and separated in 1774 by Karl Wilhelm Scheele, but it was not until 1811 that chlorine was proved by Humphrey Davy to be an element and given the name it now holds. Iodine was discovered by Bernhard Courtois in 1812, when he was investigating the various products obtained from the mother liquors of seaweed. Gay-Lussac showed that it was an element in 1815. The elemental nature of bromine was discovered and named by Antoine Ballard, who first isolated it from Mediterranean sea salt in 1826.

That all these halogens form light-sensitive compounds with silver was discovered very early on in the investigations of their properties. Of the three known halogens, chlorine had fairly widespread use as a bleaching agent. At the time the daguerreotype process was introduced, neither iodine nor bromine had any practical use aside from a few experimental medicinal uses. None of these halogens is found in elemental form in nature, and they are difficult to handle. Chlorine is a corrosive gas. Bromine is a vile red liquid that gives off an equally vile gas at room temperature. Iodine is not much better, but it is crystalline at room temperature and therefore is easier to handle than either bromine or chlorine. None of these halogens could be easily obtained either by purchasing them or by actually extracting them from other compounds. The addition of other halogens to the routine practice of daguerreotypy, an obvious next step, presented obstacles that needed to be overcome.

Bromine was the first halogen to be added to the daguerreotype process. This step was accomplished by a particularly important early group working on the daguerreotype process in Philadelphia. The daguerreotype took Philadelphia by storm, as it did most other places. The difference was that the Philadelphians were quietly working and making great strides with the process without the fuss and publicity associated with workers in other American cities. The Philadelphians working on the daguerreotype process—Joseph Saxton, Dr. Paul Beck Goddard, Robert Cornelius, Martin Hans Boye, Dr. Walter Rogers

Johnson, among others—were loosely connected through their common interests in scientific pursuits. Their skill at the new art was unsurpassed anywhere or by any other workers at the time they were active—roughly the early fall of 1839 through 1842.

Joseph Saxton (1799–1873) was a mechanic and instrument maker employed at the United States Mint (fig. 4.2). Around September 25, 1839, the day Daguerre's process was first published in Philadelphia,[4] he produced a view of the Central High School, one of the first daguerreotypes made in America (fig. 4.3). And sometime in October or November 1839 Dr. Walter Rogers Johnson (1794–1852), who at the time held the chair of chemistry and natural philosophy at the Pennsylvania Medical College, made a portrait of Dr. Ezra Otis Kendall. Little is known about the work of this important pioneer of photography except that he performed systematic experiments with the process and worked on methods for portraiture. He lectured on and demonstrated the process at several public meetings in the chemical lecture room of the medical department of the Pennsylvania College at the end of 1839 and the first few months of 1840. These lectures were also accompanied by examples of the daguerreotypes made by those already working in Philadelphia.

Early in February 1840 the Philadelphia papers published a notice commenting on the first public lecture in New York City by François Fauvel Gouraud, Daguerre's appointed representative sent to teach the art in the United States, and claiming that Gouraud had produced the first perfect daguerreotypes taken in America. The notice reported that many perfect specimens of the art had already been made in Philadelphia and credited them to Dr. Johnson. Johnson declined to take all the credit for Philadelphia's excellence in this line, saying that credit was also due to Joseph Saxton, Robert Cornelius, Dr. Paul Beck Goddard, and

Fig. 4.2. Portrait of Joseph Saxton taken in 1872 by Mathew Brady. Courtesy of the National Portrait Gallery, Smithsonian Institution, Washington, D.C.

James Swaim. (It should be noted that Gouraud visited Philadelphia for one night and never returned. He remained in New York City and then moved on to Boston teaching the art. Perhaps the Philadelphians did not need his instructions.)

Robert Cornelius (1809–93) was involved in his father's business of manufacturing plated goods, especially brass lamps, chandeliers, and candlesticks. He had been educated at private schools and had also studied chemistry with the Dutch-born mineralogist Gerard Troost. At the time the daguerreotype was introduced, the economic depression that plagued most of the civilized world had also slowed work at the Cornelius manufactory, and the thirty-year-old Robert took up this new art. A practical metallurgist in the modern sense of that term, he was skilled in the process of plating silver and other metals by cold-roll cladding and in the intricacies of polishing and finishing metal goods. It is likely that Cornelius supplied the early workers in Philadelphia with the silver-plated copper sheet needed for daguerreotype plates. Around October 1839 he began experimenting with the daguerreotype process with the initial assistance of Joseph Saxton. Sometime in October or December he made an extraordinary self-portrait (fig. 4.4). Unlike any other example of daguerreotype portraits taken at that time, Cornelius's self-portrait is not stiff and stilted.

Cornelius and Dr. Paul Beck Goddard (fig. 4.5) began working together on the daguerreotype process by early December 1839. At the time Goddard (1811–66) was an assistant to Dr. Robert Hare, professor of chemistry at the University of Pennsylvania. Applying his chemical knowledge to the practice of daguerreotypy, he and Cornelius evidently began to experiment with the use of bromine in the process that December. They began taking portraits during the late fall of 1839 but not as a commercial

Fig. 4.3. View of the Central High School, Philadelphia, taken by Joseph Saxton on September 25, 1839. This is the oldest extant American daguerreotype. Courtesy of the Historical Society of Pennsylvania, Philadelphia.

Fig. 4.4. Self-portrait of Robert Cornelius taken in November or December 1839. Source: Private collection.

Fig. 4.5. Copy photograph of a daguerreotype of Paul Beck Goddard taken in December 1839. Julius F. Sachse made this copy and published it in the *American Journal of Photography* in 1893. The original daguerreotype is lost. Source: Private collection.

venture. The Cornelius-Goddard experiments led to decreased exposure times and made possible the opening of Cornelius's daguerreotype studio, the first daguerreotype portrait studio in Philadelphia and the second one in the United States, in the late spring of 1840. While Goddard and Cornelius made no formal announcement about the adoption of bromine in their daily practice, Cornelius clearly recognized its obvious commercial advantage because it "prompted him with Dr. Goddard to purchase all the bromine in the eastern cities as they did [begin taking portraits]; and to keep their process a secret."[5] In addition to using bromine, Cornelius also used a large diameter lens and a lighting system of mirrors on pivots to concentrate light on the sitter, easing the glare by placing a large, purple glass disk from Dr. Hare's laboratory above the sitter.[6] This lighting system was similar to that used by Wolcott and Johnson in New York, and Cornelius adopted the method after visiting their studio in the spring of 1840.

The credit for the first use of bromine has most often gone to another Goddard—John

Frederick Goddard (1795–1866), an English lecturer in optics and natural philosophy hired by Richard Beard to improve the daguerreotype process.[7] Beard owned the English patent rights on the daguerreotype process, which he had purchased from Daguerre. In the December 12, 1840, issue of the *Literary Gazette*, Goddard published a note entitled "Valuable Improvements in Daguerreotype." He referred to recent experiments with bromine that increased the sensitivity of his plates, but he gave no details on his method.[8] He offered no subsequent publications about bromine or its use in daguerreotypy.

About the same time that Goddard's report appeared a few notices were published in Vienna about the use of bromine and chlorine. The first, a letter from Franz Kratochwila, a government employee, described his use of chlorine and bromine since September 1840 to shorten the exposures for daguerreotypes.[9] Kratochwila also used a Petzval lens with a Voigtländer camera. A few months later Josef Berres, an anatomy professor at the University of Vienna, published a letter claiming that the Natterer brothers could record "living streets" by using iodine and chlorine sensitization when making their daguerreotypes.[10] This is an obvious reference to the earliest daguerreotype street scenes, which were eerily devoid of people or other signs of life.

Another claimant to the discovery of accelerated exposures was Antoine Claudet (1797–1867). Claudet, originally from France, had opened a portrait studio, the Adelaide Gallery, in London in June 1841. In May 1841 he presented a paper to the Royal Society entitled "New Mode of Preparation of the Daguerreotype Plate, by which a portrait can be taken in the short space of time of from five to fifteen seconds, according to the power of light, discovered by A. Claudet in the beginning of May 1841." He described a multiple-step sensitization process that used chlorine in addition to the usual iodine. Claudet had also used bromine but preferred the effect of chlorine and the avoidance of the horrible stench of bromine. The same report was sent to l'Académie des Sciences in a letter from Nicholas-Marie Paymal Lerebours, an optician and early practitioner of daguerreotypy, to Arago in late June 1841.[11]

Others reported similar procedures for multiply sensitized daguerreotypes in 1841. F. A. P. Barnard,[12] a professor of mathematics and natural philosophy at the University of Alabama, reported a method similar to Claudet's in a letter to the *American Journal of Science* dated July 1, 1841. He refers to "artists in Atlantic Cities" who were also successful in producing accelerated daguerreotype plates by a method different from his. He goes on to say:

> I suppose that I am acquainted with the mode of preparation which they employ; but as it was communicated to me under an injunction of secrecy, before I had discovered it myself, although I had actually employed it unskillfully, and therefore without complete success before, I can say nothing of it here. It will, without doubt, soon be made public, if it is not already known.[13]

It is impossible to know what experiments or which Atlantic cities Barnard was referring to. He had been a student at Yale and maintained associations with that institution. He had taught in New York City before going to Alabama and learned to make daguerreotypes under the tutelage of Samuel F. B. Morse. Presumably, he would have also known Dr. John Draper[14] and Wolcott and Johnson. He may also have known someone in the Philadelphia circle, but no connection has been made. There was also a story that an assistant to Dr. Hare, who had been present during some of Cornelius and Goddard's early experimental work, had run away to New York and, it was feared, revealed the secrets of the Cornelius studio sometime in 1841.[15]

At any rate, Barnard's reference is clearly

to those using chemical means to accelerate exposures. At the time of his article contemporary published reports of portraiture using the daguerreotype process in "East Coast cities" were all concerned with improvements in camera and lens design and lighting schemes. Indeed, a resume on the state of American daguerreian practice that appeared at the end of 1840 states that the apparatus had been improved but that the process was essentially the same as introduced except for the adoption of gilding.[16] This article was written by W. H. Goode, who had been a chemical assistant to Dr. John Draper at the University of New York and at the time of publication was associated with the Medical College in New Haven. Goode was very familiar with the work of Draper and Morse, as well as that of Wolcott and Johnson. All these men were able to make portraits because of physical modifications to their cameras and optical systems and not through chemical modifications of the daguerreian system.

After the Barnard report there were no more American reports on the use of bromine or chlorine for almost six months. In France, however, Hippolyte Fizeau (fig. 4.6) published a short note in *Comptes Rendus* in response to Claudet's announcement.[17] Fizeau (1819–96) said that since it was known that both silver chloride and silver bromide were more light-sensitive than silver iodide, their use in photography had been only a matter of time. He asserted that experiments along this line had begun almost as soon as the announcement of the daguerreotype process and referred to the successes in England (probably John Frederick Goddard) and in Germany (most likely, he was really referring to the Austrians Kratochwila and the Natterer brothers). Fizeau added that he himself had thought of using bromine to accelerate exposures but out of respect had waited for Daguerre to publish his announced accelerated process.

Arago reported that he had received a let-

Fig. 4.6. Portrait of Hippolyte Fizeau. Source: Figuier, *Les Merveilles de la Science*, fig. 13.

ter from Daguerre about a new accelerated process that used "electricity" and shortened exposures to such an extent that they were instantaneous.[18] Daguerre visited Arago and assured him that he would make known the details of this process as soon as he completed his move to the country to Bry sur Marne.[19] A week later Daguerre sent a sketchy letter to Arago stating that an error had been made in the previous report and that he had not yet been able to determine if "electricity" really had an effect on plates. Daguerre wrote that the whole story was yet to be published and that he was devoting all his efforts to his experiments until he could deliver a complete set of results. Arago felt that this response was not going to satisfy the curiosity aroused by the first mention of the new process.[20]

Daguerre's method was never perfected, and his description of the process never came. Experimenters who had waited in deference to Daguerre before publishing new methods to accelerate plates went ahead with their papers. Fizeau finally felt free to describe his method of using bromine in daguerreotype processing, and Marc Gaudin published a method using a combination of iodine and chlorine. Daguerre's silence was greeted with indignation, as reflected in the following account from the *Athenaeum* after Arago's last announcement:

A complete stagnation of a branch of trade which has grown into a considerable one in Paris, is stated to be the consequence. Notwithstanding M. Daguerre's assurance that his new system demands no change in the disposition of the apparatus, the instrument-makers will not venture upon the manufacture of the old model, which may, they apprehend, be rendered useless, by the production of the new one—neither will their customers buy them. Every one is waiting for M. Daguerre, daguerreotype artists and savants alike; and, for the last fortnight the sale of photogenic drawings, themselves, is at an end—the purchasers of this popular species of merchandise looking forward, like others, to the results of the amended process. Under these circumstances, the parties injured are clamourous for the realization of M. Arago's announcement, and the publication of M. Daguerre's. They argue that he is not in the position of a private speculator, who may produce his discovery when he pleases; but that having been paid beforehand for his invention, by the national provision made for him, he has no right to keep his improvement concealed for a single day.[21]

Daguerre never communicated another word about the daguerreotype. In many ways he appears to have been cast in the role of a scientist by the well-meaning Arago, but Daguerre was not a scientist and could not deliver on the very difficult problems he was expected to solve.

The earliest printed report of Paul Beck Goddard's use of bromine appeared in the proceedings of the American Philosophical Society in the minutes of the meeting for January 21, 1842.[22] Even though according to Goddard he and Cornelius had been using bromine since December 1839, this first public disclosure was prompted only after he had received news that a similar method to accelerate daguerreotype exposures had been presented to l'Académie des Sciences in Paris (possibly Fizeau's or Claudet's report).[23] According to Goddard's report, multiple sensitization of the daguerreotype plate had been part of his and Cornelius's routine practice from the time of the studio's opening in the spring of 1840, but they had elected to keep their method secret because of their business interests in Cornelius's studio.

The importance of the discovery of multiple sensitization or accelerated plates has often been underplayed or misunderstood. Even during the daguerreian era some people, like Fizeau, believed the use of bromine and chlorine was so obvious that the question of who had first used it was hardly worth worrying about. However, the use of bromine was the key to practical portraiture and ultimately made possible the daguerreotype's great success. The assignment of priority for the discovery almost always comes down to one of the Goddards.

From Paul Beck Goddard's comment made in 1842, at the time John Frederick Goddard conducted his first experiments using bromine, it would seem that Cornelius had been using bromine for at least nine months. Because Paul Beck Goddard and Cornelius chose to keep their discovery to themselves and because it was made public through a late and relatively obscure notice as part of the minutes of a meeting rather than as a formal scientific paper, the usual test of first publication of results, they cannot be given the priority of discovery. This fact has tended to confound the problem of assigning priority for the discovery of multiple sensitization. Neither Cornelius nor Paul Beck Goddard seemed concerned with their future position in the history of photography and, as a result, did not make their own case for the discovery. Further,

both Cornelius and Goddard had basically stopped making daguerreotypes by 1843. Cornelius returned his attentions to the family's business, now the firm of Cornelius and Baker, because Philadelphia had begun using gas as a source of heat and light. The firm became the major American lamp manufactory during the nineteenth century, and Cornelius no longer had time for side ventures such as daguerreotypy. Goddard turned his attentions to his medical practice and research.[24]

Protecting the Daguerreotype Image

Another drawback to the daguerreotype process was the mechanical fragility of the image after it had been made. Daguerre tried coating these images with various common varnishes such as damar, copal, wax, and Indian rubber. Other experimenters made other suggestions for suitable varnishes, including water-soluble coatings such as dextrine solutions and gum arabic. All were unsatisfactory because they altered the appearance of daguerreotypes and eventually seemed to cause the images to disappear. Very few varnished daguerreotypes have survived, but the Joseph Saxton daguerreotype view of the Central High School in Philadelphia is one example.

Another resolution of the daguerreotype's mechanical fragility was to package the plate so that it would be encased and the image itself would not have to be handled. There was already a tradition of packaging painted miniatures, and the methods used for those items were adapted for daguerreotypes. Larger daguerreotypes were placed in the same types of frames used for other prints and other small works of art.

The largest number of daguerreotypes was produced in the United States, and certain American conventions developed around the preparation and packaging of these images. The common, small booklike cases with a hinge and clasp were particularly American. The conventional package included the daguerreotype, a mat or spacer, and a cover glass all bound together with a paper tape. A flexible brass tape called a preserver was bent over the edge of the package to cover the paper tape underneath and to fit the daguerreotype package snugly into the case. An entire industry grew out of manufacturing the items needed for packaging daguerreotypes in this way.

A chemical means to improve the daguerreotype's mechanical fragility was discovered in 1840 by Hippolyte Fizeau, just twenty years old at the time and part of a circle of young science students in Paris whom Arago encouraged to work on the daguerreotype process. Fizeau found that if a daguerreotype was treated with a solution of gold chloride and sodium thiosulfate and heated slightly the image was both mechanically stronger and had a more vigorous appearance. This process followed the removal of unexposed silver iodide. He stated that this process was a way to "fix" the daguerreotype image to the daguerreotype.[25] The French word he used was *fixer*, meaning to fasten or affix. Fizeau recognized that this process affixed the image to the plate in a more complete way than by stopping the daguerreotype processing after the removal of silver halide. The process, called *gilding*, was the precursor to what is now called gold toning, but it is not the same as gold toning.

An image that had been gilded could be handled without fear that the image would be wiped off by the slightest touch. The gilded image also had a luminous quality that ungilded images lacked. The highlights of the image were whiter, and the shadows were apparently darker. Fizeau's process was so effective that it was universally adopted as soon as it had been made public.

Other aftertreatments for gilding daguerreotypes involved making the daguerreotype one of the electrodes in an electric battery filled with a suitable metal-salt solution such as gold chloride. When the current was allowed to flow in the battery, a thin layer of gold was deposited on the daguerreotype plate surface. While this method of gilding was discussed in the literature and used by some experimenters, it does not appear to have been widely practiced and after the 1840s was rarely mentioned. However, a patent for a variant of this method was granted to Charles L'Homdieu, a daguerreotypist in Charleston, South Carolina, at the end of 1852.[26]

Colored Daguerreotypes

Gilding and multiple sensitization were the only two changes made to the daguerreotype process. Other shortcomings of the daguerreotype process were addressed in different ways and were more intractable. Daguerreotypes do not reproduce the actual colors of a scene. While the modern eye has become used to black-and-white images made with a camera, for those who first saw the daguerreotype this lack of coloration was a drawback, especially once portraiture had become practical. Daguerreians began to add color to their daguerreotypes—a little green here, red in the cheeks, maybe a touch of color added to a dress or prop in the daguerreotype. The addition of color varied from mere hints of color to dramatic painting that almost took away the presence of the daguerreotype itself.

Color could be added to the surface of a daguerreotype in many ways.[27] The simplest was to paint a transparent varnish or adhesive onto the daguerreotype surface and then apply ground pigments to the surface while the adhesive was still tacky. Another way was to grind pigments up with a powdered water-soluble adhesive such as gum arabic and then apply this pigment onto the plate. The pigment could be set in place by breathing on the plate or by applying a slight amount of heat to the back of the plate. Pigments were usually applied with camel's hair brushes or by sifting out powdered pigment from a shaker onto a daguerreotype protected by a stencil. One English patent granted to Richard Beard for coloring daguerreotypes used glass stencils and heavily painted backgrounds with clouds and atmospheric effects.[28] Another feature adopted by some practitioners was to paint jewelry and metallic items in portrait daguerreotypes with gold paint or sometimes to use a pointed awl to make an indented spot on the plate that would appear jewellike and glittery in the light. A coloring box used for daguerreotypes is shown in figure 4.7.

Another more elaborate method of coloring was a variant on the electrolytic method of gilding plates. An American patent was granted to Daniel Davis, Jr., and assigned to the daguerreotypist John Plumbe, Jr., for a method of coloring daguerreotypes based on this scheme. In this case other metal salts, usually copper and silver, were used in addition to gold chloride in making up the electrolyte baths for the galvanic battery. These salts varied according to the color desired. Since coloring was the aim of these treatments, the electric field had to be directed so that only those areas of the plate surface that were meant to be colored would be affected. A wire was attached to the positive pole of the battery and used like a brush directing the action of the bath. Plumbe used this process, but it was fraught with problems because a daguerreotype could cloud over if the operator was not careful.[29] A similar method was patented by Warren Thompson and assigned to Montgomery P. Simons in 1843. The major difference between this later patent and the Plumbe patent was that all areas of the daguerreotype that were not to be colored in one electrolytic bath were protected by a combination of gum arabic and any greasy substance. Once the desired portions of the daguerreotype were colored,

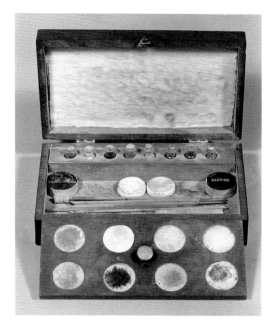

Fig. 4.7. Coloring outfit for daguerreotypes from the 1840s with a label stating that it is the "daguerreotype retouching box of Isaac McNeely." Source: Collection of Matthew R. Isenberg.

the daguerreotype was boiled in lye to removed all traces of grease. The process was repeated for each color.[30]

These electrolytic methods were very complicated, and hand coloring daguerreotypes with pigments was far easier and more direct. Even so, some experimenters hoped to be able to produce color daguerreotypes directly. The two men who had major success in this area were Claude Felix Abel Niépce de Saint-Victor, the nephew of Nicéphore Niépce, and the American Levi L. Hill. Neither man, however, produced a commercially viable process.

To understand the work of Niépce de Saint-Victor, one must step back and look at the work of Edmond Becquerel. Becquerel (1820–91) was the second of three generations of French physicists. He, like Fizeau, was part of Arago's circle of young scientist-students encouraged to investigate the daguerreotype. His first scientific paper, published when he was twenty, addressed that topic.[31] Becquerel worked for several years on the effect of spectral radiation and also electricity on the daguerreotype plates. As part of this work, he observed that if an iodized daguerreotype plate was given a normal camera exposure and then exposed to yellow (or red) radiation by being placed under a colored glass, an image would appear that was the result of the action of the "continuing rays" of yellow light. This effect required no mercury. Today this is called the Becquerel effect, and among daguerreotypists it is referred to as Becquerel development.[32] Another observation about this effect is that daguerreotypes made in this way are colored—not strongly colored but nonetheless exhibiting the colors seen in nature. He went on to discover that if a plate was chloridized in a solution of copper sulfate and sodium chloride while it was the positive pole in an electric cell, it could be used to record the colors of a solar spectrum after it was removed from the sensitizing bath.[33]

Niépce de Saint-Victor (1805–70) took up these researches (fig. 4.8). He devised a process in which chlorinated daguerreotype plates were placed in a camera obscura and given very long exposures (i.e., on the order of several hours). He determined that

there was a direct relationship between the color a chloride salt produced in a flame and the color that light would produce on a daguerreotype plate treated with that salt. Using water baths saturated with chlorine and a chloride salt to treat daguerreotype plates, he produced copies of colored engravings using this process, which he called *heliochrome*. Some landscapes were also produced. He did not produce daguerreotypes, and images made using his process could not be fixed and would gradually turn gray if left in the light.[34] In one effort to preserve these images, he tried coating them with transparent coatings, but these were not very effective. A few of these plates have survived and have been exhibited at various times. However, Niépce de Saint-Victor's work along these lines did not result in a practical process.[35]

The person who has achieved the most notoriety for work on a color process derived from the daguerreotype is Levi L. Hill (1816–65). Hill, a retired Baptist minister in Westkill, New York, had taken up daguerreotypy because he believed that breathing the chemical fumes associated with the process improved his health. Acute bronchitis had earlier forced him to give up his ministry. In late 1850 and early 1851 he announced that he had devised a color process, *hillotypes*, which was widely reported in the popular press and in photographic journals. This announcement set in motion a black comedy unlike anything seen in photography before or since.

In his autobiography Hill recounted his euphoria upon producing the first successful example of his process:

Never, before or since, did I experience such overpowering mental excitement as when I saw this result. Wearied and worn with the toils of three long years, I was, as it were, suddenly ushered into a place of repose and beauty. My brain reeled and staggered under the mighty fact that I had reached the goal of my hopes, and I shouted, like a Methodist—

Fig. 4.8. Portrait of Claude Felix Abel Niépce de St. Victor. Source: Figuier, *Les Merveilles de la Science*, fig. 22.

Eureka! Eureka! It seemed to me that the house was too small to hold my suddenly expanding thoughts, and I made my way to a clump of "willows," near a running brook, where, Ophelia-like, I soliloquized all manner of sentimentalism. . . . Indeed, I was, for the only time in my life on the verge of insanity. Suddenly, the thought struck me that I must make a desperate effort to *remember* how I had secured that picture. . . . I accordingly locked myself in my room, and went to work deliberately to *construct a formula*. In the nature of the case this effort was the keystone to my future success.[36]

After this powerful experience Hill and his wife pledged an oath to show examples of his process to no one until it had been perfected. Even so, he made his announcement, which was given lavish publicity in the photographic journals. The *Daguerreian Journal* enthused:

We are led to believe that the Hillotypes will supercede the Daguerreotypes, as the former will be altogether preferable. The discoverer *has* produced copies of colored engravings "true to the tint. . . ." They are unlike a Daguerreotype as they can be seen in any light, and possess beauty that no artist can paint, while at the same time they present Nature as she is.[37]

As a result, the daguerreotype business dwindled; people waited for Hill's new process rather than have regular daguerreotypes made. A committee was formed of three members of the New York State Daguerrean Association to check the veracity of Hill's announcement, but with the religious conviction of his oath Hill refused to let them see examples of his work. The leader of this group, D. D. T. Davy, a daguerreotypist from Utica, bullied Hill and was referred to as a desperado in Hill's autobiography. Hill was denounced as a quack, and threats were made on his life. He had to borrow a revolver and a guard dog, and his neighbors organized a warning system should Hill and his family find themselves in dire peril. Hill did show examples of his hillotypes to Samuel F. B. Morse, John Whipple, Marcus A. Root, and Jeremiah Gurney, all men of considerable stature in the photographic community, and they gave positive testimony to his having produced colored images. His application for a patent was turned over to a Senate committee to investigate the veracity of his discovery. Hill gave testimony before this committee and showed examples of his process. The committee found the discovery of obvious originality and merit but felt that the existing patent laws would not afford Hill protection because the process was "strictly chemical."[38] He was never granted a patent for the hillotype.

The hillotype process was never made into a commercial venture. In 1856 Hill published *A Treatise on Heliochromy*, which sparked the controversy all over again. The photographic community was indignant that this man, a quack, dared to sell a book at the cost of twenty-five dollars on a process that had been declared a hoax. Legal actions were taken by D. D. T. Davy to prevent the book's distribution on the grounds that it libeled the New York Daguerrean Association. The books were sold for paper rags, and only a few copies survived. A facsimile edition was printed in 1972.

The controversy over whether Hill actually produced colored images has raged since his first announcement. His work has been dismissed in many standard histories of photography; however, sixty-two examples of hillotypes are preserved in the Museum of American History of the Smithsonian Institution. The hillotypes are not daguerreotypes, even though they were made on daguerreotype plates. They have color like that seen in nature—that is, red is red, blue is blue, and so on. These colors are not saturated, and they are set against a beautiful pearly white background. Hillotypes of line drawings or engravings record the design in a characteristic deep magenta-purple, rather than black. The process was recently reproduced from Hill's instructions by Joseph Boudreau, an artist working outside New Haven, Connecticut.[39] Boudreau's specimens look very much like Hill's. While reproducing the process is a vindication of sorts, however, there has yet to be any scientific investigation of what Hill was doing or why his plates were colored.

Reproducing Daguerreotypes

The search for color daguerreotypy is far more exciting than the rather mundane problems associated with reproducing daguerreotypes. Obviously, since a daguerreotype is on a piece of opaque silver plate, it is not possible to make multiple copies of a daguerreotype in the same way that one can use a negative to make positive photographic prints. Daguerreotypes must be copied either by daguerreotyping the original daguerreotype or by using the daguerreotype as a printing plate or as a cartoon for preparing a printing plate.

A single sitting for a daguerreotype portrait very often included a number of daguerreotypes. In large studios where prominent people came to have their portraits taken, it was not uncommon for the daguerreotypist to keep and display one of the daguerreotypes made during the sitting. These daguerreotypes were sometimes copied and sold to others to display elsewhere.[40] If preserving the visual quality of a daguerreotype was the aim, the only way to accomplish this was to daguerreotype the original. Copy daguerreotypes can often be identified because the edges of the original can be seen daguerreotyped on the plate. Copy daguerreotypes may also have a lack of contrast that is different from daguerreotypes made from life, although it is not necessarily easy to perceive (fig. 4.9).

Another way to produce multiples of a daguerreotype is to obtain prints of the

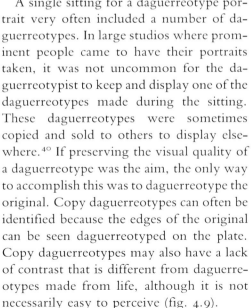

Fig. 4.9. Copy daguerreotype of an unidentified woman. Courtesy of M. Susan Barger.

image. Daguerreotypes were transferred to both lithographic stones and copper engraving plates. The first and one of the most famous examples of this process was the book *Excursions daguerriennes: vues et monuments les plus rémarquables du globe*, published in two volumes in 1841 and in 1843 by the optician Lerebours in Paris. This was one of the first illustrated travel books, and the illustrations were engraved from daguerreotypes taken in Greece, Russia, Italy, Egypt, and the Holy Lands as early as the fall of 1839. A few lithographs made from daguerreotypes were also included in *Voyages Pittoresques*, the running series of lithographs of topological views printed in France from the 1820s and into the 1870s.

In the United States the first venture into mass production of prints made from daguerreotypes was *The National Plumbeotype Gallery*, the 1846 series of engravings produced by John Plumbe, Jr. The foremost practitioner of making lithographic reproductions from daguerreotypes was Francis D'Avignon, a French-born artist raised in Russia who came to the United States in

1842. D'Avignon is best known for *The Gallery of Illustrious Americans*, his collaborative effort with Mathew Brady; he also worked with other daguerreotypists and reproduced many of the daguerreotype portraits of the day.[41]

One attractive feature of reproducing daguerreotypes as prints was that the daguerreian image could be copied directly onto a plate or stone without being reversed because the image was already reversed in camera. Printing was, in fact, an easy way to reverse images so that they read as seen in nature. (Prisms were used in some daguerreotype cameras to reverse the image as it was being taken, but for the most part daguerreotypes really are mirror images. One such prism is shown in figure 4.10.)

Attempts were made to use the daguerreotype plate itself as a printing plate—which, of course, was initially what Nicéphore Niépce had in mind—by etching the plate or sometimes by electroplating the plate. Alfred Donné published a method of making engravings from daguerreotypes in 1839 in which he claimed it was possible to pull prints from the daguerreotype's silver surface.[42] The first practical method of etching daguerreotype plates from printing was published by Josef Berres in 1840. He varnished a finished daguerreotype plate and then held it over a steam of weak nitric acid. The daguerreotype was then coated with gum arabic and finally immersed in nitric acid.[43] He noted that daguerreotypes made on silver plate made better engraving plates than those made on pure silver. William R. Grove, the pioneer of electric batteries, published a method for electroetching daguerreotype plates to form printing plates.[44] Fizeau, Donné, and Claudet all did some work in the area of electroetching daguerreotypes. None of these methods seemed to work very well, and they were not widely adopted.[45]

Another interesting but unsuccessful method of reproducing daguerreotypes was the *tithonotype*, invented by Dr. John W. Draper. In this method the back and edges of a gilded daguerreotype were varnished, and the daguerreotype surface was electroplated with copper. After twelve to twenty hours the copper layer would split from the daguerreotype plate, carrying the image with it. This tithonotype could then be used as a printing plate, or it could be transferred to other surfaces.[46] This was a variant of a procedure reported several years earlier in which mucilage or isinglass could be spread on the surface of a gilded plate and also split from the surface to make a transparency of the daguerreotype's image. This method was unsuccessful because it was almost impossible not only to get the adhesive layer to come up in one piece and but also to prevent permanent marks on the daguerreotype.[47]

This brief discussion of the kinds of methods suggested to multiply daguerreotypes gives an idea of the number of ingenious ways in which people tried to enhance the usefulness of daguerreotypes.

Fig. 4.10. The "Cathan's Reflector and Lense" with its original box. This is a reversing prism for producing daguerreotype images that are laterally correct. Source: Collection of Matthew R. Isenberg.

Fig. 4.11. A typical European daguerreian outfit showing the equipment needed for producing daguerreotypes: the camera (*a*), a plate holder (*b*), sensitizing boxes (*c*), mercury bath (*d*), plate level and clamp (*e*), plate box (*f*), gilding stand (*g*), dish for hypo (*h*), and buffing paddle (*i*). All the pieces are made to fit in the large box. The bottles along the side of the large box are for chemicals and polishing compounds. Source: Hunt, *Encyclopaedia Metropolitana*, bound into the back of the book.

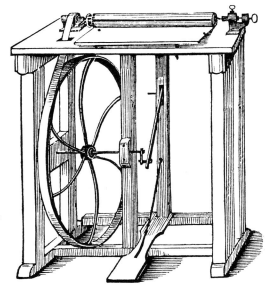

Fig. 4.12. Treadle-operated buffing wheel invented by C. C. Harrison, a camera manufacturer from New York City. Source: Snelling, *A Dictionary of the Photographic Art*, p. 19.

Standard Practice for Daguerreotypes

By the mid 1840s standard methods of practice for the daguerreotype had evolved, and commercial equipment such as that shown in figure 4.11 was readily available. The process had expanded from that which had been set out by Daguerre, and to produce a daguerreotype one followed these steps:

1. A plate was cleaned and polished to a mirror finish. Polishing methods were quickly improved over those suggested by Daguerre. Large studios even devised machines to mechanize the polishing of plates; one of these is shown in figure 4.12. The plate might also have been given a thin coating of electroplated silver over the commercially prepared daguerreotype plate. Two types of batteries used for electroplating daguerreotype plates in the studio are shown in figures 4.13 and 4.14. This process was first used in the United States and dates from sometime in the mid-1840s. The name comes from the American daguerreotypist Warren Thompson's French reference to this practice—*le procédé d'Americain*. This method of plate prepa-

ration was also referred to as *galvanizing*. During the nineteenth century this term was used both for plates manufactured by electroplating and for plates electroplated by the users. This term also carries a very specific meaning today, referring to steel or iron that has been treated with a protective coating of zinc to prevent corrosion. The authors have chosen to refer to this method of plate preparation as *American process* to avoid confusion over the various uses of this term.

2. The plate was sensitized by placing it first over iodine vapor and then over some other halogen vapor in sensitizing boxes like those shown in figures 4.15 and 4.16. A fairly standard treatment was to expose the plate to iodine until it had turned a yellow-rose color and then place it over bromine until the plate changed to the next interference color—i.e., to steel blue if the initial color over iodine was yellow-rose. Finally, the plate was put back over iodine in the dark for some fraction of the original iodizing time. The second and succeeding sensitizing steps were called acceleration, and the materials used were called accelerators or quicks.

The standard method of using three coatings was first suggested by Abbé Edmond Laborde, a lecturer in physics and chemistry in the diocesan seminary at Corbigny in Niévre, France.[48] He found that this method of sensitization made it easier to control the highlight regions of a daguerreotype, ensuring that those areas of the image were pure natural whites instead of cold blue whites. Samuel van Loan and John Johnson introduced this method to the United States early in 1844, and it was universally adopted and recommended in most daguerreotype manuals. The authors have adopted the term *multiple sensitization* as a general term to cover all the treatments that use more steps to sensitize a plate than Daguerre's original process. The methods and apparatus for applying these sensitizers im-

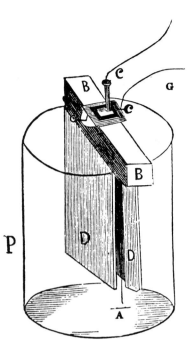

Fig. 4.13. A Smee's galvanic battery for generating electricity for electroplating daguerreotype plates. A sheet of "platinized" silver (*A*) is separated from a sheet of zinc (*D*) by a wooden beam (*B*). The binding screw (*C*) is soldered to the silver plate, and wires are connected to this metal binding to carry the current generated by the battery. When these two plates are immersed in a solution of sulfuric acid held in the glass jar (*P*), a current is generated by virtue of a flow of electrons from the silver plate to the zinc plate. Source: Snelling, *A Dictionary of the Photographic Art*, p. 90.

Fig. 4.14. A Coad's patented graduated galvanic battery, found in a Philadelphia daguerreotypist's studio. Source: Collection of Matthew R. Isenberg.

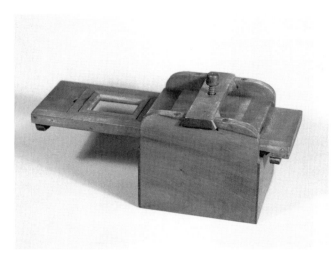

Fig. 4.15. The contents of a halogen coating box. The bottom of the open box is marked (*a*), and the glass jar liner that holds the halogen is marked (*c*). The underside of the sliding cover of the box (*d*) is shown with the ground glass cover (*e*) that seals the glass jar liner when the box is closed and the opening in the slide (*f*) where the plate is laid face down so that it can be fumed with halogen vapor. A second spring-loaded (*i*) cover (*h*) is held in place by the bar (*g*) so that halogen vapor does not escape when the box is not in use. Source: Snelling, *A Dictionary of the Photographic Art*, p. 49.

Fig. 4.16. American quarter-plate sensitizing coating box with reducing masks to accommodate sixth and ninth plates. Source: Collection of Matthew R. Isenberg.

proved so that repeatable, even coatings could be obtained easily.

3. The sensitized plate was placed in a camera and exposed to light. The exposure time varied according to the weather and time of the year, the type of multiple sensitization used, and the camera-lens combination used by the daguerreotypist. A typical daguerreotype camera is shown in figure 4.17.

4. The exposed daguerreotype was placed over hot mercury until an image appeared. No change was made here, except that the apparatus used for this step was improved over the years. In particular, mercurizing pots were made with thermometers so that the temperature of the mercury could be closely monitored, as can be seen in figures 4.18 and 4.19.

5. The remains of the unexposed light-sensitive layer were removed from the plate usually by washing the plate in a solution of sodium thiosulfate (hypo). Cyanide solutions also were used for this purpose. Another method suggested by Draper in 1840 was electrolytic fixing. In this method the daguerreotype was placed in a solution of salt water, and a zinc rod was touched to

one corner, causing the colored halide layer to be swept away from the plate surface in a wave of color.

6. The plate was gilded by heating it gently after a solution of sodium thiosulfate and gold chloride had been washed over its surface. This process was not changed after its introduction by Fizeau in 1840.

The daguerreotype process was changed very little, as noted. Changes were made in the niceties of producing plates—for instance, better polishing methods were devised—but these were not changes in the actual process itself. Daguerreotypists and others who took up this art had very quickly pushed the process to its limits, and it remained intractable to change. That did not stop the tremendous commercial business that resulted from the introduction of the daguerreotype. The daguerreotype process remained the only practical method of producing photographs until the introduction of the wet collodion process in 1851.

The Daguerreotype Industry

As Arago had hoped, the new art had opened up new areas of inquiry. A new industry had grown up around the daguerreotype, and new jobs had been created not just for daguerreotypists but also for the people required to produce the supplies

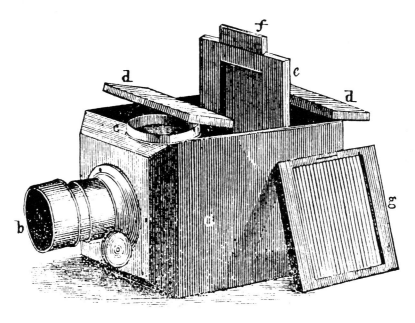

Fig. 4.17. A photographic camera showing the dark box (*a*), the lens tube with a knob for focusing (*b*), the lens cap (*c*), the lid to the camera box (*d*), the plate holder (*e*) with the dark slide (*f*), and the ground glass (*g*). The scene is focused on the ground glass, and the glass is removed and replaced with the plate holder. When the photographer is ready to begin an exposure, the dark slide and the lens cap are removed. Once the exposure time has elapsed, the lens cap is replaced. Before the exposed plate is removed to the mercurizing bath, the dark slide is returned to its position in the plate holder. Source: Snelling, *A Dictionary of the Photographic Art*, p. 29.

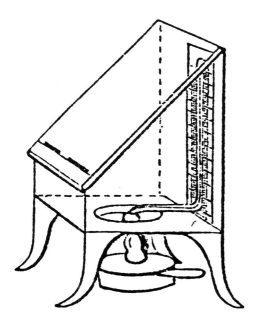

Fig. 4.18. Mercury bath made of wood with a cast-iron bottom with a depression in the center to hold mercury. The thermometer bulb is situated in the mercury dish; its scale can be viewed on the outside of the bath. An alcohol lamp is used to heat the mercury. The exposed daguerreotype is placed in a groove under the slanted, hinged lid. The legs of this bath can be folded under to make the bath more portable. Source: Fisher, *Photogenic Manipulation*, p. 19.

Fig. 4.19. Whole-plate mercury bath and reducing masks to ninth plate with thermometer and alcohol lamp. Source: Collection of Matthew R. Isenberg.

needed to practice the art. The more important aspects of the industrial response to the daguerreotype are related to plate manufacture, the production of the necessary chemicals, and the manufacture of the apparatus and accessories required for the art.

The Need for Plates

The most immediate need when the daguerreotype came on the scene was to find sources for silver-plated copper to make daguerreotype plates. Daguerre gave the following instructions for the plates to be used for his process:

> The drawings are made on sheets of silver plated copper. Although the copper serves principally to support the sheet of silver, the combination of these two metals contributes sensibly to the perfection of the effect. The silver should be of sufficient thickness to maintain the planimetry of the plate, so that the images should not be deformed; but too much thickness should be avoided on account of the weight that would be added to the apparatus. The thickness of the two metals together should not exceed that of a thick card.[49]

This brief statement was all people had to go by in choosing their plates. The early experimenters working on the daguerreotype process took these few brief instructions and turned to the traditional suppliers of silver plate. At the time silver plate generally meant copper onto which a layer of silver had been attached by rolling the two metals together in a rolling mill. This kind of plate, called *Sheffield plate*, was produced on a large scale in England and in continental Europe and to some extent in the United States. It was available as a stock item in hardware stores and other types of supply houses. Common silver plate was usually made of coin silver (900 parts silver per 1000 parts total alloy) and was not found to be satisfactory for use as daguerreotype plates. In most cases early daguerre-

otype plate manufacturers were those already in the business of making silver plate who adjusted their methods to accommodate the needs of the new market.

Sheffield plating is also called *plating by fusion* or *cold-roll cladding*. Basically, this process is accomplished by taking a piece of silver foil and binding it to a clean copper ingot using an iron wrapping. This package is placed in a muffle (a special oven) and heated until just below the melting point of silver. This step was very important. If the heating was carried on too long, the silver and copper layers would be alloyed; if it was too short and the silver was insufficiently liquated, the silver layer would not adhere properly to the copper after rolling. This package might then be cooled or rolled under light pressure, reheated, and then cooled. At this point the iron wrapper was removed, and the silver-copper ingot was cleaned. It was then rolled back and forth between two workers as the gap between the rollers in the rolling mill was gradually closed. The plate increased in length as a result of rolling. At certain times the plate would be annealed to remove strain built up in the metals during the rolling process. This process was repeated until the plate was of the desired thickness. For daguerreotype plates a modification was made to ensure the finest silver surface. In the last passes through the rolling mill two pieces of plate were rolled face to face. The silver surfaces were chalked to prevent them from sticking together. Once the rolling was completed, the plate was cut into smaller pieces and planished, or worked with a hammer to remove the curvature introduced by the rolling process.

This process was carried on by eye and by feel. It required skilled platers, and often the need for platers was an almost insurmountable problem. The best platers came from apprenticeships in Sheffield and Birmingham, England. The special requirements of silver plate destined for daguerreotype plates increased the need for these craftsmen and was a real obstacle to getting into this line of work. Silver plate for daguerreotype plates had to be free of flaws and smears, and the silver layer had to be well adhered to the copper substrate. The companies taking up this new product line had no systematic body of knowledge to fall back on, and often their customers were the only source of information for improving their product. There were also great problems connected with attaining the necessary machinery and equipment to make plates of high quality. The best rollers for this kind of work came from England, and the technology was not necessarily open for export.

Two other methods were used to make silver coatings on base metal, but these were minor compared to the amount of goods made by Sheffield plating. The first of these was called *Paris plate* or *French plating*. In this process hot silver leaf was burnished onto a prepared substrate. This method was often used to repair smears and flaws in Sheffield plate. The other minor method involved dipping or boiling formed base metal in solutions of silver or silver salts. Neither method was applicable to the daguerreotype because the silver layer had to stand up to the rigorous polishing needed to make the required mirror finish necessary to the production of good blacks. In the case of French plating, a flaw covered in this way might not be visible to the eye, but it might show up as a spot during halogen fuming as a daguerreotype was being made.

The daguerreotype plate market became an important area of business venture. Plates were made everywhere, but French and American plates were reputed to be the best available. From the outset the French silver plate industry was required by law to add plate marks. The convention was that these marks would include a number indicating the proportion of silver to copper

present, a design denoting the maker, and, usually, the word *doublé*, meaning plated. In the United States only the state of Maryland had laws requiring the use of plate marks, but American manufacturers adopted plate marks for daguerreotypes to show that their plates were as good as those from France.[50]

The largest American manufacturer of daguerreotype plates was the J. M. L. and W. H. Scovill Company of Waterbury, Connecticut. It had been in the plating business at the time of the introduction of the daguerreotype and adapted its work to meet the needs of the new photographic market. Several other American daguerreotype plate manufacturers started out with Scovill, including Pemberton, who had been hired by Scovill after an apprenticeship in England, and Israel Holmes and David Hayden of Holmes, Booth and Hayden. All these men established businesses along the Naugatuck River Valley in Connecticut, the same locale as Scovill.[51] The daguerreotypist Joseph Pennell reported to his former partner, the daguerreotypist Albert Southworth, that Scovill was producing approximately 1,000 plates per day during the winter of 1848.[52]

Another method for making silver plate was *electroplating*, developed during the same period as the daguerreotype. As mentioned earlier, American process daguerreotype plates were roll-clad plates that were electroplated by the user as part of plate preparation. The early history of electroplating is very complicated and confused. Various claimants to the priority of discovery of an electroplating method railed against each other in the press for more than twenty years during the same time as the popularity of the daguerreotype. However, the key to practical electroplating of precious metals, British Patent 1840/8447, granted March 25, 1840, is central to the issue of manufacturing daguerreotype plates by this method. This patent was granted to two brothers, George R. and Henry Elkington, who took out this patent on behalf of John Wright. Wright had made the discovery that it was possible to make coherent electrodeposited films of precious metals using cyanide plating baths. This discovery was the basis of successful silver and gold plating.

The Elkingtons had a button and ornament plating business in Birmingham, England, and in 1839 they met up with Wright in a hotel while in London to patent a method for electroplating. They found that Wright was in London on a similar mission and that his method was considerably more practical and promising than theirs. In exchange for his patent rights, they offered him a position in their plant to do further research. Wright accepted this offer but died in 1844 without having made further progress with his research. The Elkingtons also obtained the patent rights on cyanide plating baths in most European countries, thus protecting their privileged position in establishing the silver electroplating industry, and in other countries their agents were able to promote their new process without loss of Elkington control.

Electroplating is relatively straightforward but requires special attention to the details of each step. The greatest problems are producing adherent deposits that will not peel or flake away from the substrate and producing bright (i.e., shiny) coherent deposits. The electrochemical mechanisms of electroplating were not fully understood in the nineteenth century and are not today.

General scientific interest in electricity and its related phenomena promoted the rapid empirical advancement of electroplating and electrodeposition during that period. These processes require simple equipment: a battery and a glass or earthenware plating vessel. Early plating solutions were made up of boiled silver nitrate or silver chloride in a saturated solution of potassium cyanide. The object to be plated

and a silver anode of the same size were put in the plating bath and hooked up to a battery until the silver layer on the object being plated was of sufficient thickness. Unlike the Sheffield plate process, this method of producing silver plate was simple and cheap and could be done at home or in the laboratory without a great deal of capital investment. The silver layer produced by electroplating under the best conditions was finer and of purer silver than could be obtained by Sheffield plating; initially, however, electroplated silver was not considered as wear-resistant as roll-clad Sheffield plate.

Even so, it was only a matter of time before daguerreotype plates were manufactured by electroplating. This happened in the early 1850s, when the French firm Christofel et Cie. began to produce its famous "scale plates." French law required the use of plate marks to denote the quantity of silver present in silver-plated goods. A "scale" plate mark indicated that the silver plate was produced by electroplating and not by cold-roll cladding.[53] Christofel et Cie., the holder of the Elkington patent in France, commercially exploited electroplating in France.

Two obstacles seemed to mitigate against the introduction of electroplated daguerreotype plates until the early 1850s. The first involved a bitter patent dispute between Christofel et Cie. and Henri Catherin Camille Roulz, a chemist who had been employed at Christofel, regarding the provisions of the Elkington patent. In 1842 Christofel et Cie. purchased a set of plating patents that had first been filed with the French government by Roulz. The Roulz patent was for cyanide plating baths and was filed in France around the same time as the Elkington patent was filed in England. Roulz sold his patent rights to the dyer Guillaume Edouard Chapée in 1840; he later repurchased his patents and sold them to Christofel. Christofel hired Roulz to work for the commercial exploitation of electroplating in 1842, the same year that they became the French agents for the Elkington patent. Roulz remained at Christofel until 1845, and upon his departure the company turned to the Elkingtons for help in continuing the portion of their business that involved electroplating. This arrangement did not sit well with Roulz, who felt that the company had violated his patent rights by conferring with the Elkingtons. This problem caused considerable embarrassment to Christofel et Cie., which brought suit against Roulz to establish the Elkington patent priority in France, to establish the Christofel position as the Elkington agents in France, and to stop Roulz's efforts to set up a business in competition with theirs. The suit kept all parties in the courts until the early 1850s, when the court ruled in favor of the Elkington priority and the Christofel position.[54]

The second obstacle to the production of electroplated daguerreotype plates was a technological one. Silver plate manufactured by means of the earliest type of cyanide plating baths was porous and could not withstand the polishing required of daguerreotypes. In 1847, however, some of Elkington's workers discovered that by adding a small amount of carbon disulfide to plating baths a bright silver deposit could be produced that required no polishing after plating.[55] In addition, silver deposits from brightened baths were more coherent than those from simple cyanide baths. These two characteristics made daguerreotype plates manufactured by this method commercially viable. Early in the 1850s Christofel et Cie. offered electroplated daguerreotype plates that became exceedingly popular because of the inherently improved quality of electroplated surfaces. The popularity of the Christofel plates may have caused other companies to adopt this method of plate manufacture. For instance, the Scovill Extra, an American plate offered for sale

around 1850, reportedly was made by electroplating. The Scovill Manufacturing Company had begun experimenting with electroplating about 1847, but it did not make progress in this area until it hired a skilled electroplater in 1850. The decline of the daguerreotype after the early 1850s as a result of the introduction of other photographic processes ended the further development of the plate market, both for clad and electroplated plates.

The Need for Chemicals

Just as there was a need for plates, there was an equal need for the chemicals necessary for the daguerreotype process. Since chemicals are often the raw materials for other items of commerce and not mere ends in and of themselves, charting the sources of specific chemicals can be difficult. After the introduction of the daguerreotype, most pharmaceutical and other scientific journals began to publish notes on the best methods for the synthesis of specific chemicals needed for producing daguerreotypes and other types of photographic materials. These notes included synthesis for items such as hypo-sulphite of soda (sodium thiosulfate), chloride of gold (gold chloride), and the like[56] and also included notices on the fabrication of polishing compounds. Eventually, chemicals were bought in bulk by wholesalers who then repackaged these materials and produced them for sale to the daguerreian market. Scovill branched out from plate manufacture and produced or sold an entire line of materials necessary to produce daguerreotypes.

Sources for most of the chemicals needed for the daguerreotype were fairly direct. However, this was not the case for iodine or, especially, bromine. All halogens are usually found in nature as salts of alkali metals such as sodium and potassium. Unlike sodium chloride (common salt), which is fairly abundant in nature, the incidence of natural iodine and bromine salts is widespread but not highly concentrated. Both iodine and bromine are extracted from certain salt brines or the ashes of certain kelp plants.

Until the beginning of the nineteenth century the source for most of the alkalis needed in manufacturing operations such as glass making and soap making was the ashes of certain plants. Ash obtained from wood was called potash. Sea plants, another source of ashes, were divided into two categories: those obtained from the Mediterranean plant *salsola soda*, which was called barilla, and those from Scottish, Irish, or Norwegian seaweed, which was called kelp.[57] By the late eighteenth century kelp burning had developed into an important industry, especially in the British Isles, because of demand caused by European timber shortages and the attendant shortage of potash on the Continent. Even though the alkali content of kelp was lower than that of potash, the kelp industry in the British Isles expanded greatly throughout this period because of protective tariffs placed on other types of alkali ashes imported into Britain.[58]

The natural alkali market was adversely affected beginning in September 1791, when the French chemist Nicholas Le Blanc patented a method to manufacture soda from common salt. Soda obtained by the Le Blanc method was much cheaper than that obtained from kelp or potash. After 1806, when Le Blanc's patent expired, his method was generally adopted, and as its use spread the British kelp industry was almost brought to a standstill. The Napoleonic Wars brought a temporary but transitory revival of the kelp industry because foreign sources for natural alkalis dried up during the continental blockade. The final revival of the kelp industry came after the discovery of iodine in 1812. Kelp was the primary source of this element, and the lack of European and American import duties on it allowed this necessary raw material to

be sold for high prices. In the United States a few iodine plants were set up in New England during the 1820s but did not remain in business very long. The New England kelps were very low in iodine, and this fact, coupled with the limited market for iodine, meant that the plants could not produce enough iodine to make a profit.

The bromine story is similar to that of iodine, except that the bromine market was negligible when compared even to the meager market for iodine. Since both bromine and iodine were found in the waters of spas, the logical use for these elements was in medicine. However, only their salt forms—iodides and bromides—were thought to be useful for medicinal purposes. Thus, when the daguerreotype was introduced, neither of these elements was widely available or easily obtainable, although iodine was the easier of the two to obtain.

At the outset of the daguerreian era and throughout that period, the primary source of iodine was France, where the local kelps were rich in that element. In 1841 Andrew Ure set up a successful iodine manufactory in Glasgow, Scotland.

Bromine was another story. As mentioned previously, Paul Beck Goddard and Robert Cornelius had tried to buy up all the supplies of bromine in the eastern United States after they had devised a process to use bromine in daguerreotypy. Their initial experiments were done in Dr. Hare's laboratory, and they probably had some supply of bromine or a way of extracting small amounts of bromine from salt brines. At that time small quantities of bromine were produced in Germany near the salt mines of Stassfurt, the major source of bromine in Europe. German bromine had a reputation for quality, but it is unlikely that the first bromine used by Goddard and Cornelius came from German sources. Their bromine more probably came from western Pennsylvania. The bitters left from the salt works along the Allegheny River near the towns of Freeport, Tarentum, and Natrona, Pennsylvania, were found to be rich in bromine.

The first company set up exclusively for bromine manufacture was established by Dr. David Alter (fig. 4.20) and his partners, Drs. Edward and James Gillespie, around 1845 in Freeport, Pennsylvania. Alter (1807–77) had been educated at the Reformed Medical College of the United States in New York City, and after completing his studies he returned to western Pennsylvania, settling in Freeport in 1843. In addition to his medical practice, Alter was passionately involved in various scientific pursuits. He invented an electric telegraph and an electric clock, he actively observed the stars, and he was one of the first in the area to take up the practice of daguerreotypy. Alter and the Gillespies were interested in bromine primarily because they found it and its various compounds useful in the treatment of "Neuralgia, Tubercular Disease of the Lungs, and as an external application in Erysipelas."[59] The Alter-Gillespie bromine works was not only the sole producer of bromine in the United States but also the major world producer. Not until 1866, when Rosengarten and Sons of Philadelphia built a plant in Tarentum, Pennsylvania, was there another bromine works in the United States. Alter obtained two patents for bromine manufacture, one in 1848 and the second in 1867.[60] Samuel D. Humphrey, editor of *The Daguerreian Journal*, commended the Alter product, saying:

We were not aware that Dr. Alter was the only one who made Bromine in America, and when we spoke in favor of this article we did not say enough—experience has taught us that there is nothing equal for Daguerreotype purposes, bearing the name of Bromine, to that manufactured in America and by *D. Alter*. We have heard Daguerreians declare that the *German* Bromine was the best and they would

Fig. 4.20. Dr. David Alter at the age of sixty. Source: Means, "Dr. David Alter," facing p. 224.

have none other. The secret was that the [*sic*] said to be *German* was made in the United States. Let us say to all interested in the art, If you want *good Bromine* buy the American.[61]

Around 1856 the bromine market crashed. This industry was tied to the daguerreotype, and as other photographic processes became popular there was no longer a need for bromine. The market did not pick up again until Dr. Robert Bartholow initiated the widespread therapeutic use of bromides in the early 1860s. During the Civil War bromine was first used as a disinfectant and as a discharge agent in calico printing. These new uses helped revive the industry.[62]

Other Equipment

The practice of daguerreotypy also required a source of cameras, lenses, fuming or sensitizing boxes, mercury pots or baths, buffing paddles and the other equipment needed for polishing daguerreotype plates as well as the materials needed to case daguerreotypes. Initially, those experimenting with the daguerreotype obtained cameras and lenses from opticians and instrument makers or made their own equipment. Once it was clear that there was a market for photographic supplies, companies went into the business of manufacturing all these materials. As with the manufacture of daguerreotype plates, this business was quite successful. Since the majority of daguerreotypes were made in the United States, it is not surprising that the largest suppliers of daguerreotype supplies were located there. The two largest companies in the United States were Scovill Manufacturing Company in Waterbury, Connecticut, and E. and H. T. Anthony Company of New York City. Many other companies got into the business throughout the 1840s and 1850s, and a portion of these remained in the business of supplying photographic materials even after the daguerreotype no longer reigned as the principal method of producing photographs.

The daguerreotype and its success as a new art and an item of commerce more than fulfilled François Arago's hopes. Arago had recognized that this process was not just a child's game but a very practical discovery and that its ramifications would be far reaching. The daguerreotype was quickly made into a practical process that could be used for portraiture and as a commercial operation. This process, in turn, brought about changes in the manufacture of goods such as silver plate and made possible new industries such as the bromine industry.

WILLIAM HENRY FOX TALBOT's report on François Arago's announcement of the daguerreotype process was presented to the Ninth Meeting of the British Association for the Advancement of Science. It was clear to most natural philosophers from the start that the daguerreotype was not going to be an easy mystery to solve; it was going to be a great challenge and, they hoped, a boon for science. Before going to the French Parliament to ask for pensions for Daguerre and Niépce, Arago had Daguerre attempt two experiments to show that the daguerreotype would have applications outside the realm of art. To fulfill this request Daguerre made a daguerreotype of the moon and recorded the solar spectrum as a proof that scientific work could be conducted using his process. As head of l'Observatoire, Arago had special hopes for the application of the daguerreotype to astronomical research and the study of light; these applications reflected those aspirations.

In spite of great enthusiasm and activity, however, the technology of the daguerreotype sped far ahead of the science needed to describe how the process worked or how the phenomenon of recording images occurred. At first, scientific questions occupied anyone who worked on the process, but very soon the daguerreotypists and the scientists went their separate ways. For the daguerreotypists operating theories that let them keep track of their day-to-day operations were sufficient. Often these theories were derived from intuitive knowledge gained from practice with the process. For the scientists these theories were insufficient to derive a set of ideas or a science adequate to describe the process.

There is a vast difference between using a process and understanding why the process works. In the case of the daguerreotype and, indeed, all photography, the issue was complicated by the fact that photography offered a way to measure light, but the trick

5

Scientific Interest in the Daguerreotype during the Daguerreian Era

M. Arago stated to the Institute that the sciences of optics and chemistry united, were insufficient in their present state, to gain any plausible explanation of this delicate and complicated process. If M. Arago, who had the advantage of being six months acquainted with the secret, and therefore of considering its nature in all points of view, was of this opinion, it seemed as if a call were made on all the cultivators of science to use their united endeavors, to the accumulation of new facts and arguments, to penetrate into the real nature of these mysterious phenomena.

Talbot's remarks on Daguerre's process[1]

to interpreting such measurements required an understanding of how light formed the photographic image. The important scientific problems revolved around these questions: How does light cause any image to be formed? Why is what a photograph records different from what our eyes see? What is the role of mercury in forming the daguerreian image? And what could these new processes reveal about the nature of light? These problems were multilayered and interconnected. Moreover, the solutions and the problems were interdependent in a tautological way. Looking at one part of the problem meant looking at all parts of the problem. A further restriction was that the tools available to study the daguerreotype process were quite limited—the optical microscope, crude spectrometers, observation, and the effects of the process itself. A final complication was the conflicting theories about the nature of light and its behavior.

The Role of Light in Image Formation

In the theoretical framework of the time, light could act either as a chemical agent, chemically altering a body, or in a physical way, mechanically altering the surface of an irradiated body. Further, radiant energy was divided into different elements—luminous rays, which were limited to the light that could be seen by the human eye or, to use nineteenth-century terminology, "impressed upon the retina"; calorific rays, or heat; and chemical rays, or only those rays that caused a compositional (or molecular) modification of a surface. According to some, electricity was another type of radiant energy. All these rays made up the class of weightless substances called *imponderables*. The workings of any phenomenon, such as the behavior of light in the daguerreian system, would be determined by which part of the theoretical construct

was espoused by the individual experimenter.

Since the French had the advantage not only of Daguerre's presence but also of Arago's position and enthusiasm, it would seem logical that they would have led the way in the scientific search for the solution to the daguerreian mystery. They did at the beginning, but as Arago pointed out several years later his own experiments on the daguerreotype were postponed because the one camera obscura at l'Observatoire was occupied and it took some time before a new one could be built for photographic pursuits.[2] With his infectious enthusiasm he did set in motion a group of young student-scientists, and scientific daguerreian activities began to move in Paris. The English had a problem beginning their research on the daguerreotype because Daguerre's patent restricted those who could use the process. There were no such strictures in the United States.

The Americans began their scientific work on the daguerreotype with the enthusiasm and diligence that characterized their embrace of the entire process and all that it entailed. Almost as if he had been waiting in the wings, Dr. John William Draper (fig. 5.1) began a passionate pursuit of the science of the daguerreotype as an addition and continuation of his already active research on the nature of light. In May 1839 Draper had become a professor of chemistry at the University of the City of New York. He had learned firsthand about the new process from his colleague at the university, Samuel F. B. Morse, who had met Daguerre and had seen examples of his process in March 1839. Draper had already been testing the formulas published for photogenic drawings published by Talbot, but his involvement with photochemistry and, by extension, photography dated from before 1839.

John William Draper (1811–82) was the son of an itinerant Methodist minister in

marily concerned with the chemical effects of light, the application of physics and chemistry to physiology, and the extension of science education in America.

In 1837 Draper published a series of papers in the *Journal of the Franklin Institute* entitled "Experiments on Solar Light."[3] In this series he described experiments showing that chemical absorption of light was not restricted to violet rays, as was commonly believed. He stated that solar radiation is made up of three types of rays—calorific, chemical, and colorific—and that these rays are particulate with one half of the particles in each ray polarized at right angles with respect to the other half. (In spite of his description of radiant energy being particulate, Draper did not dismiss the idea of the aether.) When solar radiation is passed through some medium, according to Draper, a part of the radiation is absorbed depending upon the constitution of the medium. Chemical rays cause some chemical effect in the medium, and calorific rays cause the medium to become heated. These two types of rays are not as easily observed as colorific rays, which may be easily seen by the absence of certain colors in the solar spectrum.

Draper demonstrated in his experiments that absorption depends on color and not on other characteristics of the transporting media. According to Draper, absorption was "material loss from the ray" and not a change in the motion of the aether. He demonstrated these effects on various light-sensitive media, including metal salts, particularly silver chloride and silver bromide. In his discussion he also considered the formation of vapor images on glass. For instance, if camphor is placed at the bottom of a vacuum jar and exposed to light, crystals of camphor will form on the illuminated side of the jar. Draper showed that this phenomenon was not limited to camphor vapors but was also the property of other vapors, including mercury and water.

England. Educated by tutors, he spent two years at the University of London. After his father's death the family came to the United States at the invitation of his mother's relatives in Virginia in 1832, when he was twenty-one years old. Draper studied medicine at the University of Pennsylvania from 1835 until 1836. During his time in Philadelphia, he worked as an assistant in Dr. Robert Hare's laboratory at the university and also at the Franklin Institute with the chemist J. K. Mitchell. Upon returning to Virginia, Draper taught chemistry at Hampden-Sidney College until he assumed his position at the University of the City of New York in 1839. He would have left Virginia sooner had it not been for the depression of 1837 and his waiting to see what effects it would have on the proposed medical school at the university. While in Virginia, Draper began publishing scientific papers, partly to establish himself among European scientists and also to spread his reputation. His work was pri-

Under all conditions he found that he could form vapor patterns on the illuminated portions of the vacuum jars. He postulated that electricity might have something to do with these images and that it might be carried in solar radiation, although he offered no proof for radiant electricity.

When the first news of the details of the daguerreotype process reached New York, Draper, Samuel F. B. Morse, and Alexander Wolcott all began working to produce portraits. Like their counterparts in Philadelphia, they began by trying to get more light on the subject or into the camera. Wolcott in New York won this race by designing his camera with a reflector lens, which has been described earlier. Draper may have succeeded in producing a portrait of his sister, Dorothy, as early as December 1839, but this date is not firm.[4] He sent the daguerreotype of his sister and a paper describing how he had obtained it to the *Philosophical Magazine* in England and asked the editor to forward the daguerreotype to Sir John Herschel.[5] Dorothy Draper's portrait may have been the first daguerreotype portrait to reach Europe, but it was definitely not the first ever made.

Before he had sent notice of his successful portrait, Draper had already published a summary of his 1837 work in the *Philosophical Magazine* to establish his place in the scientific pursuit of the mysterious effects of light and photography.[6] In his notice of the application of the daguerreotype to portraits, he amplified the importance of his earlier work, and thus began a long stream of papers on the daguerreotype and daguerreian theory. In that second notice, he also made some statements as to how the daguerreotype process worked. Draper found that mercury in the daguerreotype process behaved in the same way as the other vapor images he had already described and which had been described previously by the Swedish chemist J. J. Berzelius. Since silver iodide characteristically blackens in the light and the daguerreian image is white, not black, Draper believed that the key to understanding the process is to determine what mechanism stops the action on an image before it turns black. Further, he concluded that the reappearance of an image when a used plate is polished and put back over mercury is proof that the daguerreian image is dependent upon some mechanical alteration of the plate surface caused by light. Thus Draper cautioned that those working on the daguerreotype should be careful to separate the chemical from the mechanical effects of light. Although both Daguerre and Herschel had spoken of the necessity of using achromatic lenses when taking photographs, Draper claimed that these were not essential if the sensitive plate or paper was placed at the point where blue or indigo rays came to a focus at the camera back. At the blue focus the photographic effect would take place in less time because photographic materials are most sensitive to blue light.[7] Of great significance, Draper also reported that he had succeeded in daguerreotyping the moon. These early papers clearly established Draper as one of the important scientists working on the daguerreotype process.

From the beginning Sir John Herschel was actively interested in the science of photography. Herschel was one of the most important English scientists of the day, holding a position similar to that of Arago in France. Like Arago, he had a background in astronomy and research on the nature of light. As Talbot's friend and as a member of the English committee sent to see whether the daguerreotype was a hoax, he was very early acquainted with photography and the study of photographic phenomena. Herschel had discovered the solvent effects of sodium thiosulfate on unexposed silver halides in 1819 and suggested its use to both Talbot and Daguerre. He also was the first to use the term *photograph*, and other commonly used photographic terms such as *negative* and *positive* were adopted upon his suggestion. Herschel became involved in making specimens of the new art from the beginning and may have sent the first photographs seen in Germany to his aunt, the astronomer Caroline Herschel.

Beginning in mid-1839 Herschel presented to the Royal Society the results of his own and other investigators' spectrum studies using various light-sensitive materials, including photographic materials. His own most important paper on photography was published late in 1839,[8] and that work was summarized and published in many scientific journals.[9] Herschel discovered that certain portions of light from the solar spectrum, notably the red rays, caused a formed (i.e., visible) image to become bleached or faded. This phenomenon is now known as the Herschel effect.[10] Always open and supportive with his research, he shared his results widely, actively corresponding with others working on similar problems. It was natural that Draper would have chosen to correspond with Herschel during this period.

In France several scientists applied their energies to disentangling the daguerreotype. Early in 1839 Jean Baptiste Biot (1774–1862), an old friend of Arago and one of the scientists who helped maneuver the bill of support for Daguerre and Niépce through the Parliament, began experiments concerned with the study of the solar spectrum using photographic materials. He, like Draper, was especially interested in this particular application of the new discovery because it removed the study of solar radiation from the limitations of merely what is visible to the human eye. In the first report of these studies Biot mentioned that he was aided with intelligence and zeal by "the son of one of our confreres, M. Edmond Becquerel."[11]

Just nineteen when he began his work with Biot, Edmond Becquerel (fig. 5.2)

Fig. 5.2. Portrait of Edmond Becquerel. Source: Figuier, *Les Merveilles de la Science,* fig. 27.

quickly established himself as an up-and-coming researcher in his own right. In November 1839 Becquerel reported that when two plates of silver, platinum, or gold are immersed in an acid or alkaline bath and light is allowed to shine on one of the plates, an electric current flows from one plate to the other. This effect was even more intense on silver plates treated with iodine, bromine, or chlorine.[12] Daguerre had suggested that electricity was involved in the formation of the daguerreotype image, and Becquerel's work seemed to confirm this premise or, at least, that there was some electronic change in the surface of the daguerreotype plate as result of exposure.[13] (The application of this effect to the acceleration of daguerreotype plates was part of Daguerre's supposed improved daguerreotype process that caused so much trouble in 1841; see chapter 4.) Becquerel continued his studies on the effects of light on the daguerreotype and discovered that if a plate was given a short image-forming exposure, an image would appear if the plate was sub-

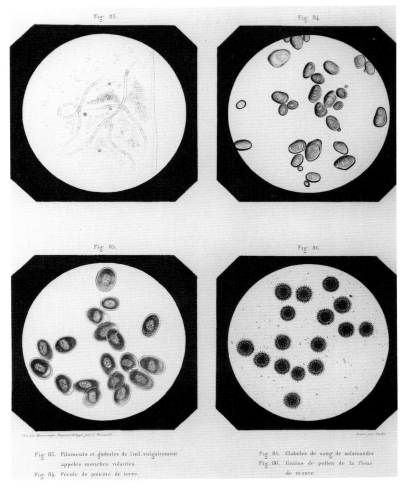

Fig. 83. Filaments et globules de l'œil vulgairement
appelés mouches volantes.

Fig. 84. Fécule de pomme de terre.

Fig. 85. Globules de sang de salamandre

Fig. 86. Grains de pollen de la fleur
de mauve.

Fig. 5.3. Plate 20 from Alfred Donné and Léon Foucault's textbook on optical microscopy for medical students. The majority of the illustrations for this book were engravings taken from daguerreotype micrographs. In this plate figures 84, 85, and 86 were made from daguerreotypes and show, respectively, potato starch, salamander blood, and pollen from the mallow flower. The text notes that the specimen of salamander blood shown in figure 85 was difficult to daguerreotype because it changed very quickly in the air. Figure 83 was made from a paper photograph and shows nasal mucus. Source: Donné and Foucault, *Cours de Microscopie Complementaire des Études Médicales*, pl. 20. Courtesy of the International Museum of Photography, at George Eastman House, Rochester, N.Y.

sequently given a longer exposure to red light.[14] He called the red light of the second exposure *continuing rays* because they seemed to continue the initial exposure. Today this phenomenon is called the *Becquerel effect.*[15]

Alfred Donné (1801–78), a member of the faculty at l'École de Médécin in Paris, was among the first to study the changes in the daguerreotype surface during image formation using an optical microscope. As a microscopist, he was interested to see what daguerreotype images and the process of image formation looked like up close. According to Donné, a sensitized plate is uniformly covered with a homogeneous, crystalline layer of silver iodide. Light exposure causes the adherence of this layer to break down, and the plate looks powdery wherever light has struck. Mercury can attack the plate surface wherever the silver iodide layer is disturbed. Thus, the highlights of the image are formed of mercury

globules, which are probably a silver amalgam, and the shadows only are silver.[16]

Other descriptions of image formation on the daguerreotype based on observations using a microscope were published about the same time as Donné's paper. Golfier Bessyre claimed that the light causes movement in the silver iodide layer, resulting in a molecular modification proportional to the intensity of exposure. In the mercury bath mercury condenses on the modified silver iodide surface, forming yellow mercury iodide, which leaves the plate surface and condenses on the surfaces of the mercury bath. According to Bessyre, silver iodide had two functions: (1) to receive and retain mercury and (2) to form a strong chemical reaction with the silver of the daguerreotype plate.[17] Similarly, Augustus Waller (1816–70), a medical student in Paris, reported that the basis of photography was light acting upon silver halides in such a way that the action can be fixed by mercury vapor. He felt that silver should not be the only metal that is suitable for the process.[18]

Donné was also interested in the application of the daguerreotype to recording the images seen through an optical microscope. With the assistance of Léon Foucault (1819–68), Donné made micrographs of various medical samples on daguerreotype plates and published the first atlas of microscopy made using photography as the source of the illustrations (fig. 5.3).[19] Similarly, Waller and Bessyre made micrographic medical daguerreotypes. This medical application of the daguerreotype spread, and soon there were reports of daguerreotype micrographs being made in Germany,[20] the United States, and elsewhere.

In 1841 Robert Hunt (1807–87), another of the English photographic pioneers, wrote and published *A Popular Treatise on the Art of Photography, including the Daguerreotype and all of the new method of producing pictures by the chemical agency of light,* the first general manual of photography.[21] Hunt was a self-educated chemist from Cornwall and worked as a pharmacist while carrying on his scientific research in Devonport. Like Draper, Hunt carried on a lively correspondence with Herschel. In his *Popular Treatise* Hunt had carefully included all the experimental work published to that time on all types of photographic processes and had provided a wealth of detailed information about how each process worked, with hints for performing each operation. He always carefully credited his sources and sometimes included long passages from previously published papers. For instance, the first edition of his book included a long section on daguerreotype portraits, and Hunt acknowledged, "This very interesting application of Daguerre's discovery has been perfected by Dr. Draper."[22]

In 1841 Draper published what is probably his most important contribution to photochemistry.[23] In his investigations of the solar spectrum, Draper found that the amount of chemical change caused by light was dependent upon whether light was absorbed by a sensitive surface. He showed that specific effects could be attributed to light of specific wavelengths. By iodizing daguerreotype plates until they took on the characteristic colors of silver iodide layers—yellow, yellow-rose, red, blue, and so on—he showed, first, that these colors were directly related to the thickness of the silver iodide layer. More important, he found that the colors affected the sensitivity of the plate. From this he was able to show that if light is not absorbed by a sensitive surface, no matter how much light is present, it will not cause a chemical change in the surface. This observation is called the Grotthuss-Draper Law.[24]

Draper concluded that the sensitivity of a surface depends on its chemical nature; however, the quantity of chemical change depended upon the optical quality of the surface—that is, its interior and exterior molecular arrangement. On the basis of his observations he tried to explain the dark lines in the solar spectrum by saying that they were caused by absorption of components of the aether so that in those portions of the spectrum there was nothing to reach and react with a sensitive surface.

Draper, like everyone else, had a theory of image formation in daguerreotypes, and he used this paper to discuss his ideas; these were, to use his own words, "what appears to be the proper theory of the Daguerreotype."[25] He determined that both mercury and iodide must cover the entire surface of a daguerreotype once it has been made and before it has been fixed. Using a paper saturated with starch, which turns blue in the presence of iodine, Draper concluded that iodine is not evolved, or given off as a gas, during exposure because these treated papers did not turn blue when laid on an iodized plate during exposure. He determined that light mechanically alters the plate so that in the mercury bath exposed portions of the plate are more susceptible to mercury vapor. Iodine is liberated by these image portions when they come in contact with mercury vapor. However, the iodine immediately corrodes the underlying plate surface so no iodine is ever lost until the plate is treated with sodium thiosulfate. Once an image has been fixed, the shadow portions are composed of uncombined mercury, the highlights are silver amalgam, and the midtones are some combination of the two. Once again, Draper reminded his readers of the daguerreotype's similarity to the vapor images he had described in his earlier papers.

A Scientific Disagreement

Up until this point, the discussion of the daguerreotype among scientists was fairly open and collegial. This situation changed sharply in 1842, when Draper perceived that some of his colleagues had failed to properly acknowledge his previous work and took great umbrage at this fact. During the summer of 1842 a new process was communicated in a letter to Sir David Brewster in England from the German astronomer Friedreich Wilhelm Bessel at the University of Königsberg.[26] Ludwig Moser (1805–80), a physics professor at the same institution, had found that it was possible to make images on black plates of polished horn or agate using various types of vapors. He also found that different vapors behave in a way analogous to the light of different colors. Moser likened these images to the image on a daguerreotype. This phenomenon came to be called *Moser-bild*.

Bessel reported that Moser's images were "nearly as good as those produced by Talbot's process." Brewster said that these images must be due to some sort of thermal effect, but Bessel protested; Moser found that light affected vapors in different ways and that these vapors could yield "latent light," which formed these images. A discussion of the role of hypothesis in pushing forward a new area of study followed in which Herschel commented that it was a shame to encumber research on photographic phenomena with speculations associated with the emissive and undulatory theories of light. He went on: "[No] true philosophy, without a certain degree of boldness in guessing; and such guessing, or hypothesis, was always necessary in the early stages of philosophy, before theory has become an established certainty; and these bold guesses, in their proper places . . . should be encouraged, and not repressed."[27] These reports from Königsburg were greeted with great curiosity and interest.

Later translations of Moser's papers gave more details of his experiments.[28] Moser believed that light, physical contact, and condensation of vapors have the same effect on all bodies. Light and physical contact mechanically modify materials in such a way that after exposure or contact, vapors can condense differentially on any polished surface. This fact, according to Moser, was the basis of Daguerre's discovery; the daguerreotype was merely a specific case of this general property of matter. Through later experiments Moser found that even physical contact was not necessary for bodies "to reciprocally depict each other," and thus he concluded that all bodies must be "self luminous" or possess latent light. This property of matter held even in complete darkness, Moser contended. He also concluded that his so-called latent light was not the same thing as heat or phosphorescence. To make these vapor images, an object needed to be placed near but not touching a polished surface and left in the darkness. After an appropriate time the polished surface could be treated in a vapor to reveal the image. These images were called, variously, *vapor images, hauchbild, Moser-bild, thaubilder,* and *roric images.* Moser carefully demonstrated that these images are the result of mechanical action, not chemical action.

Harking back to Draper's work on the varying sensitivity of the silver iodide layer on daguerreotype plates according to their color, Moser believed that the latent light associated with vapors was also colored. For instance, the latent light of mercury vapor was yellow because it continued exposure just as Becquerel's continuing rays did. Likewise, iodine, bromine, and chlorine all possessed either blue or violet latent light because all these substances were capable of destroying the effects of the camera exposure on daguerreotype plates in the same manner as blue light.

Draper immediately responded to the publication of the reports of Moser's work with a letter to the editor of the *Philosophical Magazine* that began:

If there be a thing in which I have a disinclination to engage, it is controversy of a personal kind with scientific fellow-labourers.

But, as you well know, it ordinarily happens that there is no other gain to philosophers beyond the *mere credit* of their discoveries, they may be forgiven for reluctantly endeavoring to secure this their only reward.[29]

Draper reported that not only had the discovery of vapor images been credited to Moser by the English but also it had been reported in France as a new and important discovery. He reminded the editor and the readers that he had reported these facts several years earlier and had pointed out their similarity to the daguerreotype phenomena.

Along with this letter he enclosed a daguerreotype with a solar spectrum that he requested be forwarded to Herschel. He had made this particular spectrum in Virginia and claimed that the same spectrum could not be made in New York or in England. This point became part of the groundwork for another theory that would soon be delineated. Draper believed that whatever caused the photographic effect on daguerreotype plates (and other photographic materials, for that matter) was not exactly the same as visible light. He reported that there are six types of rays that act in concert to form an image. These rays include, in part, protecting rays, rays that whiten, rays that blacken, and rays that whiten intensely, and their action varies according to the geographic location. For this reason the same person could not obtain the same solar spectrum results on daguerreotype plates produced in different locales.

In the December issue of *Philosophical Magazine* Draper explained that his research led him to conclude something quite remarkable. In spite of the general supposition that photography and photochemical effects are caused by light, as their names imply, they are in reality caused by something distinct from light—an imponderable much like light, heat, and electricity, which were already accepted. He said that this fourth imponderable was much like heat and light, yet it was quite different from them (fig. 5.4).

Draper named this new imponderable *tithonicity*, after Tithonus, a figure in Greek mythology. Aurora, the goddess of the dawn, fell in love and married a beautiful youth named Tithonus. At her request the Fates made Tithonus immortal but failed to grant him eternal youth. So, unlike his goddess wife, he became weak and helpless

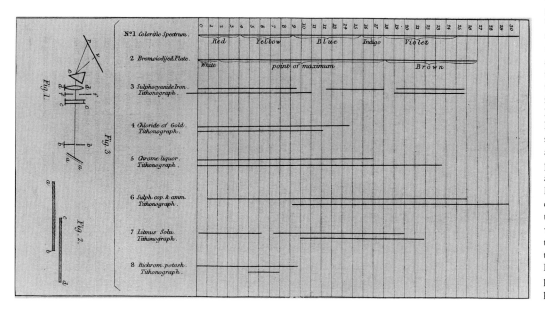

Fig. 5.4. Drawing of absorption spectra and their corresponding tithonographs obtained by John William Draper using the apparatus shown in figure 1. Sunlight is reflected off a heliostat mirror (*a*) through a slit (*b*) and then through a cell (*c*) containing various colored salt solutions as indicated. The light then passes through a second slit (*f*), a lens (*d*), and a prism (*e*) that projects the spectrum onto a receiving screen (*rv*). Figure 2 shows Draper's experimental setup for determining the existence of what he called specific tithonicity, which he likened to specific heat. Source: Draper, "On a new Imponderable Substance," pl. 1.

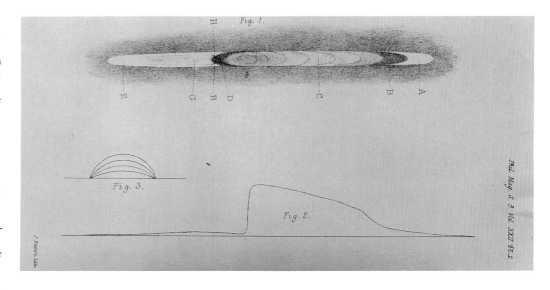

Fig. 5.5. A lithographic copy of the solar spectrum recorded on a daguerreotype plate by John William Draper in Virginia that accompanied Herschel's description of that plate. The spectrum measures approximately 3.8 inches long and 0.8 inches wide. Figure 1 shows the spectrum with letters to indicate regions of different colors. Figure 2 shows the profile of the intensity of the spectral response recorded by the daguerreotype from one end of the spectrum to the other. The lateral depth of the image is shown in figure 3. The daguerreotype from which this illustration was made is in the collection of the Science Museum, London. Source: Herschel, "On the Action of the Rays of the Solar Spectrum on the Daguerreotype Plate," pl. 2.

after a time and had to be taken care of like a baby. Aurora took pity on her husband in this state and metamorphosed him into a grasshopper. This story, Draper proposed, was analogous to what happened in photography: compound radiant beams when exposed to a sensitive surface are separated so that chemical rays sink into the surface and light rays have no effect. Observed photographic phenomenon is the result of escaping chemical rays altered by the sensitive material. Draper likened the loss of photographic effect (what is now called latent image fading) to the loss of strength experienced by Tithonus. This effect would be called *tithonicity* and related terms would follow, such as *tithonoscope, tithonometer, tithonography, tithonic effect,* and so on. According to Draper these words are musical to an English ear, and the rest of his paper was devoted to proofs of these effects.[30] After this paper was published Draper always used his new words to describe photographic phenomena. These terms have confused some historians, who have regarded them as indicating entirely separate processes.

It did not take the scientific community long to respond to Draper's suggestions. In the next issue of *Philosophical Magazine* Herschel described in minute detail Draper's daguerreotype spectrum, which had been forwarded to him by the journal's editor (fig. 5.5).[31] Herschel found that this spectrum was in no way different from any other spectrum obtained on silver iodide regardless of the support material (paper or a silver plate) or the geography of the location where it was taken. The differences in the Draper spectrum and other spectra recorded on the same medium were due to variations in the chemical composition of the prism used to break the solar beam into a spectrum and the use of achromatic lenses rather than true differences in the light of different venues. Herschel cautioned that spectral studies had to be conducted using a prism and not colored glass filters, for "the present state of our knowledge of the absorptive powers of such glasses, serves only to confuse or mislead."[32] Herschel suggested that Moser's work, for example, needed to be analyzed more carefully because much of it had been done using filters, and this, he felt, had confused the Moser results. Herschel's paper was so long that he did not have space to comment on Draper's new imponderable, but he warned that giving an old idea a new name was something that should not be done lightly.

Many people took up Moser's line of experiments and investigated vapor images. Robert Hunt, for one, had set about to repeat Moser's experiments. While Hunt believed there was evidence for the existence of latent light,[33] he felt that Moser's experiments demonstrated that vapor images were caused by heat and did not confirm the possibility of latent light.[34] In the case of the daguerreotype, according to Hunt, light exposure caused the destruction of the silver iodide layer, resulting in the liberation of iodine and the production of finely divided silver on the daguerreotype plate surface. This iodine liberation could be monitored by placing a polished metal plate in close contact with an exposed daguerreotype plate and noting any etching of the opposing metal surface.[35] The etching would produce a vapor image on the facing metal plate that was caused by the chemical action of the iodine, not latent light.

Another Englishman, Horatio Prater, suggested further causes of vapor images on the basis of his experiments. Vapor images were the result of "chemicomechanical" action or catalytic effects.[36] Hippolyte Fizeau offered yet another origin for vapor images, proposing that they were merely the result of slight traces of organic material left on surfaces.[37] Draper simply dismissed Moser's line of reasoning, saying that latent light was nothing more than the reradiating action characteristic of tithonic rays.

In a sharp reply to Hunt and Draper, Moser argued that both men had inadequately represented his results. Hunt had not gone far enough with Moser's experiments, thus leading to erroneous conclusions. Even though a material body becomes luminous when heated, according to Moser, this was merely a visible manifestation of the latent light already present in that body. Moser felt that his experiments had already shown that light is radiated from bodies at all temperatures and that, as

the temperature of a body is increased, this light increases in intensity while at the same time becoming less refrangible. Light and heat are not the same even though they may share some characteristics. As proof Moser explained that while heat is emitted from a heated body, once light has created the change necessary for vapors subsequently to condense on a surface, there is no such corresponding reradiation of that surface. The action of the light in Moser's reasoning becomes "extinct" and no longer has any effect. He then went on to dismiss Draper's work:

[I]t is impossible not to be astonished at a physicist who conceives he has discovered a new force—the tithonic, and asserts that it radiates without even having made one single experiment to prove this. Dr. Draper knows no more than what every person who has been engaged in experiments on light is aware of, that the image on an iodized silver plate, it comes from the camera obscura, disappears after a time, and can no more be made to appear in the vapours of mercury. Does it hence follow that the image radiates from the plate? . . . Dr. Draper is so convinced of the radiation of these images, that he pretends to preserve them by covering them, and in this scientific manner has discovered specific light analogous to specific heat![38]

This debate over what is now called latent image fading is an important argument because it shows the difficulty of sorting out what happens during image formation in a daguerreotype when the tools for examining the process were limited and the theoretical constructs inadequate to describe and bolster experimental observations.

Not wanting to be dragged into a personal dispute, Hunt responded to Moser in his usual manner, praising Moser for opening a new and important area of scientific inquiry into the nature of matter and then going on to say that, nonetheless, people were free to draw their own conclusions from other work. Reiterating his own po-

Fig. 5.6. Line spectra recorded by Edmond Becquerel. Figure 1 shows a drawing of the visible spectrum with the Fraunhofer lines indicated by the letters. The same spectrum recorded on a daguerreotype plate sensitized with iodine and either bromine or chlorine is shown in figure 2. Figure 3 shows the chemical spectrum as recorded on an iodized daguerreotype plate, and the effect of the continuing rays on the same daguerreotype plate is shown in figure 4. The effect of the solar spectrum on silver bromide paper is shown in figure 5. Figures 3, 4, and 5 were recorded in a camera. The same paper used to record the spectrum in figure 5 was used to obtain the spectrum in figure 6 except that the second was recorded by projection instead of in a camera. The spectra shown in figures 7 and 8 were obtained by projection on papers coated with potassium dichromate, guaiacum resin and silver nitrate (fig. 7) and guaiacum and chlorine (fig. 8). The chemical spectrum drawn from a composite of all the spectra is given in figure 10. Figures 11 and 12 show the phosphorogenic spectra of barite and calcium sulfate. In these last two spectra the upper set of lines is what Becquerel calls the luminous spectrum and the lower set is the chemical spectrum. Source: Becquerel, "Memoir on the Constitution of the Solar Spectrum," pl. 9.

sition that light modifies the surface of a body, Hunt noted that "the molecular change which bodies undergo, under the most trifling of circumstances, is certainly one of the most curious matters which photography has brought us acquainted [*sic*]."[39] Hunt clearly stated that the solar spectrum is made up of light and color, calorific rays, and chemical rays. Light is the element that allows for sight, and anything that does not activate sight is not light. Further, Hunt claimed that vapor images are not caused by light but are most likely caused by heat or something analogous, such as "voltaic electricity."

Draper was not sitting idly by while his new theories were being mocked by his foreign colleagues. He chose first to attack Hunt by pointing out the historical errors in the second edition of Hunt's *Researches on Light*. He protested that Hunt had not given him the credit for taking the first portraits from life and had misrepresented his experiments. Draper said Hunt's book gives the "impression that the science of photography is the intellectual creation of Mr. Hunt" and that the "air of vanity and curious errors which pervade this book will impose on some and amuse others."[40]

The real problem for Draper was that no one wanted to take the idea of tithonicity seriously; worse yet, they thought his beautiful name was silly. He pounded away that the effects of tithonicity had been discovered during the eighteenth century and that since then chemists and physicists had been at a loss to explain their observations on photochemical effects. He believed not only that tithonicity could explain the phenomena associated with photographic images but also that he had contributed a great service to the community by defining the principle:

The idea that they [these effects] constitute a fourth imponderable of the same rank as light, heat, and electricity, belongs to me. . . . I have the same undoubted right to impose on that imponderable a name, which Davy exer-

cised in giving currency to the word chlorine and its derivatives. The cases are parallel. Has not the universal receipt of that name settled the law?[41]

Phosphorescence

To complicate the issue further, Draper suggested that there must be another imponderable—the phosphorogenic rays. This suggestion came as result of work published by Edmond Becquerel and by Joseph Henry that Draper felt showed phosphorescence was another class of radiant energy. Phosphorescence is the characteristic emission of light from certain materials not obviously caused by heating. At this time this phenomenon, long a fascination of scientists, could not be adequately explained and, indeed, had to wait for substantial advances in both the physics of solids and the preparation of ultrapure materials. Certain phosphorescent materials, like the Bologna Stone (barium sulfate or barite), had been used by early experimenters searching for a way to capture the images in a camera. Even Daguerre used it for experiments done in 1824, before he had heard of Niépce.[42] The introduction of photography sparked new interest in phosphorogenic phenomena because the photographic effect and the cause of phosphorescence were perceived to be related. The study of phosphorescence actually became the life work of Edmond Becquerel.

Becquerel began his study of phosphorescence as part of his research on spectral studies on solar radiation. His father, Antoine Becquerel, had been interested in and had written about this phenomenon, so it was natural that Edmond would also take it up. Becquerel used a spectrometer of the usual prism, slit, and lens arrangement to separate different types of solar rays so they could be compared and studied individually. He was the first to do an extensive, correlative study of spectra obtained from many materials.

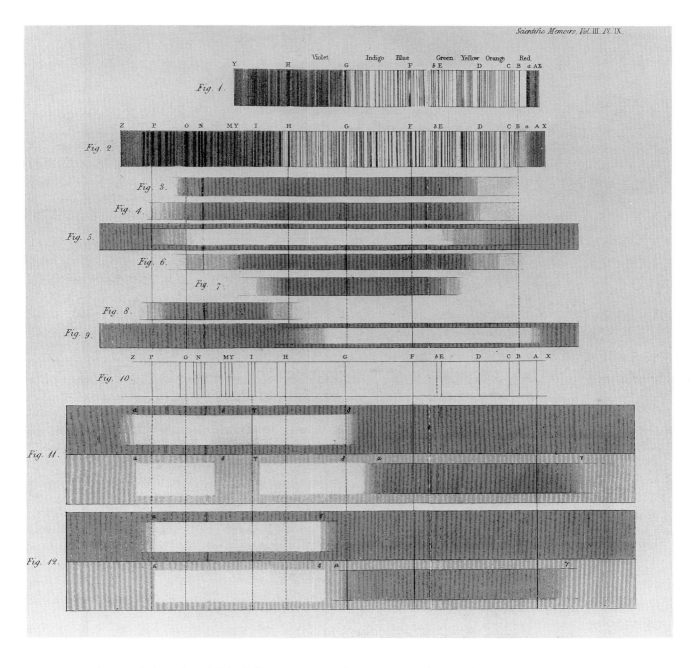

Scientific Memoirs, Vol. III. Pl. IX.

Becquerel recorded and published line spectra which were obtained on a variety of light-sensitive materials—daguerreotype plates prepared in different ways; silver bromide paper; papers coated with potassium dichromate solution; gum guaiacum; and several phosphorescent materials—and described the differences and similarities of these spectra to the Fraunhofer spectrum, the simple solar spectrum (fig. 5.6). These various features were due to diversities in the action of the components of solar radiation: the phosphorogenic spectrum, the chemical spectrum, and the luminous spectrum. While these components have the same physical properties—that is, reflection, refraction, polarization, and interference—they are caused by different effects. The chemical spectrum is manifest when certain molecules that are weakly bound together are broken apart. The phosphorogenic spectrum is manifest when luminescent materials are "impressed" with the motion of the phosphorogenic rays, thus causing the emission of the "electricities" needed to maintain molecular equilibrium. Finally, the luminous spectrum is composed only of the rays that cause the vibrations of the aether perceived by the retina of the eye.[43]

Joseph Henry (1797–1878), then a professor at Princeton University, performed experiments exposing daguerreotype plates to solar light filtered through both luminescent and nonluminescent materials. He had difficulties attaining consistent exposure conditions because of variations in his plate preparations and diurnal fluctuations of the light. However, on the basis of an afternoon's experiments, Henry concluded that there was evidence that "the Phosphorogenic emanation and the chemical are as distinct as the luminiferous and the calorific."[44] Taking into account both Becquerel's and Henry's work and his own experiments, Draper concluded that phosphorescence must be another class of forces operating on matter. Thus came about the fifth imponderable.

At this point Hunt had had enough. To Draper's last attack he replied that even though Draper felt he had not been given the proper recognition, "I am not aware that he [Draper] has devised a single photographic process; I cannot discover that he has even improved one."[45] The object of Hunt's book was "to show the remarkable powers exerted by solar radiations" and not the mechanical arrangement of cameras. Hunt did not dispute the likelihood of a fourth imponderable; indeed, he thought it a distinct possibility. On his own he had previously suggested that the force that causes photographic images be called *energema, energia, helioplaston,* and *metamorphia.* The objection to the name *tithonicity* was

simply its fanciful origin. Ever the gentleman, Hunt concluded:

I am not ambitious to deprive Dr. Draper of any of the fame that he covets, but I hope to be allowed to pursue my investigations, which I enjoy from their own exceeding great reward, without disturbance. I shall always feel it my duty to express my opinions on matters which have come under my investigations with that freedom which I hope others will use towards me.[46]

The Role of Mercury in Image Formation

A purely chemical approach to image formation was presented in 1843 at the time of the initiation of the dispute over Draper's fourth imponderable. Marie Charles Isidore Choiselat (1815–58) and Stanislas Ratel (1824–1904) published two papers detailing the action of light and mercury in the formation of the daguerreian image.[47] Their theory was based on the premise that light changes the silver iodide on the daguerreotype plate to a silver subiodide in proportion to the intensity of light exposure. In highlight areas the subiodide and some iodine set free by exposure attack the silver of the plate, while shadow areas remain unaffected. During subsequent exposure to mercury, a mercury protoiodide is formed, and when it comes into contact with silver subiodide both red mercury iodide and free mercury are formed. Some mercury in contact with the silver subiodide is converted to a mercury protoiodide, and silver is released. In this scheme the highlights of the daguerreotype are a silver amalgam formed by the double decomposition of silver subiodide and mercury protoiodide into silver and mercury, which form the amalgam and some mercury subiodide.

Choiselat and Ratel found that if a plate is exposed to light and fixed but not put over mercury, an image composed of a white, insoluble powder of silver subiodide is left on the plate surface. In their scheme the highlights of a daguerreian image are composed of a powder of silver amalgam. When this powder is silver-rich, rather than mercury-rich, the tones of the highlights are more lively. The shadow regions are composed of a large quantity of mercury with very little silver. They found that accelerators shorten exposure times because small amounts of bromine and chlorine tie up, or combine with, silver iodide so that iodine is not liberated during exposure. Thus, iodine is protected by the accelerator so that it may be used to form the silver subiodide necessary for creating the image.

Choiselat and Ratel did not believe that the chlorine or bromine in accelerators combines with the silver in the plate surface. They reminded their readers that silver iodide must be present in order to form a daguerreotype but that neither silver chloride nor silver bromide alone could make daguerreotypes.[48] This theory or a variant of it was held for many years by those who felt that the production of a daguerreotype must be solely a chemical process, not a physical process or some hybrid of the two.

The theories about the daguerreotype process that seemed to be based more closely on observation and therefore were more satisfactory were more along the lines of the mechanical theories favored by experimenters investigating vapor images. The major difference between the daguerreotype and other photographic processes is the vapors involved in making the image. The key to understanding the formation of those images seemed to be discovering how the vapors worked. Another attempt to sort out the role of mercury vapor in image formation was put forth by Augustus Waller.

Waller (1816–70) had first been involved in daguerreotypy as a medical student at l'École de Médicin in Paris. His first con-

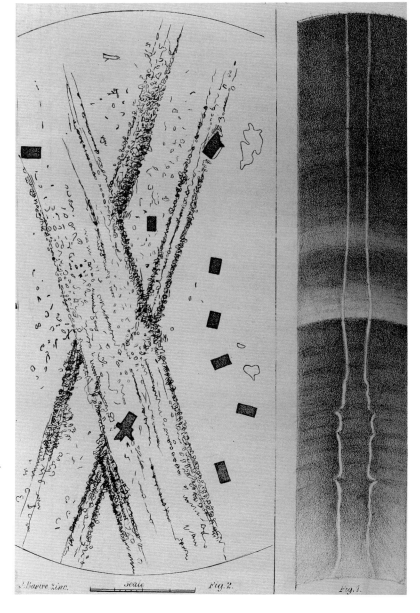

Fig. 5.7. Figure 2 is a drawing of the crystals described by Augustus Waller that are formed as a result of drawing through films of certain salt solutions. This figure was made to demonstrate the fineness of the lines formed by this rapid crystallization reaction. Each large division in the scale bar is equal to 0.1 millimeters. Source: Waller, "Observations on Certain Molecular Actions of Crystalline Particles," pl. 3.

tribution to the attempts to understand these compelling images came in 1840, when he reported that halogen vapors in contact with silver have a remarkable property—the subsequent action of light on silver treated in this way is "fixed" by mercury vapor.[49] Waller finished his medical degree and returned to his native England before he made another offering in this area. It had long been known that certain salt solutions rapidly crystallize if a film of the solution is spread on a glass or polished surface; also, if a pointed object is drawn through the film, lines of crystalline precipitate will be left in the wake of the tracings (fig. 5.7). This fact can be demonstrated using a solution of sodium phosphate and ammonium carbonate mixed with a solution of magnesium sulfate. Waller observed that a similar effect occurs when a volatile liquid is poured on a polished surface, except that the tracing is effervescent.

While these phenomena may seem far removed from the daguerreotype, in each case, according to Waller, a mechanical disturbance in the surface of the substrate causes the formation of a particulate image. Similarly the vapor image produced by placing an object on a polished surface and then breathing on the surface once the object is removed is also the result of a mechanical disturbance of the substrate. In this case the simplest contact alters the surface so that droplets of water from the breath will adhere to the places of contact. The more capable a vapor is of wetting a surface and depositing minute globules of liquid, the more likely that vapor can produce an image. From Waller's observations vapors from liquids capable of forming permanent images are those composed of bubbles or vesicles of liquid that because of their volatile nature crystalize as they are deposited on a surface.

Waller found a direct analogy between the components of the daguerreian system and these phenomena. In his view "chemical rays" of light act on the daguerreotype plate to cause a "molecular disturbance," which differentially alters the plate so that certain portions will be wetted by mercury vapor. The image formed in this way is a crystalline amalgam. Waller discovered that the image on a newly made daguerreotype is composed of groups of hexagonal amalgam crystals.[50] This was a sophisticated twist on Draper's and Moser's vapor images because it introduced the idea that vapors crystalize as they are deposited and that this crystallization is what fixes the image to the surface.

At this time the discussion about the daguerreotype among scientists began to change subtly. N. M. P. Lerebours communicated findings concerning the bleaching of an image exposed to red light. This was the same observation that Herschel had made in 1840 using silver chloride papers, now called the *Herschel effect* (see note 10). Using daguerreotype plates prepared with iodine and bromine, Foucault and Fizeau took up this line of research and showed that red light could have both positive and negative effects on the daguerreotype.[51] Now the discussion of how light affects the daguerreotype was confounded by two new types of light—the positive rays and the negative rays. Becquerel felt compelled to reply to this work because it seemingly contradicted his own observations about "continuing rays." He felt that Foucault and Fizeau had confused the issue by using plates sensitized with iodine and bromine so that the experiment now had two variables—the colored light and the mixed halides—which needed to be separated in order to interpret the real results.[52] This discussion would not have been complete without Draper, and he added his thoughts that different regions of the spectrum have different effects on the daguerreotype— some have no effect on the plate, some produce the whites of the highlights, and some produce the blacks characteristic of solarized or overexposed daguerreotypes.

In "Researches on the Theory of the Principal Phenomena of Photography in the Daguerreotype Process," a summary appearing in 1850 of the scientific work during the first decade of photography, Antoine Claudet pointed out that all aspects of the daguerreotype process had yet to be explained.[53] The phenomena still to be explained included the action of light on the surface of the prepared daguerreotype plate, the role of mercury vapor in forming an image, "which rays impart to the chemical surface an affinity for mercury," and the difference between the visible and photogenic focus in achromatic lenses. Moreover, no satisfactory ways of measuring the photographic effect had been devised. This summary was very discouraging, and, indeed, these scientific efforts had petered out by the end of the 1840s.

The diversity of observations about the science of the daguerreotype was related to the variation in the workers' experimental settings. For instance, spectral studies on daguerreotypes that used spectrometer instruments made variously of colored glass filters, lenses and prisms of assorted types of glass, and diffraction gratings generated results that could not be rigorously compared one with another. Further, there was no adequate way for all the participants in this exchange to compare their results in a quantitative way. As mentioned, there were several sets of axioms about the nature of light but no agreed-upon way to apply those axioms to the interpretation of findings. In spite of the tremendous efforts made by these intelligent and dedicated scientists, they were unable to answer the questions of what made the daguerreotype process work. The failure to achieve a scientific explanation of the daguerreotype was definitely not a failure of those who had embraced this subject, for they went on to work that was more amenable to discoveries and breakthroughs.

Draper, in spite of the caricature that he became over his fourth imponderable, made many important contributions, especially in defining the Grotthuss-Draper Law. He became his own greatest promoter, and because of his own interpretation of his role in the pioneering stages of photography he has gained almost universal but erroneous acknowledgment for having taken the first portrait. This perception is being corrected, but Draper's important contributions to photographic theory and early astronomical photography have not been given the prominence they deserve. Moser remained at the University of Königsberg but published very little subsequent to his papers on the daguerreotype and vapor images. For Becquerel the daguerreotype had been a vehicle for recording spectra, and the failure to solve its mysteries did not stop his continued studies of luminescence and other radiation phenomena. Hunt moved on to other methods of photography and pursued his various researches into all aspects of science; he was elected a fellow of the Royal Society during the 1850s. Foucault and Fizeau worked together at the Bureau des Longitude in Paris on various aspects of optics and photometry until a personal falling out in the late 1840s put an end to their joint work.

6

The Daguerreotype as a Scientific Tool

It appears at first sight most convenient to make use of sensitive papers and a darkened chamber; but we soon find that this is not the most advantageous method. No well-defined results are obtained, and we always remain in uncertainty, at least as far as regards the more delicate phenomena. The method of Daguerre is the only one which can be advantageously adopted. The operations proposed by him are calculated to produce a perfect artistical representation, and they answer their purpose effectively. . . .

Ludwig Moser, "On Vision and the Action of Light on All Bodies"[1]

LUDWIG MOSER'S ASSESSMENT of the advantages of the daguerreotype over other modes of photography for scientific data collection was the commonly held opinion of most who were interested in using photography as a tool of science. While the scientific issues surrounding the exact nature of the daguerreotype process remained unresolved and at the time were unresolvable, the application of the daguerreotype as a tool of science did move forward.

It was apparent to François Arago that the daguerreotype had great potential in science, and he used this argument as part of his justification for singling out the daguerreotype and its discoverers and granting them monetary support by the French Parliament. However, the daguerreotype's usefulness was not immediately obvious to scientists unless they were already engaged in a branch of science that relied heavily on visual observation and the communication of those observations through verbal descriptions of the appearance of phenomena or through drawings, prints, and similar media. The daguerreotype offered even experimenters with poor drawing skills a way to record visual descriptions exactly and with fine detail. This very idea had been an important motivation for those searching for a reliable tool that would use sunlight to imprint images obtained with a camera. The question that needed to be answered was, How could the daguerreotype be assimilated into the repertoire of tools available for science?

Initially, because of the limitations of the daguerreotype process itself, the most logical applications of scientific photography were in geology, for recording monuments and geological sites for purposes of exploration, defining national boundaries, and supplying military information; in archaeology; in medicine; and in astronomy. The lack of sufficient light sensitivity impeded the daguerreotype's usefulness in medicine until the process had been altered and practical portraiture was possible. The daguerreotype also found application in newly emerging fields such as anthropology and in fields now no longer considered sciences, such as phrenology.

Moser's mention of the artistic merit of the daguerreotype in the same breath that he discussed its suitability for scientific recording was no doubt intentional. The exacting detail characteristic of the daguerreotype process, not possible from the first paper-based photographic processes, meant that a daguerreotype contained a complete visual description of the total appearance of its subject. The daguerreotype image was strikingly similar in its appearance to the tradition of Dutch art dating from the seventeenth century.[2] As such, the daguerreotype falls into the art-as-description tradition already established by the Dutch two centuries before. Much like the eye and the retina, the camera and the daguerreotype could capture all the components of a scene seemingly without bias and provide powerful testimony to scientific truth and the wonders of nature.[3] In a few scientific fields, such as botany, the daguerreotype was not the preferred means of pictorial or photographic recording. For botanists and others engaged in the systematic descriptions of shapes of plant or animal parts, photogenic drawings offered sufficient detail for the accurate recording of important information. Here, the daguerreotype offered too much detail and a quality that took away from the scientific purpose for recording such images (fig. 6.1).

Another impediment to the adoption of daguerreotypy (and, for that matter, all photographic methods) as a scientific tool was the considerable skill required to make daguerreotypes. Initially, experimenters like Draper, Becquerel, Fizeau, Donné, Moser, and their colleagues produced their own daguerreotypes. In part, they did so because in the early days of the daguerreian era daguerreotypy and photography were

the frontier of science. The daguerreotype frenzy seized the imagination of people from all walks of life, and scientists were in the thick of the action. However, the daguerreotype process itself lost its charm as an object of study because its secrets were impenetrable. And as time went on the standard of recognized quality for daguerreotype images developed to such an extent that scientists found it necessary to hire professionals to produce daguerreotypes that could convey the accuracy they required in their observations.

Within a few weeks of the August 1839 announcement of the daguerreotype process, groups of artists were sent by N. M. P. Lerebours, a Paris optician and publisher, to daguerreotype the monuments of the Levant, Italy, Corsica, Spain, and France for *Excursions Daguerriennes*, the first publication that used daguerreotypes as the source of its illustrations.[4] Between 1839 and 1844 some 1,200 daguerreotypes were commissioned, shown, and sold by Lerebours in his shop on Place du Pont Neuf.

Fig. 6.1. Photogenic drawing from "Photogenic Drawings of Plants Indigenous to the Vicinity of Philadelphia," an album of 233 botanical specimens made by Mathew Carey Lea in 1841 using the process of Mungo Ponton, a photographic process based on potassium dichromate. Courtesy of The Franklin Institute Science Museum, Philadelphia.

Of these 111 were published in various issues of *Excursions Daguerriennes* between 1841 and 1843. The daguerreotypes made for these editions were made not for scientific purposes but to cater to the general artistic taste for engravings and lithographs of topological views and for travel views. Even though these daguerreotypes were often stylized and romantic, they heralded the possibilty of using daguerreotypy and photography as tools for field recording and as methods of systematically logging structural and contextual information about geological and archaeological sites. Both geology and archaeology were just emerging as separate scientific fields at midcentury, and the daguerreotype's usefulness to those fields was initially made apparent by publications such as *Excursions Daguerriennes*.

Geological Applications

The necessity of having artists as members of expeditionary teams was understood from the outset of the era of exploration, during the fifteenth and sixteenth centuries. The onset of the nineteenth century magnified this need because of the growing interest among scientific circles in geology and systematics. Obviously, the Europeans did take up the daguerreotype for this type of recording, as can be seen by the secondary themes in *Excursions Daguerriennes* and the work of European travelers such as Girault de Prangey.

Americans were also quick to see the advantages the daguerreotype offered exploratory expeditions over simple drawings, even those made using a camera obscura or a camera lucida. Words and drawings could not convey the vastness of the western regions of North America to those who had not been there. The commercial exploitation of the West for trade and the establishment of the railroads depended upon reports brought back by the great exploration surveys of the midcentury. The daguerreotype offered a superior way to present the magnitude and potential of the West to the congressmen and businessmen who provided the funds for western expansion. Whether as an intermediary for the engraver or as a quick method for accurate recording, the daguerreotype brought a great deal of promise to explorers and entrepreneurs alike. Even so, it took a while for the daguerreotype's results to match the expectations for the new art.

The first use of photography as a surveyor's tool came in 1840 as a result of the long-standing border dispute between England and the United States over the exact location of the northeastern U.S. border between Maine and New Hampshire and New Brunswick, Canada. The border commission appointed by President William Henry Harrison was headed by the architect-engineer James Renwick, then an engineering professor at Columbia University. Renwick regarded the gathering of both historical and newly made pictorial evidence on the border's location as one of the commission's primary supportive aims. Thus, it was very important to include expedition members skilled in the newest methods of visual recording to help the surveyors and metrologists, the core members of such expeditions. To that end Renwick decided to hire Edward Anthony, a former student, as a daguerreotypist on the surveyor's team. Anthony had been taking lessons in daguerreotypy from Samuel F. B. Morse and was also a civil engineer; this combination of skills made him uniquely qualified to be the first American photographic surveyor.[5] The pictorial evidence provided both by daguerreotypes and by drawings made with a camera lucida during the survey was the most powerful substantiation of the border's location. The Anthony daguerreotypes have not survived; however, drawings made from the daguerreotypes remain with the records of Renwick's commission report.[6]

The great expeditions to the western regions of North America that took place throughout much of the nineteenth century began to use daguerreotypy and photography as recording media almost as soon as these tools were available. Many photographers, such as Robert H. Vance and Carlton E. Watkins of California and Thomas Easterly of St. Louis, took western views that were exhibited and offered for sale. In California many such views were exhibited after the discovery of gold in 1848 set off the Gold Rush. Even though these daguerreotypes had much in common with those taken on the western expeditions, they were taken for different reasons and used for different purposes.

John C. Frémont was the leader of five expeditions to the West between 1842 and 1853.[7] He himself tried without much success to make daguerreotypes to record the places he visited during the first expedition, in 1842. He wished to make daguerreotypes because they would testify to the possibility of travel to the West much more convincingly than any number of drawings and words. The cartographer Charles Preuss, who traveled with Frémont on three of his expeditions, described Frémont's first attempts in this way:

Frémont set up his daguerreotype to photograph the rocks; he spoiled five plates that way. Not a thing was to be seen on them. That's the way it often is with these Americans. They know everything, they can do everything, and when they are put to the test, they fail miserably.[8]

A few months later Preuss added, "Today he [Frémont] said the air up here is too thin; that is the reason his daguerreotype was a failure. Old boy, you don't understand the thing, that is it."[9] Preuss was not so scornful of Frémont's attempts at producing images that he did not express disappointment when Frémont failed to get a particularly beautiful view. Daguerreotypes from Fré-

mont's first expedition did not survive, and it was not known until the discovery of Preuss's diaries in 1954 that daguerreotypes had even been attempted.

On the third Frémont expedition, Edward Kern was employed to record the topology of the trail using a camera lucida. Since no daguerreotype apparatus was taken along, Frémont evidently hoped that the drawings made using a camera lucida could compete with daguerreotypes. He evidently was reluctant to repeat the poor results of his previous attempts at making daguerreotypes in the field.

The next time that Frémont attempted to use daguerreotypes in his explorations was during his last expedition, in the winter of 1853–54, to investigate a route for the railroad. Frémont hired Solomon Nunes Carvahlo, who had been taking daguerreotypes since sometime in 1850, and placed him in charge of all visual documentation during the expedition. Carvahlo's often quoted descriptions of the hardships he suffered to make daguerreotypes under extraordinary conditions during the Rocky Mountain winter are testimony to both his skill and his endurance. His most striking narrative is that of making a particularly difficult and arduous climb with Frémont in order to take a daguerreotype: "After three hours' hard toil we reached the summit and beheld a panorama of unspeakable sublimity. . . . Plunged up to my middle in snow, I made a panorama of the continuous ranges of mountains around us."[10] The conditions during this last expedition became dire, and to prevent the loss of his men Frémont had to cache all items that were not absolutely necessary for survival. The daguerreotypes were kept by Frémont and carried to New York, where they were copied by Mathew Brady; some were made into paintings and engravings. Frémont's public letter to Congress tried to convince lawmakers that Carvahlo's daguerreotypes demonstrated the overall feasibility and usefulness of the railroad even during harsh winter weather.[11]

By the time Frémont and Carvahlo set out in 1853, the United States Topographical Corps already employed several daguerreotypists. While the daguerreotype was the medium of choice for Americans during this period, by 1850 Europeans were using waxed paper negatives and paper prints instead of daguerreotypes to do the same type of recording. Europeans never acquired the skill and facility with the daguerreotype process characteristic of American operators, especially during the 1850s. Americans who made expeditionary daguerreotypes included Edward Kern and John Mix Stanley. Kern, the artist who had been employed on the second Frémont expedition, learned to make daguerreotypes from Edward and Henry T. Anthony. He was employed as the daguerreotypist on the North Pacific Exploring Expedition under the leadership of Cadwalader Ringgold. It is recorded that Kern purchased 432 plates before the expedition's departure.[12] Stanley made daguerreotypes for Gov. Isaac I. Stevens's survey of the Washington territory for the Pacific railroad during 1853.

In 1852 Eliphalet Brown, Jr., was hired to make daguerreotypes for Commodore Perry's first expedition to Japan. Since Congress had not expressly granted the authority for Perry to hire artists, Brown had to petition Congress for payment for his services after he returned to the United States. He claimed:

As the employment of artists was so very essential to the success of an expedition like that in charge of Commodore Perry, the failure on the part of Congress to confer the authority may be safely charged to inadvertence rather than design. Commodore Perry foresaw, what he supposed, must be the evil results of this inadvertency and engaged artists to join the expedition. They were enlisted as master's mates, with the understanding and expectation that they would seek a suitable compensation from Congress after the return of the expedition.[13]

Brown served on several ships for Perry between March 1852 and December 1854, working both as a daguerreotypist and a draftsman. He furnished his own daguerreotype apparatus and took more than four hundred daguerreotypes, which all became the property of the United States government. Engravings made from these daguerreotypes were used to illustrate Perry's reports (fig. 6.2). Brown sought and received $1,500 per year for his services during the time he served the Perry party.

Anthropological and Archaeological Applications

The first use of the daguerreotype for anthropological research came in October 1839 when John Lloyd Stephens and Frederick Catherwood set out on an expedition to the Yucatan. Stephens was already famous for his writings on travels in the Levant and for being the first American to visit Petra. Catherwood was an English

Fig. 6.2. Lithograph by T. Sinclair (Philadelphia) from a daguerreotype by Eliphalet Brown, Jr., taken during the Perry expeditions to the Pacific region, 1852–56. Source: Hawks, *Narrative of the Expedition of an American Squadron to the China Seas and Japan Performed in the Years 1832, 1853, and 1854*, plate facing p. 419.

painter and an engraver known for his paintings of the Levant. The two met during one of Stephens's trips through London because of their mutual interest in the Mideast. When the two set out in 1839, only three historical sites were known in the Mayan area of the Yucatan and Central America. They returned to the United States in 1840 and left on a second trip to the Yucatan at the end of 1841. At the close of their second trip they had found and described forty-four Mayan sites; only one, Uxmal, had been known previously. Dr. Samuel Cabot, a physician from Boston, was the third member in their party on their second expedition.

During their second trip Stephens and Catherwood invited another traveler, Baron Emmanuel von Friedrichsthal, to make daguerreotypes while he was traveling in the Yucatan,[14] but his daguerreotypes were later lost. However, Stephens and Catherwood had also brought daguerreotype apparatus, and along with Cabot the three made daguerreotypes more or less successfully. Catherwood was the first person to use the daguerreotype process extensively for archaeological research. He had hoped the daguerreotype would simplify the recording of the buildings and ruins in the archaeological sites they visited, but his "results were not sufficiently perfect to suit his ideas."[15] The three were plagued with problems ranging from difficulties with the daguerreotype process itself to unfavorable weather and sickness among the members of the party. The daguerreotype process could not adequately record the contrast range of scenes that included the extremes of light—from bright sunshine to deep shade—and capturing the contrast and detail in the buildings and ruins they discovered was impossible. Thus, Catherwood resorted to the use of a camera lucida to make additional drawings to go along with the daguerreotypes, while Stephens and Cabot concentrated on making da-

Fig. 6.3. Engraving of ruins at Chunjuju from daguerreotypes by John Lloyd Stephens and Samuel Cabot and camera lucida drawings by Frederick Catherwood made during their second expedition to the Yucatan in 1841. Source: Stephens, *Incidents of Travel in Yucatan*, pl. 134.

guerreotypes. Their faith in the reliability of daguerreotypes over an artist's drawings was attested to by Stephens in this report after the expedition: "[I]n order to insure the utmost accuracy, the daguerreotype views were placed with the drawings in the hands of the engravers for their guidance."[16]

The Stephens-Catherwood daguerreotypes and all the artifacts that they had collected during their second expedition were exhibited and destroyed in a fire on July 31, 1842, within weeks of their return from the Yucatan. The content of the daguerreotypes was preserved because engravings had been made as soon as the travelers arrived back in the United States. Stephens's book, *Incidents of Travel in Yucatan*, first published on March 23, 1843, contained eighty-five engravings from views taken on their last expedition (see fig. 6.3).

Just as the daguerreotype was employed to record monuments and land forms, as soon as portraiture was possible the daguerreotype was used to record living forms for anthropological, ethnographical, biological, and medical study. The daguerreotype portrait allowed a scientist to make careful scrutiny of subjects for purposes of taxonomy and for making systematic catalogues. The underlying belief in the veracity of the camera added a potent new political tool to midcentury scientists who were laying down the fundamentals of both modern science and modern society. Central to the scientific applications of the daguerreotype portrait are various aspects of craniometrics, sciences based on skull and brain measurements.

Phrenology

With regard to the daguerreotype, the principal among these sciences was phrenology. While this "science" is largely ridiculed today, during the nineteenth century it was an influential system that colored many other branches of science. The science of phrenology dates from the end of the eighteenth century and the early nineteenth century, when Franz Joseph Gall, an anatomist who specialized in the head and the brain, and his follower Johann Christoph Spurzheim defined a system to relate the physical attributes and appearance of the head and brain (i.e., its size and shape) to mental qualities and capacity. Spurzheim became Gall's student in Vienna and moved with him to Paris in 1807. Shortly afterward, the two split up, and Spurzheim became the first propagandist for phrenology, the name he attached to Gall's system. He lectured widely in England and France and died while on a lecture tour in Boston in 1832.

Another early promoter of phrenology was the Scottish anatomist George Combe. Initially very critical of phrenology, Combe made a deeper study of it after being favorably impressed by Spurzheim's skill in dissection at a private meeting during one of the latter's lecture tours in Scotland.[17] Thereafter, Combe became one of the great popularizers of phrenological science both in Europe and the United States. His brother, Andrew, became involved in phrenology while completing his medical studies in Paris between 1817 and 1819. The two Combes founded the Phrenological Society in 1820, and in 1823 they began to publish the *Phrenological Journal*.[18] This interest carried over into the United States, where the great proponents of phrenology were Orson S. and Lorenzo N. Fowler. These two brothers became acquainted with Spurzheim's ideas while students at Amherst College in Massachusetts in 1834. Phrenology came to the United States in a big way in 1838, when George Combe began a two-year lecture tour in America. That same year the Fowler brothers founded the American Phrenological Association and began the *American Phrenological Journal*.

In essence, phrenology was based on the following premises: The skull corresponds to the shape of its interior parts, that is, the shape of the brain. The faculties and powers of the mind are seated in specific localities in the brain, and thus specific enlargements or diminutions of parts of the brain correspond to changes in the mental capacities of a particular person. The mind of a specific person can be analyzed and treated, if necessary, by analyzing the shape of the various portions of the skull and head. While analyzing bumps on the head is now regarded as fairly silly, phrenologists were the first to espouse the idea of the brain as the seat of the mind and to suggest that different functions are associated with specific areas of the brain. The implication that changes in the outward shape of a person's head accompany changes in his or her behavior resulted in the evolution of theories about treatment of various physical and mental conditions. Initially, at least, many phrenologists were physicians, anatomists, and scientists of various kinds. At that time, while phrenology did have many detractors, one could have sound, scientific reasons for espousing that area of inquiry. In addition, phrenological thought became tied to the teachings of many social reformers and influenced many popular movements during the period. As popularized through the various phrenological journals and books, phrenology affected prevailing scientific points of view espoused by those outside the scholarly and scientific elite.

What does phrenology have to do with daguerreotypes? This science was based on both visual and tactile information about head shapes, and daguerreotypes provided records of specific "types" that could be studied and reproduced in journals. The nearly simultaneous popularization of phrenology by Combe and the Fowler brothers and the introduction of the daguerreotype wedded the two professions together for some practitioners. Indeed, some American daguerreotypists were also practicing phrenologists.

The American edition of Marmaduke B. Sampson's book *Rationale of Crime and its Appropriate Treatment*,[19] a treatise devoted to the idea that criminals have some kind of faulty brain development or "disease" that can be corrected through phrenological treatment, was illustrated with engravings made from daguerreotypes. First published in England in 1843, Sampson's ideas were quite influential and sparked a great deal of interest and following among those involved in prison reform. The champion of this book in the United States was Eliza Farnham, the matron of the women's prison at Sing Sing in New York. She wrote a set of practical notes to accompany the American edition and also decided to include a set of illustrations to bring out Sampson's views. Farnham hired Mathew Brady, an unknown daguerreotypist at the time, to make the daguerreotypes used in her edition of Sampson's book (see fig. 6.4). Brady's first daguerreotype studio, at the corner of Fulton and Broadway in New York City, was just a block away from the Nassau Street location of the Fowler brothers' phrenological enterprises. Farnham

knew the Fowlers, and perhaps they suggested that she contact Brady, who later became involved in the phrenological movement.

In keeping with the general interest in physiognomic ideas, many American daguerreotype studios displayed daguerreotypes of customers and others both as a kind of advertising and also as a form of propaganda for "the wonders of nature" and the general scientism of the age. This latter quality is seen in some daguerreotypes that carry messages beyond the simple depiction of the sitter. For instance, the daguerreotype portrait made by an unknown daguerreotypist around 1847 of Laura Bridgeman and her teacher, Sarah White Shaw (fig. 6.5), is a testimony to the triumph of Dr. Samuel G. Howe's modern scientific approach and his Perkins Institute for the Blind in training the blind to lead productive lives. Bridgeman was the first person blind and deaf from early life to be educated, and her case was well known and widely celebrated. Charles Dickens wrote about Bridgeman in *American Notes*,[20] and later Helen Keller was strongly influenced by Bridgeman.

Fig. 6.4. Engraving of Blackwell's Island inmate "S.S.," made from a daguerreotype by Mathew Brady. Eliza Farnham commissioned this and other daguerreotypes to show the phrenological characteristics of typical criminals. Source: Sampson, *Rationale of Crime and its Appropriate Treatment*, p. 157.

Fig. 6.5. Daguerreotype portrait of Laura Bridgeman and her teacher, Sarah White Shaw, made by an unknown daguerreotypist around 1847. Bridgeman was brought to Boston in 1837 and placed under the care of Dr. Samuel G. Howe at the Perkins Institute for the Blind. Courtesy of Paul Katz.

Fig. 6.6. Portrait of Renty, a Congo-born slave from Edgehill, the plantation of B. F. Taylor in Columbia, South Carolina. This series of portraits was commissioned by Louis Agassiz to show the characteristics of the "African race" and was taken by the daguerreotypist J. T. Zealy of Columbia. Courtesy of the Peabody Museum, Harvard University.

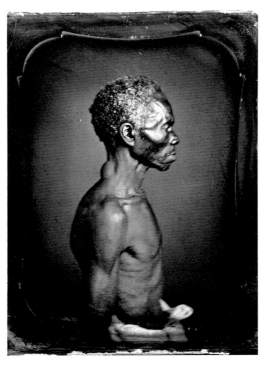

Polygeny

In more conventional scientific circles many scientists were involved in the classification and cataloguing of all manner of natural systems. One important group of classification scientists was those who believed in polygeny, the idea that the various races are separate, unrelated species. This belief held currency in many fields of biological inquiry and was used to provide scientific support for the continuation of slavery within the United States and elsewhere. Dr. Samuel Morton, a prominent anatomist from Philadelphia, had come to the conclusion, through his measurements of the cranial capacity of hundreds of skulls collected from many locations in both the New World and the Old World, that there were many races, each created by God to be specially adapted for life in particular geographical regions. He published two influential books, *Crania Americana* (1839) and *Crania Aegyptica* (1844) about his discoveries, which were read widely and both de-

cried and lauded.[21] Belief in polygeny was sufficiently prevalent that it was considered characteristic of the "American School" of anthropology. As with phrenology, many of the exponents of polygeny used craniometrics to demonstrate the superiority of the Caucasian race over all other races.[22]

Another important and influential proponent of the idea of polygeny was Louis Agassiz, also known as the father of American biology. He was brought from Switzerland to Harvard College in the mid-1840s to help raise the recognition of science in America and establish the Museum of Comparative Zoology at Harvard (now the Peabody Museum of Archaeology and Ethnology). Almost immediately upon his arrival in the United States, Agassiz encountered black-skinned people for the first time, and his revulsion of them was immediate and complete. This emotional reaction cemented his beliefs, not grounded on any empirical evidence, and influenced his subsequent writings and teachings on the subject of the origins of humankind. In addition, Agassiz became acquainted with Morton, his followers and their ideas on the subject. Agassiz publicly affirmed his support of polygeny through statements made during the 1850 meeting of the Association for the Advancement of American Science in Charleston, South Carolina.[23] Before he returned to Harvard, he visited plantations near Columbia, South Carolina, to examine African-born slaves for further proof of his theories on separate races. He commissioned a series of daguerreotype portraits of the slaves he examined to illustrate the characteristic features of the black race (see fig. 6.6). These daguerreotypes were taken by J. T. Zealy and sent back to Agassiz at the Museum of Comparative Zoology. Polygeny and its attendant beliefs were dealt a deathblow by the publication of Charles Darwin's *Origin of the Species* in 1859 and the effects of the Civil War; however, the daguerreotypes were stored away and remained forgotten until they were rediscovered in 1977.[24]

Systematics

Taxonomists and anatomists engaged in the classification of plants and animals into systematic hierarchies also used body shape and measurement to accomplish their work. The teachings of men such as Carolus Linnaeus, George Louis Leclerc, Compte de Buffon, and Jean Baptiste Lamarck during the eighteenth and early nineteenth centuries ultimately led to the theory of evolution as expressed in *The Origin of the Species*. In all these classification schemes, illustrations played a major role in conveying important portions of the scientific arguments upon which the systems were based. In this field, too, daguerreotypy and photography offered important and seemingly impartial tools for providing evidence for new theories. For instance, the copy of the daguerreotype of a laboratory at the Philadelphia Academy of Natural Sciences (fig. 6.7) is a rather formal portrait of various laboratory workers along with part of a collection of vertebrate skeleton specimens. Daguerreotypes such as this one impart a great deal of information about the layout of laboratories and the types of specimens found there, as well as provide portraits of the workers associated with a particular laboratory.

Medical Applications

"Medical" daguerreotypes assumed a variety of guises. Like other types of scientific daguerreotypes, they often seem awkward and stylized because there were no conventions for posing subjects for medical photography. Some doctors began to use daguerreotypes of patients afflicted with various diseases or disease symptoms as teaching devices as soon as the daguerreotype appeared. The daguerreotype in figure

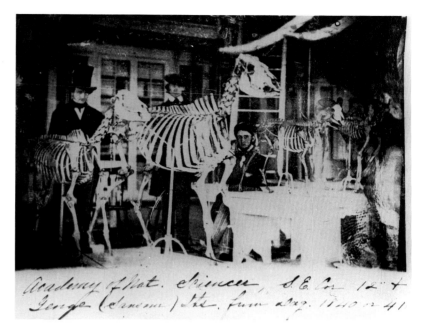

Fig. 6.7. Gelatin copy photograph made before 1900 of an early daguerreotype taken in a laboratory at the Academy of Natural Sciences of Philadelphia. The daguerreotype, made about 1840, is attributed to Paul Beck Goddard. The figures are unidentified; however, it has been conjectured that the central figure is Joseph Leidy, who became a leading American paleontologist. The seated figure on the right may be Edgar Allan Poe, who worked in Philadelphia assisting Thomas Wyatt of the Central High School in writing books on natural history. Source: Library, Academy of Natural Sciences of Philadelphia, coll. 9, folder no. 7, acc. no. 144 (neg. no. 000696).

Fig. 6.8. Portrait of a woman after her nose was reconstructed using a flap of skin taken from her forehead. This daguerreotype, taken by John Adams Whipple, is from the teaching collection of pathology specimens that belonged Dr. Henry J. Bigelow, Massachusetts General Hospital. Courtesy of Fogg Art Museum, Harvard University, Cambridge, Massachusetts; Loan—Massachusetts General Hospital.

6.8 is an example from the collection of teaching daguerreotypes commissioned by Dr. Henry J. Bigelow of Massachusetts General Hospital during the 1850s. It shows an unidentified woman's nose after it was reconstructed with a flap of skin taken from her forehead. Nevertheless, for some physicians the daguerreotype itself may not have been as satisfactory for the depiction of various pathologies as colored drawings or lithographs, for it lacked the nuance and subtlety of color and tone that could be achieved in a drawing. As today, certain types of medical illustrations are drawn because a drawing more effectively conveys the most pertinent visual information filtered from information that might be considered extraneous or distracting.

Daguerreotypes were used to record or commemorate medical events such as the first use of anesthesia in surgical operations, documented by a series of daguerreotypes taken by Albert Sands Southworth and Josiah Johnson Hawes at Massachusetts General Hospital in Boston. These daguerreotypes were taken sometime after the actual surgical operation, which was the first performed using ether anesthesia and was held for public view on October 16, 1846. Figure 6.9, one of the series, shows Dr. William T. G. Morton, the dentist credited with the discovery of ether anesthesia (in the fancy waistcoat), and Dr. John Collins Warren, the chief surgeon of Massachusetts General Hospital (seventh from left), who performed the operations. These daguerreotypes were taken from the gallery of the operating theater; thus, they have an aerial angle of view reminiscent of the operating theaters seen in engravings and paintings of the Dutch realists.

Fig. 6.9. Daguerreotype of a reenactment of one of the first surgical operations performed using ether anesthesia. The surgeon who performed the operation was Dr. John Collins Warren, pictured third from right, and Dr. William T. G. Morton, the dentist who introduced the use of ether is pictured at the center dressed in the fancy waistcoat. Courtesy of Fogg Art Museum, Harvard University, Cambridge, Massachusetts; Loan—Massachusetts General Hospital.

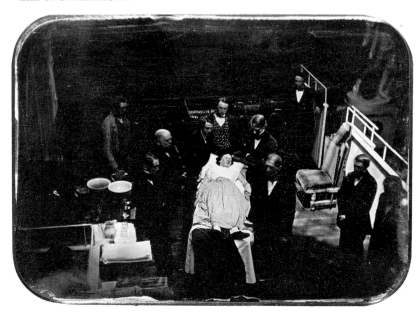

Diseases and deformities were recorded on daguerreotypes in a variety of ways. Some were made directly for medical records. Some were daguerreotypes taken as records of freaks, such as the portrait of the unidentified man with no limbs, probably the result of some sort of birth defect (possibly interuteroamputation) (fig. 6.10) or that of an unidentified man afflicted with leprosy (fig. 6.11). And some were daguerreotypes taken as postmortem portraits (fig. 6.12). During the daguerreian era, daguerreotypes of the latter two categories were not taken for medical records but have been used in subsequent times in studies of disease pathologies.

An impression of the practice of medicine during the midnineteenth century can be gleaned from daguerreotype portraits of the physicians themselves. Some of these portraits are very formal and much like all other depictions of the human visage. Some, like the occupational daguerreotype portrait of a dentist with a patient strapped in his chair (fig. 6.13), convey some idea of health care practice during the 1840s and 1850s.

As mentioned in the previous chapter, Alfred Donné was the first to make daguerreotype micrographs of biological tissue specimens. By attaching a plate holder to the eyepiece end of a microscope tube,

Fig. 6.10. Portrait of an unidentified man born without limbs, possibly because of interuteroamputation, taken by an unknown daguerreotypist. Source: Collection of Paul Katz.

it was possible to record specimens on a microscope slide. Few of these daguerreotypes survive, although few may have been taken. The skill required to obtain a daguerreotype micrograph would have been very exacting because of the low light levels

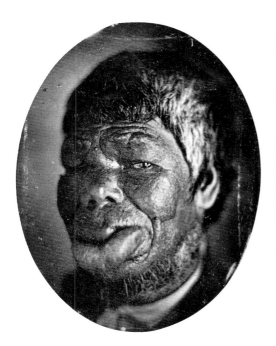

Fig. 6.11. Portrait of an unidentified man with lepromatous leprosy taken in the late 1840s or early 1850s in Saint Louis by Thomas Easterly. Source: Collection of Paul Katz.

Fig. 6.12. Postmortem portrait of a young girl. Source: Collection of Kenneth E. Nelson.

Fig. 6.13. Occupational portrait of a dentist and his patient taken by an unidentified daguerreotypist. Courtesy of Matthew R. Isenburg.

available even using a solar microscope and the blue sensitivity of the daguerreian system. The reddish or brownish coloring of many live tissue samples would have been recorded merely as black blobs by the daguerreotype. Histological staining of specimens might have helped the situation; however, until the early 1860s stains were largely limited to cochineal and carmine.[25] The red color of these dyes would have compounded the daguerreotype's problems with spectral sensitivity.

Astronomical Applications

The acme of scientific daguerreotypes were those made for astronomy. When Arago first met Daguerre and Niépce, he felt that the daguerreotype would have great potential for the study of the sun and the moon. At Arago's request, Daguerre captured the faintest hint of an image of the moon on a daguerreotype plate even before the details of the process were made public.[26] The results, Arago enthusiastically reported to the

Chamber of Deputies, raised his hopes that it would soon be possible to make photographic maps of "our own satellite" in a few minutes without the arduous and biased task of drawing what is seen at the eyepiece of the telescope.

Astronomical daguerreotypy required the mastery of a combination of both photographic and mechanical skills. To make an astronomical daguerreotype, the telescope is made into a camera by positioning the unexposed plate in a holder at the eyepiece of the telescope, much like a plate is placed on a microscope. Further, to compensate for the movement of the earth relative to the other heavenly bodies, the telescope had to have an accurate tracking mechanism or clock drive to move it during exposures; otherwise, images would appear smeared, blurred, or misshapen on the plate. The low light levels of most stellar objects required the daguerreotypist to make plates with the utmost sensitivity. In addition, inherent limitations in the daguerreotype process further restricted what could be daguerreotyped. The primary blue sensitivity of the daguerreotype presented two distinct limitations. First, it limited the selection of stellar objects that could be considered for daguerreotyping. Red stars, for instance, were beyond the reach of the daguerreian system. Second, when daguerreotypes were made using a refractor telescope or one in which light is focused by being passed through glass lenses, the plate had to be placed at the point along the telescope barrel where blue light comes to a focus. This point was called the chemical focus and is not the same as the visual focus. Because of prismatic effects due to refraction of the light by the lenses, radiation of different colors comes to a focus at different points along the telescope barrel. The difference between chemical and visual focus was a problem for every application of the daguerreotype process where light passed through a lens before exposure; moreover, the increased focal length of a telescope magnified the problem so that it could not be ignored.[27]

All these limitations curtailed the widespread use of the daguerreotype in astronomy, and despite hopes for the daguerreotype it took almost ten years before any concerted effort was made toward the systematic application of photography to astronomical observation. Isolated experimenters had individual successes, but ultimately astronomer-daguerreotypist teams made the most progress in this area because the team members could perform the tasks in which they had the most experience without having to master the skills of another discipline.

The first successful lunar daguerreotype was made by John Draper in the spring of 1840. At a meeting of the New York Lyceum of Natural History on March 23, 1840, he announced this accomplishment and described the conditions under which the daguerreotype was taken. His presentation was summarized in the minutes of the meeting:

Dr. Draper announced that he had succeeded in getting a representation of the moon's surface by the daguerreotype. A portion of the figure was very distinct, but owing to the motion of the Moon the greater part was confused. The time occupied was twenty minutes, and the size of the figure was about one inch in diameter. Daguerre had attempted the same thing but did not succeed. This is the first time that anything like a distinct representation of the moon's surface has been obtained.[28]

Draper experimented with various arrangements of heliostat, lenses, and camera throughout the winter of 1839–40, and the daguerreotype he showed and subsequently presented to the Lyceum was his best effort.[29] He also reported on his success for his foreign colleagues at the end of one of his first papers on the daguerreotype.[30] This

substantial contribution to astronomy and daguerreian art was only minimally highlighted by Draper himself. While he was never shy about claiming his undeserved priority in producing portrait daguerreotypes, he made scant mention of his lunar daguerreotypes, where he truly could claim to have had the first success. Perhaps he was not as satisfied with his lunar results as he was with his portrait of his sister.

The most obvious stellar object waiting for capture on the daguerreotypist's plate was the sun. Several experimenters, including Draper and Becquerel, recorded the solar spectrum on daguerreotype plates, but actually making an image of the sun's surface was not so straightforward. Reportedly N. M. P. Lerebours made a daguerreotype of the sun in 1842 at the behest of Arago, but the results were found to be unsatisfactory because the solar disk was overexposed and solarized. Arago then instructed Léon Foucault (fig. 6.14) and Hippolyte Fizeau, who were both working at l'Observatoire, to try to daguerreotype the sun using shorter exposure times than

Fig. 6.14. Portrait of Léon Foucault. Source: Figuier, *Les Merveilles de la Science,* fig. 95.

Fig. 6.15. Lithograph made from a daguerreotype of the sun with sunspots taken on April 2, 1845, 9:45 A.M., by Hippolyte Fizeau and Léon Foucault. The original daguerreotype is in the collection of Conservatoire National des Arts et Métiers in Paris (item 17551). Source: Arago, *Astronomie Populaire*, fig. 163.

those used by Lerebours.[31] The pair was finally successful and made a daguerreotype of the sun on April 2, 1845. A lithograph of the daguerreotype was published in *Astronomie Populaire* (fig. 6.15).[32] This daguerreotype was of particular note because it showed two groups of sunspots.

The possibility of recording eclipses was also an area where the daguerreotype could be used to record the transit of the shadow of the interposed celestial body as it blocks out the eclipsed object. The calculations for determining the times and regions for viewing eclipses were already worked out before the daguerreian era, so preparations for recording eclipses could be made in advance. The first eclipse captured on a daguerreotype was the solar eclipse of July 8, 1842, which was visible in parts of Europe, across Russia to China, and in the Pacific. The Italian astronomer Giovanni Alessandro Majocchi, a professor of physics at the Lyceo di S. Alessandro in Milan during the time of the eclipse, was able to daguerreotype the event. He needed a two-minute exposure to make his daguerreotype, which recorded the eclipse just before totality; he was unable to make any daguerreotypes of the sun's corona at totality.[33]

The Whipple-Bond Alliance

The impetus needed to push forward astronomical daguerreotypy came from an unexpected quarter. On September 1, 1849, Samuel D. Humphrey, a daguerreotypist in Canandaigua, New York, made two daguerreotypes of the moon (fig. 6.16). These showed multiple exposures of the full moon. The notations shown on the side of the image indicate the length of exposure for each image. Note that the longer exposures resulted in spread-out images of the moon because Humphrey's telescope did not have a tracking mechanism, and he was not able to trail the moon's motion. Humphrey described these daguerreotypes and their preparation in detail in *The Daguer-*

reian Journal of November 1, 1850.[34] He also sent one to Jared Sparks, the president of Harvard College.[35]

Two years previously, using funds gathered by public subscription Harvard had purchased a very expensive, foreign-made 15-inch refractor telescope, then the largest telescope anywhere. The Great Equatorial, as it was called, was the centerpiece of the newly founded Harvard College Observatory. The first director, William Cranch Bond, understood that to maintain public interest in and support of the observatory he needed to highlight programs of broad appeal and interest. As a result he opened the observatory for public observation, and he and his sons made major contributions to the important business of accurate time determination.[36]

On October 23, 1847, just four months after the installation of the Great Equatorial, Bond recorded in the observatory diary that John Adams Whipple had brought out his daguerreotype equipment to attempt to make daguerreotypes of the sun and moon using the new telescope. Whipple (fig. 6.17), a leading daguerreotypist in Boston, often publicized his studio business using photographs that demonstrated the limits of what photography (and he) could offer. It is not clear whether Bond made the initial contact with Whipple to ask for assistance in astronomical photography or whether Whipple suggested that the new observatory would be an ideal place to start such work. In any event the Whipple-Bond association was very fruitful and marks the beginning of celestial photography. Their first attempt at celestial daguerreotypy was cut short when Bond accidentally got his coat sleeve burned as the sun became aligned in the objective of the telescope and the lenses acted as a burning glass. Despite the mishap, Whipple went away satisfied that it would be possible to record celestial phenomena using the Great Equatorial.[37] Subsequently, public interest in Hum-

Fig. 6.16. Multiple-exposure daguerreotype of the full moon taken by Samuel D. Humphrey at Canandaigua, New York, on September 1, 1849. The notations on the side of the image indicate the length of exposure for each image. Note that the moon's shape is distorted for the longer exposures because the telescope used to make the daguerreotype had no tracking mechanism. Courtesy of Harvard College Observatory, item no. 6806-18G.

Fig. 6.17. Portrait of John Adams Whipple, drawn on stone by Francis D'Avignon, which appeared in the August 1851 *Photographic Art Journal.* Courtesy of the Boston Athenaeum.

phrey's daguerreotype encouraged President Sparks to lend his weight to advance the work in astronomical daguerreotypy at Harvard.

Over the next decade Whipple and his partners, William B. Jones (until 1851) and James Wallace Black (from 1856 to 1859), were regularly involved with photography at the Harvard College Observatory. As time went on Bond's son, George Phillips Bond, became the primary representative from the observatory involved in the daguerreotype experiments. Whipple's dedication to the association was remarkable in that during the day he maintained a busy and successful studio trade and at night volunteered at the observatory when weather permitted photographic work. George Bond praised Whipple and Black in a letter written in 1857 to William Mitchell, the overseer of Harvard and a prominent amateur astronomer:

To be of essential service to astronomy, it is indispensable that great improvements be yet made, and these, I feel sure, will not be accomplished without a great deal of experimenting. To do this properly we need for at least a year to come the services of the excellent artists [Whipple and Black] who have literally given us their assistance, expensive materials and instruments. They should be liberally remunerated, and feel at liberty, when the prospect is good for a fair night, to give up their day's business and come to the work fresh and fit to spend the whole night at the telescope. As matters are at present they come to the observatory thoroughly exhausted, for it generally happens that the best nights are preceded by their busiest days. They make no charge for their time, costly chemicals and instruments, and cannot, in conscience, require more of men utterly exhausted than they have done. But could we press this matter on, we should soon be able to say what we can and what we cannot accomplish in stellar photography—the latter limits we certainly have not reached as yet.[38]

The accomplishment of the Whipple-Bond team that grew out of this dedication was unparalleled throughout the world.

During 1848 experiments to make daguerreotypes at the observatory were unsuccessful. The next attempts were made near the end of 1849, a few weeks after the arrival of the Humphrey lunar daguerreotype. The December efforts, too, yielded no results, and other problems within the Harvard community ended the work for that year. On July 17, 1850, the team daguerreotyped the star Vega (alpha Lyrae), the first star aside from the sun that had been captured by the camera. The importance of this particular daguerreotype was far greater than the image itself for it demonstrated that a photographic image could be used to measure the position and distance between double stars.

The team also made some images of the moon, but these were not very successful because the original German mechanical clock drive on the telescope did not work with a smooth and constant motion and had no slow-motion controls for guiding. The team also had problems with the shimmer in the atmosphere caused by the differential refraction of light through the heated air near the earth's surface. This shimmer made the objects that Whipple and Bond tried to daguerreotype appear to waver and shake. These two problems made it extremely difficult to get clear, sharp images.

On March 12 and 14, 1851, Whipple and Bond finally achieved daguerreotypes of the moon that were "a better representation of the Lunar surface than any engraving" (fig. 6.18).[39] They were able to compensate for the faulty clock drive by monitoring their exposure so that it was made only during the times that the drive was steadily following the moon. They reported on March 22 that they had another first: they had daguerreotyped the planet Jupiter. Two of the planet's bands were visible on these daguerreotypes. The only thing that dampened these successes was that the length of time necessary for properly exposed daguerreotype plates of celestial objects discouraged further experimentation until a better clock drive could be installed on the Great Equatorial. Thus, aside from some daguerreotype activity by Whipple during the first few months of 1852, the Whipple-Bond photographic experiments were suspended for five years. Altogether, they made some seventy daguerreotypes, more than any other group.

At the Great Exposition in 1851 Whipple received a gold medal for one of his lunar daguerreotypes that was displayed at the Crystal Palace. George Bond spent the spring and summer of 1851 in Europe making the acquaintance of the directors of the major observatories in Europe, presenting each with daguerreotype copies of their successful celestial daguerreotypes. These daguerreotypes and the prize-winning daguerreotype were widely acclaimed throughout Europe, for the results obtained at Harvard were unmatched anywhere else in the world. Indeed, Whipple's prize-winning daguerreotype had a profound effect on the English astronomer Warren de la Rue, who became the most well-known early advocate of astronomical photography. De la Rue often acknowledged that the Whipple daguerreotype inspired him to advocate the establishment of photographic recording of celestial events for all observatories. The Whipple-Bond daguerreotypes were clearly the apex of astronomical daguerreotypy.

National and International Cooperation

Other experimenters also worked in the area of astronomical daguerreotypy. In 1850 the British Association for the Advancement of Science formed a committee, headed by Sir John Herschel, which issued

Fig. 6.18. An early daguerreotype of the moon, taken by John Adams Whipple and George Bond. Courtesy of the Harvard College Observatory, item no. 7908-36A.

"Suggestions to Astronomers for the Observation of the Total Eclipse of the Sun on July 28, 1851." This committee stated that

any fixed observatory within the path of the shadow [of the eclipse], which is furnished with a telescope mounted equatorially, and moved by very good clock-work (adapted in its rate to diurnal movement of the sun), it is extremely desirable that arrangements should be made for Daguerreotyping or Talbotyping the image of the sun, or of the light surrounding the moon when the sun is hidden.[40]

This eclipse, which was visible in Scandinavia, Russia, and North America, was recorded from many venues. George Bond joined the many European astronomers and the handful of Americans who gathered in Lilla Edet, Sweden, to view the eclipse. This village was chosen as a viewing site because of its proximity to the central line of the path of the eclipse. Bond made a series of drawings of the eclipse; one drawn at the totality was published in the United States.[41] Several daguerreotypes were taken of the eclipse. The one claimed to be the best was made at the Königsberg Observ-

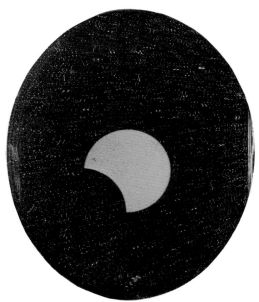

Fig. 6.19. Daguerreotype of the partial solar eclipse taken July 28, 1851, by John Adams Whipple and William Cranch Bond at the Harvard College Observatory. Courtesy of the Harvard College Observatory, item no. 6702-12D.

atory in Prussia by the daguerreotypist Berkawski at the request of the observatory director, Dr. August Ludwig Busch.[42] Berkawski's daguerreotype showed the corona of the sun at the moment of totality. Other daguerreotypes were made in Leipzig by the daguerreotypist Hermann Krone, in Warsaw by the daguerreotypist Bayer, and in Rome by Father Pietro Angelo Secchi, director of the observatory of the Collegio Romano.[43] Across the Atlantic Whipple made a series of daguerreotypes of the partial eclipse, which was all that was visible in Massachusetts (fig. 6.19).

Much like the call for an international effort to monitor the 1851 eclipse, the American Association for the Advancement of Science issued a call for national coordination of the observations of the annular solar eclipse of May 26, 1854. This eclipse was visible primarily in the United States. The report from the AAAS committee listed the items that all observers should note, including items to keep track of if one or more workers were present at an observing station. One task in the later category was to make photographs of the progress of the eclipse.[44] On the day of the event, many observing sites were overcast, and the eclipse could not be seen. However, a line of sites with clear weather followed south from New York to Washington, and at many of these sites the eclipse was successfully recorded by daguerreotypists and photographers hired by the astronomers. They included daguerreotypist M. A. Root, who worked at an observatory in New York City with professors Elias Loomis of the University of the City of New York and John L. Campbell of Robert and Lee University; daguerreotypists E. H. Olds, in Ogdensburgh, New York, and Robert M. Boggs, in New Brunswick, New Jersey, both working with Professor Stephen Alexander of The College of New Jersey[45]; and daguerreotypists William and Frederick Lagenheim of Philadelphia, who

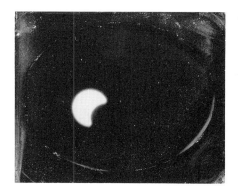

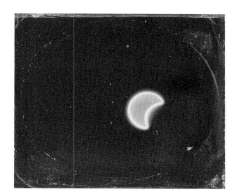

Fig. 6.20. Series of sixth-plate and sixteenth-plate daguerreotypes taken on May 26, 1854, by William and Frederick Langenheim during the annular solar eclipse. Courtesy of the Gilman Paper Company.

made an extraordinary set of daguerreotypes (fig. 6.20). Reportedly, daguerreotypes were also made under the direction of Professor W. H. C. Bartlett at West Point, New York; however, Bartlett reported that he had hired the New York photographer Victor Prevost to record the eclipse. Prevost made nineteen photographs that were reproduced in the *Astronomical Journal* by means of a tipped-in paper photograph.[46] There is no evidence that the Prevost photographs were daguerreotypes.

Ultimately, the daguerreotype was merely the herald for astronomical photography. In 1851 the wet collodion negative process was introduced by Frederick Scott Archer and gradually replaced the daguerreotype. The new process had increased sensitivity and a broader spectral range than the daguerreotype process, making it more suitable for astronomical purposes. As with the 1854 eclipse, for a while both daguerreotypes and collodion negatives were used interchangeably because it took some time for workers to become skillful in the application of the new process. Even with the introduction of the collodion process, however, astronomers still had to work with photographers because the level of skill required to record minute objects required more time to attain than most astronomers wanted to spend.[47] By the mid to late 1850s daguerreotypes were no longer used by astronomers. The daguerreotype process was revived briefly to record a singular celestial event—the Transit of Venus in 1874.

Four times every 243 years Venus appears to transit the solar disk, much like the moon does during a solar eclipse. These rare astronomical events, each called the Transit of Venus, occur in pairs separated by eight years. The last Transits of Venus were in 1874 and 1882, and the next will be shortly after the turn of the twenty-first century—in 2004 and 2012. The astronomical community perceived the significance of these events because it was thought that by timing the Transit of Venus across the solar disk from limb to limb, in several widely spaced observation sites, the exact distance between the sun and the earth could be calculated. Thus, scientific teams were sent

out during each of the last four transits to make these measurements. In every case the data gathered were insufficient to determine the distance because of limitations in the equipment available, weather problems, and other variables. The sun-earth distance has since been determined by other means, and the significance of the Transit of Venus has diminished.

Because of the advent of photography, the nineteenth-century Transits of Venus were especially important. An international effort, headed by Warren de la Rue, coordinated the photographic activities of the various investigation teams sent out by England, Germany, France, and the United States.[48] Photography enabled the members of these teams to record the Transit of Venus as the event was occurring. It was hoped that these record photographs could later be used to determine the exact time that Venus touched and let go of the edge (or limb) of the solar disk. Because it would be necessary to make exact measurements from these photographs in order to calculate the distance between the sun and earth, there was a great deal of interest in assessing the dimensional stability of wet collodion

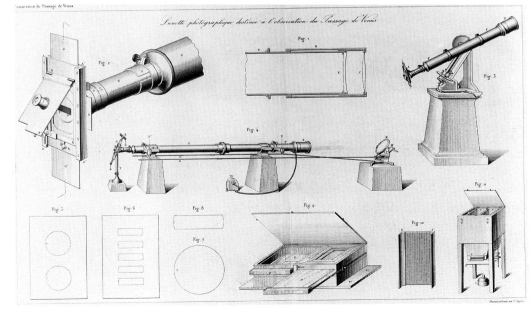

Fig. 6.21. Engraving showing the daguerreotype equipment used by Jules Janssen for recording the Transit of Venus in 1874. Figure 1 shows the focusing mechanism on the telescope. The plate holder is shown in figure 2; its placement on two different telescopes is indicated in figures 3 and 4. Different types of plates are indicated in figures 5, 6, 7, and 8. A two-chambered halogen fuming box is illustrated in figure 9 and a mercury bath in figure 11. Source: *Récuil de Mémoires.*

films before and after processing and drying. Obviously, if an image made while the collodion was wet changed during drying, measurements made on that plate would not be meaningful.[49]

One of the members of the French photographic advisory team was Hippolyte Fizeau, now the director of l'Observatoire in Paris. Fizeau suggested that the daguerreotype process was suitable for making images of the solar disk; because it was a dry photographic process, dimensional alteration would not be a concern. Fizeau taught the members of the French Transit of Venus teams to make daguerreotypes, and the process was accordingly revived for the 1874 Transit.[50] One of these daguerreotypes, taken by Jules Janssen[51] using his *revolver photographique*, a rotating camera invented in 1873, has survived. This camera used a circular daguerreotype plate that was rotated past the lens at regular intervals in order to obtain serial images of celestial events (fig. 6.21). The image is recorded in separate frames around the circumference of the daguerreotype plate (fig. 6.22). The finlike shape in the image is a portion of the disk of the sun, and the small dark circle is Venus as it makes contact with one of the limbs of the solar disk and begins its transit. The oval cutout and the notch in the inner edge of the daguerreotype plate were there to attach the plate to the mechanism that advanced it past the lens of the telescope. Janssen made this daguerreotype on December 9, 1874, in Japan, where he was sent to observe the transit.

New Photographic Recording Instruments

Daguerreotype plates and some photographic papers were also used in early photographic recording instruments employed in making measurements of physical phenomena for astronomical, meteorological, magnetic, navigational, and spectroscopic applications. These instruments were usually designed to perform tasks requiring

many consecutive observations that could not be performed adequately even by diligent human observers. Photographic recording, in theory, offered a fairly easy and inexpensive method of graphically tracking events on a continuous basis. Photographic observations were also considered to be impartial and not colored by the "personal equation."[52] Just as problems were encountered when daguerreotypes were used for recording images for other types of scientific observation, these photographic chart recorders were often plagued with problems requiring the aid of a skilled photographer.

A coordinated international effort to monitor terrestrial magnetism began in 1833 under the direction of Karl Friedrich Gauss of the observatory in Göttingen.[53] Determining variations in the earth's magnetic field was of great interest and importance primarily for purposes of navigation. Magnetic observatories were often established as part of astronomical observatories

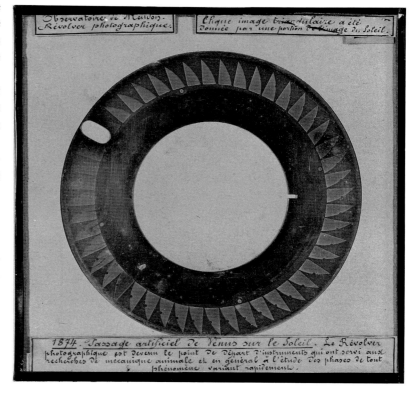

Fig. 6.22. Daguerreotype of the Transit of Venus taken by Jules Janssen using a *revolver photographique*. Courtesy of l'Observatoire, Paris.

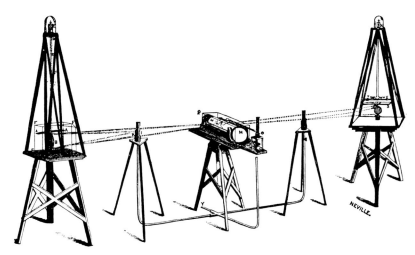

Fig. 6.23. Self-recording magnetograph used at the Royal Observatory, Greenwich. This particular magnetograph used paper on the central drum; however, an earlier model used daguerreotype plates. A gas lamp is positioned at the top of each side structures. The light from the lamp is reflected down to a mirror through a slit and focused as a beam on the central rotating drum. The line formed by the light reflected from the stationary mirror on the left is the baseline. The light from the light on right is reflected from a mirror attached to a compass needle, and the line formed by its reflection is indicative of changes in the earth's magnetic field. Source: Snelling, *A Dictionary of the Photographic Art*, fig. 78.

because of their common duties making calculations for navigational ephemerides. The British leader in this movement, Sir Edward Sabine of the British Ordnance Survey, established a series of magnetic observatories in various British colonies from the mid-1830s through the 1840s. The relation of variations in magnetic measurements to diurnal events, solar rotation, and sunspot activity was established during this period because of systematic efforts to record magnetic fields and the use of new self-recording instrumentation.

One center for geomagnetic measurements was established at Kew, outside London, and there Sabine began building photographic magnetographs around 1844. A *magnetograph* is a self-recording instrument that measures changes in the intensity of the earth's magnetic field as a function of time (fig. 6.23). Light is reflected off a stationary mirror through a slit onto a recording drum covered with some light-sensitive material. As a result, a light line is recorded on the drum corresponding to a baseline measurement. A second light is reflected off a mirror attached to the magnet of a compass through a second slit onto the same drum. The second light forms a line whose undulations follow deflections of the compass

needle and, hence, represent changes in the magnetic field. The length of a line drawn perpendicular to the baseline from any point on the curved line is proportional to the magnet's deviation from a standard position. The measured values can then be used to calculate the absolute values for terrestrial magnetism in a given place.

Sabine's first instruments used photographic papers, but he switched to using daguerreotype plates late in 1849. His 1850 instrument used a daguerreotype plate measuring three inches by twelve inches for recording. Sabine claimed that ordinate measurements made using his daguerreotype-recording magnetograph could be resolved to one five-hundredths of an inch. Photographic papers could not produce lines as fine as those recorded on daguerreotype plates because the texture of the paper fibers obscured the exact location of the lines, making absolute measurements difficult. The daguerreotype instrument was consequently more sensitive than those that used early photographic (i.e., salted) papers. Each day the plate was changed, and the light lines from the previous day were traced onto a special gelatin paper usually used to trace drawings for engravings. The daguerreotype instrument was expensive because of the cost of daguerreotype plates and the necessity for the observatory to employ both a "photographist" to make the daguerreotypes and someone to make the gelatin tracings each day.[54] Similar instruments were used in most magnetic observatories.

Sir John Herschel pointed out the value of instituting regular, photographic monitoring of sunspots in the early 1850s because of their perceived effect on geomagnetism. At his behest a photoheliograph to monitor the positions of sunspots on the solar disk was built at Kew Observatory. Initially, the instrument, designed by Warren de la Rue, was set up to make collodion negatives; however, problems with the collodion itself prompted Sabine to note in his annual report for 1856–57 that "'it has been deemed advisable to make attempts to produce positive pictures and recourse may ultimately have to be made to the daguerreotype process.'"[55] After another year of intense work the photoheliograph finally succeeded without de la Rue having to resort to making daguerreotypes. The instrument at Kew, one of the most famous and successful of these early photographic recording instruments, was used to make a daily record of sunspots from 1858 to 1872.

Various instruments to measure the intensity of light and compare the light sensitivity of different materials were built by Herschel, Hunt, and others. One of these was the *photographometer*, built by Antoine Claudet (fig. 6.24). A sensitized plate was placed at the bottom of an inclined plane behind a fixed plate with a regular array of holes. A movable plate with geometrically graded slots aligned with the holes in the fixed plate was allowed to fall past the sensitized plate, which received varying light exposure according to the size of the slots and the rate of descent of the slotted plate.

Exposure times were controlled either by the number of times the movable plate was allowed to fall down the inclined plane or by adjusting the angle of fall.

In 1856 Robert Wilhelm von Bunsen (1811–99) and Henry Enfield Roscoe (1833–1915) used a *tithonometer*, an instrument devised by Draper, in their important work on photochemical induction. They used the tithonometer to measure reaction rates in numerous photoreactive gases because it measured only the intensity of blue spectral rays (i.e., the blue, indigo, and violet portions of the spectrum), where, according to Draper, the tithonic effect was most pronounced. In order to measure both light rays and tithonic rays at the same time, Draper's instrument was used in conjunction with a photometer.[56] These now-famous Bunsen-Roscoe experiments on photochemical induction had to do with measuring the time it takes for a chemical reaction to begin once initiated by light. Most of these experiments were conducted using hydrogen and chlorine gases, which combine explosively to form hydrochloric acid when exposed to light. A portion of their experiments concentrated on photochemical induction in photography (i.e., the Becquerel effect) and were done using daguerreotype plates and photographic paper. Bunsen and Roscoe found that Draper's tithonometer was inadequate for making their measurements. Moreover, their experiments also put to rest the idea of tithonic rays, for they concluded that photographic action is merely the manifestation of the chemical rays of the spectrum and not the result of tithonic rays.[57]

As described in chapter 5, many of the first workers who investigated the daguerreotype process were trying to understand the nature of light and used daguerreotypes to record solar absorption spectra. Early experiments recording spectra on daguerreotypes confirmed the existence of the infrared and ultraviolet regions of the elec-

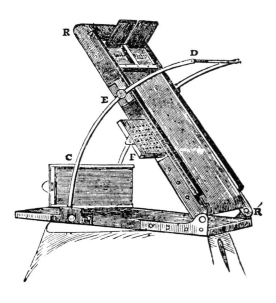

Fig. 6.24. Photographometer invented by Antoine Claudet to test and compare the light sensitivity of photographic materials. The material to be tested was placed behind the plate (F) and the slotted plate (R) was dropped down the incline plane (R,R^1). Exposure length could be controlled by changing the angle of fall or by dropping the slotted plate more than once. Source: Snelling, *A Dictionary of the Photographic Art*, fig. 81.

tromagnetic spectrum. Essentially, all spectra are formed by passing a beam of light through a slit, a lens, and a prism and then placing something to reflect or record the resulting spectrum projected from the prism.

The light and dark lines seen in such spectra result from the absorption of the light source and, possibly, absorption from the prism. The characteristic lines of the solar absorption spectrum had been mapped early in the nineteenth century by Joseph Fraunhofer, who tagged the major lines of the solar spectrum using the letters *A* through *G* and determined the constancy of these lines. The Fraunhofer notation was used universally and his lines were referred to as Fraunhofer lines. Other workers found that by placing cells filled with colored liquids or colored flames in the optical path certain portions of the solar spectrum were altered in characteristic ways. These

are all absorption spectra because the observed dark lines are the result of absorption of radiation by the transmitting media.

Work with the daguerreotype led Draper to his discovery, now known as the Grotthus-Draper law, that the recording medium's sensitivity to light affects what could be recorded. Likewise, Becquerel was able to record spectral lines on daguerreotypes in what is now called the ultraviolet region of the spectrum. Before the daguerreotype there was no easy way to detect this region of the spectrum, a spectral area beyond the sensitivity of human vision. The presence of the infrared had been detected by Herschel using thermometers before the invention of the daguerreotype. The sun was the radiation source for all these observations.

The sun is not the only source of luminous radiation. About the same time as Fraunhofer, several physicists found that colored flames could be used as sources to create spectra. They observed that these flames produced characteristic band structures that differed from the solar spectrum. Both Herschel and Talbot mentioned in papers published in 1826 that these flame spectra might be an aid to chemical analysis, but neither they nor any chemists followed up on these observations.[58] Colored flames have long been used by chemists to differentiate one compound from another, but their dispersal using a prism and the systematic study of the resulting spectra was not taken up for thirty or forty years after the observations of Talbot and Herschel. These spectra are called *flame* or *emission spectra* because the radiation dispersed by the prism is the result of characteristic emission of the compound burned in the flame. This is a very useful technique for identifying the chemical composition of unknown compounds.

The first person to publish data on flame spectra was David Alter, the bromine manufacturer in Freeport, Pennsylvania.[59] In

Fig. 6.25. David Alter's emission spectra as published in the *American Journal of Science*. Source: Alter, "On certain Physical Properties of Light," p. 55.

R.	O.	Y.	G.	B.	I.	V.	
|	|	|	|			—	Silver.
	||	|	|||			—	Copper.
|	||			|||	|		Zinc.
	|	|		|	|		Mercury.
||	|	||	|||	||			Platinum.
|	|	|	|				Gold.
	|||	||	|				Antimony.
|	|	|	||		|	|	Bismuth.
|	||	|||	|	|	|	|	Tin.
|	||	||	||||	||		|	Lead.
		|	||||				Iron.
|	||	|	|	|||	|||	|	Brass.

Fig. 6.26. Daguerreotype of Robert Cornelius demonstrating the proper technique for using a syringe-bottle in chemical analysis. Source: Robert Cornelius, Sipley Collection. Courtesy of the International Museum of Photography at George Eastman House.

Fig. 6.27. Engraving made from the Robert Cornelius daguerreotype shown in figure 6.26. Source: Booth, *The Encyclopedia of Chemistry*, fig. 18.

1854 and 1855 he published two papers in the *American Journal of Science* describing observations made while analyzing the flames emitted from various metals and gases when exposed to a powerful electric spark (fig. 6.25).[60] Alter recognized that the colors observed in these spectra were indicative of the chemicals combusted in the flame. He also reported recording spectra on daguerreotype plates. While his paper was also translated and published in Europe, it was not until 1859 that Gustav Kirchoff and Bunsen began collaborating on the work that laid down the basis for chemical spectrum analysis.

The daguerreotype even found its way into science education when James C. Booth decided to use engravings made from daguerreotypes to illustrate techniques for chemical analysis in his book *The Encyclopedia of Chemistry*,[61] the first chemistry textbook published in the United States. Booth commissioned Robert Cornelius to produce these daguerreotypes. Figure 6.26 shows Cornelius himself demonstrating the proper way to use a syringe-

bottle to wash a precipitate out of a beaker onto a filter. The engraving made from this daguerreotype is shown in figure 6.27.

Before the daguerreotype bowed to other processes, it had shown that photography would be a valuable tool for the scientist. The widespread adoption of the daguerreotype for scientific recording was impeded because of a scientific tradition that was not primarily visual, the limitations of the daguerreotype process itself, and the specialized skill required to make daguerreotypes. In spite of these impediments, the daguerreotype fulfilled François Arago's prediction: "When observers apply a new instrument to the study of nature, what they had hoped for is always but little compared with the successions of discoveries of which the instrument becomes the source—in such matters, it is on the unexpected that one can especially count."[62]

7

Scientific Interest in Daguerreotypy after the Daguerreian Era

The theory of this process is so exceedingly obscure and uncertain that at present any attempt at explanation of it would involve much that is hypothetical.

"Daguerreotype," in *A Dictionary of Photography*[1]

THE DAGUERREIAN ERA came to a close around 1860. Other photographic processes unrelated to the daguerreotype were adopted, and for all but a very few practitioners of the art the daguerreotype became an obsolete process. The primary impetus for photographers and scientists to continue investigating the workings of the process essentially came to an end at that time; nevertheless, the scientific fascination with the daguerreotype continued. There were several reasons for its continued scientific appeal. First, since daguerreotypes do not look like any other kinds of photographs, the relation of the daguerreotype's appearance and its other optical properties remained an absorbing puzzle. Second, daguerreotypy is the only photographic process in which simple silver halide is present without being suspended in a colloidal medium such as albumen, collodion, or gelatin. Accordingly, some scientists have thought that the daguerreotype process could be studied as a model photographic system. Third, once time passed and daguerreotypes came to be considered irreplaceable records of history and culture, there was a need to preserve these images for posterity. While all but the last of these questions were essentially the same ones that had been investigated during the daguerreian era, the reasons for taking up these inquiries had changed.

Even with the evolution of scientific theory and the advent of new instruments, the daguerreotype's riddles were not easy to solve. New explanations of the daguerreotype process emerged; however, the daguerreotype remained an essentially unsolved puzzle embedded in a certain amount of myth. The daguerreotype never merited the full attentions of later experimenters, and generally it was only a small part of many other research concerns. Experiments on the daguerreotype process were conducted using conventional daguerreotype plates, silver blanks, and silvered glass or film base. Some experimenters described their investigations as based on the daguerreotype process when they really used only a loose approximation of the original process; any photographic image created on a silver mirror surface or any photographic system that used a layer of silver halide on any sort of support material was called a daguerreotype.

As already noted, the study of light phenomena and optics was an active field of research throughout the nineteenth century. Indeed, François Arago saw the daguerreotype as a vehicle to extend many of his own research pursuits in the field of optics and light. Many scientists mentioned previously, including Arago, Léon Foucault, Hippolyte Fizeau, Edmond Becquerel, and John Draper, are more widely known for their work in optics than for their work with the daguerreotype. Optical problems did not fade after the daguerreian era, and the solution to some of these problems lay in the application of photography to recording fleeting phenomena for later study.

Daguerreotypes and the Search for Color Photography

Another problem that remained unsolved at the close of the daguerreian era was color photography. This question could be viewed as a photographic problem or as another optics problem. A number of scientists investigated color photography during the last half of the nineteenth century and at the beginning of the twentieth century, but the most straightforward approach was taken by James Clerk Maxwell (1831–79). By applying Thomas Young's theory of additive primary colors, Maxwell demonstrated in 1861 that it was possible to produce full-color images. Maxwell had Thomas Sutton, then an instructor of photography at King's College, make three glass-plate transparencies of a colored rib-

bon, one each through a red, green, and blue filter. When these three transparencies were projected through the filters with which each was made, the result was a full-color reconstruction of the original scene.[2] This classic demonstration proved to be the basis of modern color photography. Many early color processes, especially the photomechanical processes, used the principle of additive primary colors.

Another way to approach color photography was to find the source of color observed in some of the photographic processes already known. For instance, many early paper processes produced photographs that had a wide variety of colors— from yellows and browns to reds and purples. For the most part, the images produced by these paper processes were monochromatic.

Some daguerreotypes are colored, especially those produced using Becquerel development. In Becquerel daguerreotypes the colors are not very robust, but they are there, and they are correctly rendered—that is, red objects appear red, green objects appear green, and so forth. In this method of making daguerreotypes, an iodized daguerreotype plate is given an image-forming exposure in white light and then is allowed to remain in red or yellow light for an extended period of time until an image appears. In other experiments, Becquerel also demonstrated the appearance of spectral colors on daguerreotypes that had been sensitized in chlorine solutions or over chlorine vapor.[3] When plates prepared in this way were exposed to light from a prism, the resulting images were colored in the same pattern produced by the prism, and this fact indicated that something about this later system could provide a key to color photography. Indeed, Abel Niépce de Saint-Victor worked from Becquerel's observations when he devised his color photographic process.

Moreover, even ordinary daguerreotype

images can appear colored. The most commonly seen color is blue, although other colors can occur. Blue usually occurs in daguerreotypes that have overexposed highlights. The frequent appearance in cheap daguerreotype portraits of blue shirtfronts on men gave these daguerreotypes the name "blue-fronts."

The colors that appear on all these types of photographic materials could as easily be the result of either chemical changes in the light-sensitive materials or optical effects such as interference or scattering in the image-bearing layers of the photograph. Interference and scattering were two highly studied optical phenomena during the nineteenth century, an interest arising from the perception that understanding these two phenomena could help unravel greater problems associated with the discernment of the nature of light.

Interference refers to that set of phenomena in which the wave trains of light rays from a source or sources combine to reinforce or cancel each other out.[4] Light interference almost always is associated with color. The irridescent colors of oil films on puddles and the colored rings seen when two glass plates are pressed together sandwiching a thin film of water are examples of interference. *Scattering*, on the other hand, refers to the phenomena in which wave trains of light are deflected or diffused because of the presence of particles in the medium in which they are traveling. Scattering, too, produces color. It is very difficult through simple visual observation to distinguish colors that are the result of interference from those that are the result of light scattering. These are formidable optical problems, and the tools available during the last half of the nineteenth century were not adequate to discern the difference between these two phenomena. Nevertheless, a great deal of progress was made.

The national prize system to encourage the investigation and solution of problems of national importance for the advancement of science and technology was still a viable part of the scientific scene throughout the nineteenth century. One of these prizes set off a series of investigations that brought together all the work being done on interference phenomena using photography and, in particular, the daguerreotype. In 1865 the French Académie des Sciences announced that the 1867 Prix Bordin, a prize for physics, would be given for an experimental proof of Augustin Fresnel's theories about the propagation of light and the Fresnel vector, the vibration direction of light waves in the aether. Fresnel had worked out his theories about optics and especially interference using both observation and mathematical deductions during the second decade of the nineteenth century. Fifty years later there was still no adequate experimental verification of his work despite its importance to the entire study of light and optics, largely because making the necessary measurements of interference fringes by eye is almost impossible.

The announcement of the Prix Bordin stated that the winner of the prize did not actually have to conduct an experiment; it was sufficient merely to propose the proper experiment that would finally prove the correctness of the Fresnel vector. According to the rules, the contestants for this prize submitted untitled, anonymous manuscripts to a prize committee made up of French physicists, including Edmond Becquerel, and headed by Hippolyte Fizeau. The prize was awarded to Wilhelm Zenker, but since the manuscripts were anonymous and untitled, Zenker's name appears nowhere in Fizeau's report on the prize-winning work.[5] Not much is known about Zenker (1829–99), other than that he was a scientist working in Berlin and published papers covering a broad range of topics from astronomical phenomena to marine invertebrates to color vision and interference phenomena.[6]

Zenker described a novel, clever optical experiment that would solve the problem set by the rules for the Prix Bordin. He correctly pointed out that the real difficulty was actually observing the interference fringes that resulted from his optical setup, and he suggested that they could not be adequately seen by simple visual observation. The fringes could be recorded only by using some photographic means or a fluorescent screen, or an observing device such as a ground glass needed to be devised. One important point that Zenker made was that double-beam interference could be applied to making measurements of the diameters of stars. This observation was subsequently credited to Fizeau because he had been the communicator of Zenker's anonymous paper.[7]

The Prix Bordin was not the end of Zenker's foray into the world of interference and optics, but his other contributions were equally as obscure as his anonymous prize manuscript. He privately published *Lehrbuch der Photochromie*, a book in which he discussed photographic processes that had color, such as Becquerel plates and some of the early paper processes.[8] Previous experimenters had commonly believed that certain of the colors seen on daguerreotypes were the result of light interference from an image with a laminar, or layered, structure; that is, like the colors on the surface of a puddle covered with oil, light from the image portions of a daguerreotype passes through the image layers and is reflected from the plate surface to the viewer. The reflected light comes both from the surface of the image layers and from the daguerreotype plate; the difference in the origin of the reflection causes the various waves of light to interfere with each other, and this interference is seen as color. Zenker suggested in his *Lehrbuch* that a standing wave is formed during exposure of the plate and that it is permanently recorded in the light-sensitive layer so that

after image formation different image areas can reflect different colors. A standing wave is the result of the interference or combination of the wave motion of the incident and reflected radiation; consequently, the wave motion appears to stop or stand still. Thus, Zenker introduced the more complicated notion that standing waves, rather than simple interference, were responsible for the color seen in Becquerel daguerreotype plates.

In 1871 Carl Schultz-Sellack (1844–79), then assistant in astronomy and physics and later a professor at the University of Berlin, reported that the appearance of color in daguerreotypes was due to the presence of silver subiodides or subchlorides present in a mechanically divided (i.e., powdered) layer on the plate surface.[9] He thus discounted Zenker's claims. Later, Mathew Carey Lea (1823–97) (fig. 7.1) developed this Schultz-Sellack notion when he discovered that colored, monochrome images of various hues of red, yellow, blue, and brown could be produced using silver halide collodion and gelatin emulsions.[10] Lea became the chief proponent of the idea that color in silver-based, light-sensitive materials was due to chemical changes in the silver halide that

Fig. 7.1. Portrait of Mathew Carey Lea. Courtesy of M. Susan Barger.

could be produced either by exposure to light or in the dark. Lea felt that the colors he produced could not possibly be the result of optical effects of light interaction with particles of silver. Since he was never able to detect silver by direct chemical analysis, Lea came to the conclusion that the colors seen in his experimental plates were caused by the formation of allotropic forms of silver halides, the so-called silver subhalides.

Interference phenomena remained an active topic of research, and more sophisticated methods of making reflectance measurements and new possibilities to study the optical properties of thin films pointed up the inability of Fresnel's classic interference theory to describe new experimental observations. Fresnel's theory holds only when there is an abrupt difference in the composition of a thin film and its substrate. During the 1880s and 1890s John Strutt, Lord Rayleigh, expanded Fresnel's theories for application to thin films with periodic

structures.[11] Other scientists investigating electricity and magnetism also contributed to interference theory. Chief among these was Heinrich Hertz (1857–94), who demonstrated that electromagnetic waves had the same properties as light waves. Hertz had been able to demonstrate the existence of standing waves caused by phase changes in electromagnetic waves.

In 1889 Otto Wiener (1862–1927) tried to demonstrate the existence of standing waves in visible light by adapting the methods Hertz had used to measure electromagnetic standing waves. Wiener (fig. 7.2), one of the pioneers in the study of the physics of thin films, produced standing light waves in thin metallic films that had been deposited by evaporation on plane metal mirrors.[12] To prove that light waves have motion in materials, Wiener needed a material that would be permanently altered by light exposure, thus making a record of its exposure to light. He tried fluorescent materials, vapor images (i.e., *Moserbilder*), and photographic materials. In the literature available to him, he believed that he found evidence that early photographers had made use of standing waves without knowing they had done so. With the help of a photographer, A. Hruschka, Wiener pursued the use of photography to record his observations, and his early experiments were done using both collodion photography and daguerreotypes.

Wiener had initiated his experiments before he had heard of Zenker's previous work. In the footnote on page 230 of his 1889 paper, Wiener explained that Zenker's work had only recently been brought to his attention. He learned that Zenker had received the "Paris prize" in 1867 for his work but had not done the actual experiments described in his prize paper. Thus, Wiener felt sure that he was the first to demonstrate by experimental evidence the existence of standing light waves.

Wiener's paper described further exper-

Fig. 7.2. Portrait plaque of Otto Wiener. Source: Lichtenecker, "Otto Wiener," p. 73.

iments aimed at understanding the cause of color in thin films and, as a side issue, suggested applications of those principles for color reproduction. He used daguerreotypes because they were the ideal recording media for his experiments. The daguerreotype is essentially a polished plane metal mirror with a transparent, light-sensitive film whose thickness is small compared to the wavelengths of light and whose color is directly related to the film thickness. Just as the daguerreotypist could gauge the formation of the light-sensitive silver halide layer by the color changes on the daguerreotype plate as it was exposed to halogen vapor, Wiener had a way to gauge the effect of standing waves of light in silver iodide layers of specific thicknesses. He came to the conclusion that standing waves play a major role in determining the sensitivity of daguerreotypes.

Two years following Wiener's first paper, Gabriel Lippmann (1845–1921) discovered a method to produce color photographs using interference colors.[13] In his process a fine-grained gelatin emulsion was placed on a glass plate backed with a reflecting surface of mercury. Upon exposure, incident light is reflected back through the emulsion, and the incident and reflected beams combine to produce standing waves that are recorded in the emulsion. Once the film is developed and the image viewed in reflected light, the colors of the original scene are strongly reinforced by reflection from the various planes of the image recorded in the emulsion.

Unlike any of the previous interference-based color photographic processes, Lippmann's process had the possibility of being commercially viable. Everyone wanted to get in on the act and explain the source of the colors and present solutions for the few technical difficulties that needed to be overcome before the process could be presented for sale. These commercial possibilities also meant that the discovery was surrounded

with controversy. At the same meeting in which Lippmann's first paper was presented, Becquerel, a few months away from death, pointed out that the phenomenon of color that Lippmann observed was the same that he had seen in 1848 using daguerreotype plates.[14] Lippmann did not agree with Becquerel that the colors were the result of the same phenomena. Other pioneers in color photography such as Frederick E. Ives and the Lumière brothers also scrutinized the discovery. Even Zenker reappeared on the scene, publishing a review of his previous work.[15]

Wiener took up the task of explaining the colors in Lippmann plates, in Becquerel daguerreotypes, and in nature. He demonstrated that the appearance of color in daguerreotype plates is caused by the same phenomenon that caused the colors in Lippmann plates. In a series of ingenious experiments using a prism to control the exposure of Becquerel daguerreotype plates, Wiener demonstrated that the colors seen on those plates are due to interference and not to the formation of different chemical compounds, as claimed by Schultz-Sellack and Lea. He also stripped Becquerel image layers from their daguerreotype substrate and showed that the colors seen from the front of the image are not the same as those viewed from the back. This fact gave further weight to his conclusion that these colors are due to interference and not to absorption, as would be the case if the color was due to a chemical change.[16]

Wiener continued his experimental work on colors of thin films and especially of daguerreotypes, and in 1899 he published the last of his papers on the appearance of interference colors on daguerreotype plates. This paper briefly summarized his work on the daguerreotype and interference color problems over the previous ten years.[17] He said that he had found that the light sensitivity of daguerreotypes shows periodic increase as a function of the silver iodide film

Annalen d. Phys. u. Chem. Bd. LXVIII.   *Taf. 1.*

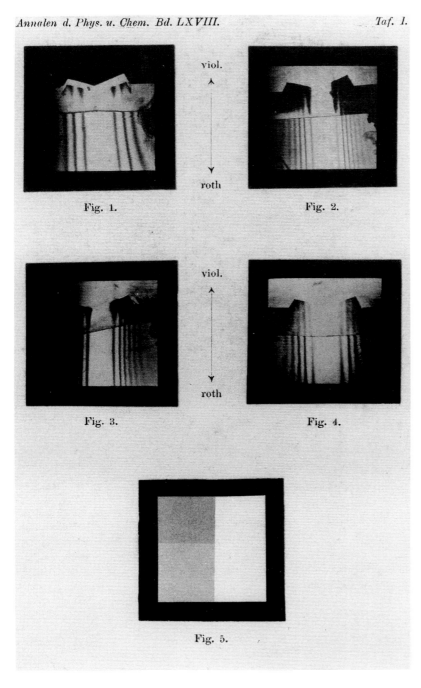

Fig. 1.   Fig. 2.

Fig. 3.   Fig. 4.

Fig. 5.

Fig. 7.3. Experimental plates showing interference bands in the silver iodide layer of iodized daguerreotype plates. Source: Scholl, "Ueber Veranderungen von Jodsilber," pl. 1.

thickness and that there are three maxima of sensitivity. These findings, he explained, had been made previously but had never been published because they were incomplete.

In particular, Wiener enumerated three problems that he felt he needed to understand before these findings could be disclosed. First, he needed to prove that the locations of the maxima of the daguerreotype's sensitivity to light and sensitivity to "electrical force" are recorded in the same dark bands on a daguerreotype plate. Second, he needed to discover whether the changing sensitivity of the silver iodide layer as a function of thickness was due to normal interference of transmitted light in the sensitive layer or whether it was the result of standing waves. Last, he needed to pinpoint the relationship between standing waves and light sensitivity. On the last point Wiener reported that he had observed a mysterious effect—only the uppermost portion of the silver iodide film on a daguerreotype plate was the most important for light sensitivity even though the ultimate source of the light sensitivity appeared to be from the silver of the plate itself. This paper was in actuality a review of the previous state of affairs and a preamble to the work of his student Hermann Scholl (1872–1923), who had solved the problems that had prevented the earlier publication of Wiener's last experimental observations on the daguerreotype.

Scholl's paper covered each of the three points that had not been adequately understood from Wiener's previous experiments.[18] Scholl performed a set of clever experiments to test the sensitivity of the silver iodide layer as a function of both the thickness of the layer and the color of the light used for exposure. His paper was illustrated with a series of plates that showed what was recorded on his daguerreotype plates as result of his experiments. By placing a glass tube at the center of his daguerreotype plates during exposure to iodine, Scholl was able to form two wedges of silver iodide that increased in thickness from the center to the outer edges of the plate. He then exposed the sensitized plates to a spectrum of light that was dispersed parallel to the wedges of silver iodide, thus allowing him to look simultaneously at the spectral sensitivity of his plates and the effect of the thickness of the silver iodide layer on sensitivity.

The results are shown in Scholl's figures 1–4 (fig. 7.3). The spectral distribution is given by the notation "violet to red" at the center of the page of figures. In the figures themselves the parallel vertical banding is the result of interference caused by standing light waves in the silver iodide layer. The thickness of the individual interference bands is related to the thickness of the iodide layer. Wiener had shown the same type of interference bands recorded on daguerreotypes in several of his earlier papers.[19]

The importance of Scholl's illustrations is his resourceful overlapping of two experiments: (1) the effect of the thickness of the silver iodide layer and (2) the effect of the spectrum on the sensitivity of daguerreotypes. The dark horizontal zones in the blue spectral region as marked in the illustration are the most sensitive areas of the plate.[20] In a second set of illustrated figures (fig. 7.4), Scholl also showed that standing waves could be seen in the silver iodide layer both before (row I) and after exposure to blue light (row II). Row III showed a considerably altered banding pattern after the exposed plates were developed; however, the banding present in the last set of illustrations is barely visible in Scholl's illustrations, as he also pointed out.

The remaining portion of Scholl's paper was devoted to a discussion of the behavior of silver iodide in the daguerreian system. Scholl studied the effect of light exposure on silver iodide films formed on silvered glass plates. In this way he could examine these films in transmitted light without mechanically disturbing the thin films. He

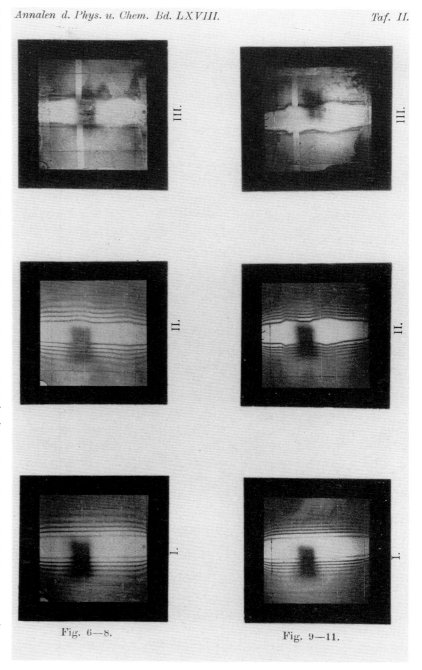

Annalen d. Phys. u. Chem. Bd. LXVIII.   *Taf. II.*

Fig. 6—8.   Fig. 9—11.

Fig. 7.4. Experimental plates showing interference bands in the silver iodide layer of iodized daguerreotype plates before and after light exposure and also after development. Source: Scholl, "Ueber Veranderungen von Jodsilber," pl. 2.

found that upon exposure the silver iodide layer becomes turbid; this clouding, he felt, was the result not of a chemical change but rather of the mechanical division of the layer. He proposed that a motion was set up in the silver iodide layer caused by the rapid interaction of silver iodide and oxygen to form silver oxide and iodine, which rapidly reform to make silver iodide and oxygen, and the cycle is repeated. According to Scholl, this rapid transference of atoms leads to a loosening and clouding of the silver iodide layer. The locations of the maxima of turbidity correspond to the locations of standing light waves in the surface of a silver iodide film where the film is in contact with the air. From these experiments Scholl concluded that he had confirmed Wiener's observation: the silver iodide surface of a daguerreotype is more sensitive than the interior portion of the film.

In the final portion of his paper, Scholl mentions that certain questions arose during the course of his work that he looked forward to investigating at some other time. These questions centered around the electrical effects in the daguerreotype system. Scholl thought that iodine moved through the silver iodide layer because iodine, oxygen, and silver could all become electrically charged, forming a sort of electrolytic solution. Scholl implies that whatever these deeper effects are, they must be related to Hallwach's effect or photoelectric excitation[21] and, therefore, had application to problems beyond the daguerreian system.

The Wiener-Scholl experiments provided empirical demonstrations of interference in the daguerreian system, in the Lippmann process, and in certain other natural phenomena. Neither Wiener nor Scholl derived any mathematical description of the physics of their discoveries, and they did not carry on their work after the 1899 papers. The chemical colors that had been studied by Schultz-Sellack and Lea and attributed to silver subhalides and allotropic silver were shown to be colors that arise from light scattered from finely divided silver particles. The scattering phenomenon was described mathematically in 1904 and 1905 by J. C. Maxwell-Garnett in his papers "Colours in Metal Glasses and in Metallic Films. I and II."[22] These papers formed the basis of what is now called Maxwell-Garnett scattering theory.

The commercial exploitation of Lippmann's process moved forward, although not through his own efforts. As early as 1903 he was nominated for the Nobel prize in physics. This nomination was based not only on his discovery of a way to make permanent, colored interference photographs but also, primarily, on his use of Fourier transforms in the mathematical analysis of the source of the colors observed in his process. Lippmann plates were touted as new instruments of research, as was the daguerreotype when it was introduced. Finally, in 1908, Lippmann was awarded the Nobel Prize, but he was the second-place nominee and his election was more a political ploy to prevent the prize from being awarded to Max Planck.[23] Zenker, Wiener, Lord Rayleigh, and Becquerel were all mentioned in the award citation, which proclaimed that Lippmann had made not only a contribution to photography but also an extraordinarily elegant illustration of the general laws of optical physics.[24] Ironically, the Lippmann process fell into disuse only a few years after he received the Nobel Prize and, like the rest of interference photography, was forgotten.

The Daguerreotype's Appearance

The next time the daguerreotype was scrutinized was during the late 1930s and early 1940s. At that time interest in the daguerreotype was derived from studies on the grain structure of photographic images. The systematic study of the relationship between exposure, the density of the silver image in photographic materials, and the resulting optical properties of the developed image dates from the work of Ferdinand Hurter and Vero Charles Driffield, first published in 1890.[25] The field of densitometry and sensitometry developed from their work; however, it was not until the 1920s that the relationship between the perception of photographic images and their image structure began to be studied.[26] This study was possible partly because of the availability of better microscopes for looking at and recording the microstructure of photographic materials; the introduction of new psychophysical methods; and the expansion of scientific research on the photographic process brought about by the founding of industrial laboratories such as the Kodak Research Laboratory.

The appearance of a photograph is dependent upon many factors, some of which are related strictly to photographic considerations, such as the nature of the photographic material, light exposure, and development conditions, and some of which are related to the visual perception of the image. Thus, on a microscopic level a photographic image is made up of a collection of grains of silver metal. A conventional photographic image appears to have light and dark areas because of the differential absorption of light in these collections of silver grains. The size and spacings of the silver grains control the amount of light that is absorbed and, therefore, the density of the image. When the shadows in an image are made of coarse grains that are not closely spaced, the image is perceived as not having saturated blacks, as being "grainy," and as lacking in resolution. Conversely, when the shadow image grains are small and closely spaced, the image is perceived as having saturated blacks and high resolution.

The photographic materials characterized as being essentially "grainless" and, there-fore, as having high resolution are Lippmann plates and daguerreotypes. For this reason it was important to measure the grain size and to determine the factors that controlled grain growth in these materials. For daguerreotypes this subject was mentioned by Lüppo-Cramer in a paper entitled "Grain Structure and Graininess," published in 1939.[27] Lüppo-Cramer did not actually measure the grain size of daguerreotypes; rather, he estimated the grain size of daguerreotypes based on notes published in Josef Maria Eder's journal *Handbuch der Photographie* and also on early accounts of making daguerreotypes and measuring image particles reported in Eder's *History of Photography*. From these sources Lüppo-Cramer determined that an individual daguerreotype grain is on the order of 0.08 millimeters (80 micrometers) in diameter, which would be considered a very large grain size given the resolution of daguerreotypes.

Heinz Jaenicke followed up on Lüppo-Cramer's paper and pointed out that elsewhere in Eder's *History* it is reported that the grain size was on the order of 0.04 millimeters (or 40 micrometers). Both measurements were derived from work by Brongniart, who had made these measurements using an optical microscope in 1839.[28] Jaenicke pointed out that even if one considered the smaller measurement correct, this was a large grain size. To check the actual grain size, he made optical micrographs of daguerreotype surfaces and used the micrographs to estimate that daguerreotype grain sizes should be on the order of 1 to 3 micrometers, more than an order of magnitude smaller than the previous estimates. Jaenicke also estimated the grain size for both collodion negative materials and Lippmann plates. He concluded by saying that looking at daguerreotypes under a microscope is interesting because they are not common and not studied since they had little theoretical value when compared with modern photographic materials.

As an aside, Jaenicke noted that while it is tempting to say that a daguerreotype is grainless or has a grain size that is not visible, his findings indicated otherwise.[29]

The appearance of the daguerreotype was also scrutinized by the Kodak Research Laboratories at the request of Beaumont Newhall, director of the George Eastman House, in 1952. Newhall felt that daguerreotypes, when properly prepared and viewed under optimal conditions, have an image quality superior to that obtained using modern photographic materials. No body of experimental data and measurements existed to substantiate his observation, so Newhall provided a daguerreotype to the physics division of the Kodak Research Laboratory for examination.[30] Because the view of the image on a daguerreotype depends on the viewer's position relative to the source of illumination, J. L. Tupper and K. S. Weaver made reflectance measurements from a highlight area and a shadow area of the sample daguerreotype using a goniophotometer to determine the best viewing geometries.

A *goniophotometer* is an instrument that allows the measurement of light reflectance from a sample as a function of angle. Accordingly, in Tupper's experiments the sample daguerreotype was positioned in such a way that it was illuminated at normal incidence (at 90 degrees to its surface), and a series of measurements was made at angles varying from 0 degree to 90 degrees relative to the incident beam. The measured intensities were then compared to similar measurements made on a perfect diffuse reflector. By comparing the reflectance measurements obtained from the daguerreotype to those taken from the diffuse reflector for each angle, it was possible to determine the angle of view at which a daguerreotype lighted at normal incidence will have the appearance of a negative rather than a positive and where it will have the appearance of the greatest density. Both are important factors to consider when determining how

best to display a daguerreotype for exhibit. Tupper found that daguerreotypes present the greatest difference between the highlight and shadow densities when illuminated normally and viewed at angles varying from 70 to 90 degrees. This viewing geometry would be unacceptable for exhibition; however, if the geometry was reversed—that is, if the daguerreotype was viewed normally while being illuminated at an oblique angle—this would be the optimal viewing geometry.

The main point of Newhall's question was to determine whether daguerreotypes have superior image quality when compared to modern photographic materials in the best viewing conditions. To settle this question, the density of the sample daguerreotype was measured using a reflectance densitometer. The lowest density value was subtracted from the highest value to obtain the density scale for the daguerreotype. After making this set of measurements, Tupper found that the daguerreotype had approximately the same density scale as matte-surface developing-out photographic paper. Tupper thus concluded

Fig. 7.5. Portrait of Irving Pobboravsky. Kevin Higley; courtesy of Gannett Rochester.

that, when compared on the basis of the reflectance characteristics, a daguerreotype was not superior to conventional photographic materials.

Pobboravsky's Spectral Studies

The most significant modern study of the optical properties of daguerreotype silver halide films is "Study of Iodized Daguerreotype Plates," conducted by Irving Pobboravsky (fig. 7.5) as part of his master's degree in photographic science at Rochester Institute of Technology in 1971.[31] Pobboravsky's thesis was a modern reexamination of John Draper's study of the source of the color observed on iodized daguerreotype plates and the relation of the colors to plate sensitivity.[32] His work was also related to the work of Otto Wiener and Hermann Scholl, although Pobboravsky was unaware of the earlier German work.

A portion of Pobboravsky's work was concerned with characterizing the colors observed in the silver iodide films formed during iodine fuming of silver plate. To do this, he produced a series of silver iodide films of different colors on daguerreotype plates and stripped these films from the silver surface using a gelatin coating. Thus, he was able to make optical measurements without having to compensate for the interactive effects of the underlying silver plate. The optical properties of these gelatin-supported silver iodide films and gelatin blanks were measured using a spectrophotometer with an integrating sphere. By varying the placement of the sample in the integrating sphere of the spectrophotometer, Pobboravsky obtained measurements for the transmittance, the front-surface reflectance, and the back-scattered reflectance from the silver iodide films. The thickness of the films was calculated using the values from the spectrophotometric results. Calculations from x-ray fluorescence measurements were used as a second method to verify the thickness of the silver iodide layers.

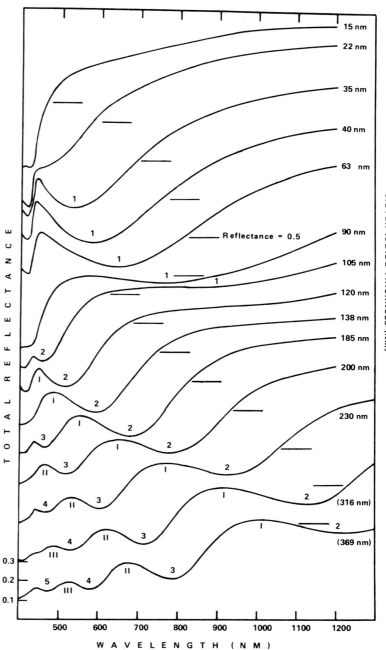

Fig. 7.6. Spectra showing the total reflectance of silver iodide of various thicknesses on daguerreotype plates. Source: Pobboravsky, "Study of Iodized Daguerreotype Plates," fig. 12a.

Finally, Pobboravsky used his spectral transmittance and reflectance data to compute the colorimetric description of the silver iodide films.

After characterizing the colors of the silver iodide films and their thicknesses, Pobboravsky went on to investigate the source of color observed in the films. His ability to make direct spectral measurements of the silver iodide films made it possible to determine whether the colors of the iodide layers on a sensitized daguerreotype were the result of interference, scattering, or absorption phenomena. He measured the total reflectance of silver iodide layers of known thickness on daguerreotype plates in the visible and near infrared regions of the spectrum (fig. 7.6). The minima and maxima in the measured curves are evidence of interference in the silver iodide films. These measured data were compared with calculated reflectance curves derived from a physical model of interference in silver halide films on a silver surface; however, the calculated and the measured curves did not match. Since Pobboravsky's model assumed that the silver iodide layer was continuous and parallel to the plane of the underlying daguerreotype plate, deviations caused by discontinuity or surface roughness in the silver iodide layer would not be indicated in the calculated curves.

The next set of experiments was devised to test the additional hypothesis that the colors in the silver iodide films might also be the result of light scattering. Pobboravsky verified that light scattering played a role in the optical appearance of the silver iodide layer. He also conjectured that because the measured total reflectance values of the films were significantly lower than his calculated values, light absorption must also be involved in the complex of optical mechanisms responsible for the silver iodide colors.

The final problem addressed by Pobboravsky was the relation of the color of the silver iodide films and daguerreotype sensitivity. Draper had found in a qualitative way that daguerreotype sensitivity was at a maximum when the silver iodide layer was yellow. The field of sensitometry—the quantitative measurement of photographic sensitivity—had developed in the time between Draper's study and Pobboravsky's work. Again, Pobboravsky extended Draper's work; he measured and compared the density of daguerreotypes that had been produced on plates with differing thicknesses of silver iodide to determine the dif-

ferences in daguerreotype sensitivity. He used Becquerel development as his primary method of image production; however, he did conduct a few experiments using mercury.

From his last set of experiments, Pobboravsky confirmed Draper's observation that daguerreotypes are most sensitive when the silver iodide layer is yellow or yellow tinged with red, presumably because silver iodide is most sensitive to blue light and a yellow layer would absorb blue light. Surprisingly, he found that the yellow of thick silver iodide layers was fairly insensitive to light, contrary to Draper's findings. According to Pobboravsky, "the fastest iodized plate can produce a good image when given an exposure of 60 seconds at f/4.5 on a bright sunny day and developed by the Becquerel phenomenon."[33] Finally, Pobboravsky found that mercury-developed daguerreotypes had effective speeds five times slower than those determined for Becquerel-developed plates and that the tonality of the resulting images varied for the two methods of development. In his final remarks he mentioned other developers used for conventional photographic materials as possible substitutes for mercury.

Pobboravsky's work has striking similarities to that of Otto Wiener and Hermann Scholl. Since all these men were interested in the colors and interference phenomena of silver iodide films, it would be interesting to compare their findings. However, such comparison is not feasible because Wiener and Scholl did not quantify their discoveries. Even though they both published photographs of daguerreotypes showing interference bands, none of these illustrations had any markings to indicate their scale. Thus, it is impossible to determine the locations of the bands or ascertain the exact thickness of the silver iodide films in which these interference phenomena occurred. If this was not the case, it might be possible to corroborate the locations of the reflectance maxima and minima seen in Pobboravsky's reflectance spectra using the earlier data and matching the three interference maxima seen in Scholl's work. Pobboravsky had an advantage over Wiener and Scholl because he had a spectrophotometer and was able to actually measure and quantify the optical properties of silver iodide films on daguerreotype plates.

The Latent Image

Throughout the evolution of the photographic process, there has been a concerted effort to discover the nature of the latent image. Obviously if the latent image is understood, improvements in photographic processes, such as increasing light sensitivity without sacrificing image quality, would be possible. Such improvements require an understanding of how light interacts with silver halide to form the latent image and, further, what methods are available to make that formation process more efficient. Of course, one major difficulty in deciphering what is going on in the photographic process is separating effects associated with light sensitivity and exposure from those associated with the actual formation of the image and development. This process is very complicated because light effects cannot be examined without developing the image and vice versa.

As noted earlier, Paul Beck Goddard and Robert Cornelius increased the daguerreotype's photographic speed and made portraiture possible by adding bromine and other halogens to the daguerreotype process. Even though other approaches were tried to increase the daguerreotype's photographic speed, nothing else was discovered that made any significant improvement. All photographic processes that came after the daguerreotype could be manipulated in a number of ways to shorten exposure times. Regardless, all the later processes had more complicated formulae and

more constituents that made it difficult to sort out all possible chemical and physical interactions. Experimenters looking for a simple system to test certain hypotheses about the photographic process often looked to the daguerreotype.

The daguerreotype has always been considered the simplest photographic system because it is based on pure silver iodide (or silver bromoiodide) with no organic binding media or other elements to complicate the reactions. The daguerreotype was often mentioned in passing by scientific workers who addressed the photographic process during the first few decades following the decline of its commercial use. However, during the 1890s the daguerreotype received careful scrutiny not only by Wiener and Scholl but also by the photographer James Waterhouse, who used the daguerreotype to investigate photographic development.

A member of the Royal Bengal Artillery and later associated with the Surveyor General's Office in India, Waterhouse (1842–1921) was a member of the Photographic Society of Bengal. He made many contributions to both the advancement of the photographic process and the advancement of photography in India. Some of his photographs were included in the monumental work *The People of India*, and he was part of the eclipse expeditions in 1871–72 and 1875 and the India Transit of Venus expedition in 1874–75. In particular, he was interested in the scientific issues of photographic processing and wrote widely on the various aspects of the topic.[34]

One recurrent question about the photographic process has been the role of "electricity" in the image formation process. Daguerre himself had mentioned that electricity must be involved in the daguerreotype process, and other experimenters such as Edmond Becquerel, Joseph Henry, John Draper and others also surmised that there must be some association between electricity and the process. In part, this conjecture derived from the observation that an electrical spark produces light and, therefore, light may also possess "electricity." However, not until James Clerk Maxwell's electromagnetic theory in 1864 and Heinrich Hertz's experimental work in the 1880s, which provided physical proof of Maxwell's theory, was there an adequate scientific language to describe electricity in materials and its relationship to light. In part, Otto Wiener's extension of Hertz's discovery of standing electromagnetic waves to show the same phenomena in light waves helped broaden the electromagnetic spectrum to include visible light.

In 1891 Waterhouse began a series of experiments on the electrochemical effects produced during photographic development. He found that when he performed the following experiment he could detect a current formed as the result of photographic development. He immersed two silver plates in bromine water and exposed one plate to light. Then he hooked the two plates to a galvanometer and immersed them in a photographic developer solution. When he used a normal developer, current flowed so that the exposed plate acted as a negative electrode and a negative image formed on the exposed plate. He further found that he could reverse the direction of the current and make a positive image simply by adding thiocarbamide solution to the developer. With much more difficulty, he was also able to produce a current using a gelatin dry plate to which he had attached electrodes of gold or silver leaf. Waterhouse concluded from these experiments that there is "electrolytic action" during photographic development in gelatin dry plates that causes the exposed silver halide in the image to be reduced to silver metal.[35]

Waterhouse continued his experiments on development and published several papers on the topic throughout the decade. Apparently, he continued working with silver daguerreotype plates as a means of monitoring developing mechanisms. In 1898 he described a set of experiments in which he used various developer solutions commonly used for wet collodion plates or gelatin dry plates. These developers included both acid iron–pyrogallol developer and the alkaline developers usually used for gelatin dry plates. He sensitized his daguerreotypes by fuming them with iodine and bromine vapors; then, following camera exposure, he developed these plates in the various developer solutions and noted that the images produced in this way were somewhat different from those typical of daguerreotypes.

Waterhouse had two primary objectives for this set of experiments. The first was to see if he could produce something like Lippmann photographs directly on the grainless, light-sensitive silver halide film of a daguerreotype. This was not exactly the same objective that Wiener and Scholl had, but it was strikingly similar to some of their experiments. The second purpose was to produce images on metal plates that could be etched and used for photomechanical printing. This later object was greeted with high praise as having great commercial promise. Waterhouse seems not to have been aware of the fact that fifty years earlier Biot, Fizeau, and Draper, among others, had attempted etching daguerreotype plates in order to use them for printing plates but had rejected this approach as impractical.

In 1899 the honor of presenting the second annual Traill Taylor Memorial Lecture to the Royal Photographic Society was given to Waterhouse, and he chose as his title "Value of Daguerreotype as a Teaching Process."[36] He began his lecture by stating that even though the daguerreotype process had once been regarded as unique and unlike other photographic processes, both Mathew Carey Lea and he had shown that this was not the case. Most especially, his own work of the previous year had shown that daguerreotype plates could be processed just like conventional gelatin dry plates: the sensitivity of the plate could be enhanced using various dyes and halogen acceptors, images could be produced using conventional developers, and these images could be strengthened using conventional methods for intensification. For Waterhouse, the only difference between more conventional photographic systems and the daguerreotype was that in the daguerreotype silver halide is present without any organic materials to complicate the photographic reaction. In his Traill Taylor lecture, Waterhouse described some of his recent work on daguerreotypes; however, in general, the paper was a review of work on topics related to the daguerreotype and photographic chemistry, particularly the interaction of the daguerreian system with light.

Waterhouse had conducted a series of experiments in which he tested the light sensitivity and electrical properties of simple silver metal. He found that he could produce weak images on silver plate or on silvered glass simply by exposing the silver surface to light. Like Moser, who had made similar observations before him, Waterhouse was unable to determine what changes in the silver caused these images. Correspondingly, he found that light-exposed silver plates could be immersed in water (or other liquids) and if hooked up to a galvanometer would produce weak currents. Such plates also took on an altered surface color in areas that had been exposed to light. These results suggested that the role of silver metal in both the daguerreotype and conventional photography was greater than supposed by most scientists and photographers.

Adding iodine to his experimental system, Waterhouse studied the behavior of silver iodide in the presence of an excess of silver. He determined that silver appeared to darken more when there was an excess

of silver and ceased to darken when there was an excess of iodine. Citing the work of Scholl, Waterhouse corroborated the observation that no iodine appeared to be set free when an iodized daguerreotype plate was exposed to light. This conclusion indicated that upon exposure certain particularly light-sensitive compounds are formed that darken readily in the light, while iodine is driven deeper into the underlying silver surface. Waterhouse posited that this iodine may even act as a sensitizer for silver iodide.

Much of the rest of Waterhouse's lecture was taken up with a review of the scientific work done by others concerning the daguerreotype or related photographic phenomena. He paid particular attention to Scholl's work and its relation to previous work. The work of both men seems to overlap at several interesting points, a fact that Waterhouse did not seem to recognize. First, both used either the Becquerel phenomenon or conventional photographic developers to produce their images. The images produced using conventional developers were noted by both men as differing somewhat in appearance from daguerreotypes produced using mercury. Second and most important, both men reported observing electronic effects as a result of light exposure. Scholl had even made the observation that the daguerreotype might be an example of Hallwach's effect, as mentioned earlier. Subsequently, other scientists made the same observation.

In the 1880s Heinrich Hertz and Wilhelm Hallwach (1859–1922) independently discovered the basis of what is now called the *photoelectric effect*. Essentially, they found that certain negatively charged materials lose their charge when irradiated with ultraviolet light. In 1898 J. J. Thompson and P. E. A. Lenard expanded the understanding of this phenomenon when they both discovered that the loss of charge occurred because light is absorbed by the charged surface and electrons are ejected as cathode rays, now known as photoelectrons. They also determined that the velocity of these ejected photoelectrons is independent of the intensity of the light but that the velocity increases as the irradiating wavelength decreases. In 1905 Albert Einstein proposed a model of this phenomenon in which the light acts as discrete bundles, or quanta, that are absorbed singly by individual electrons. The energy of these quanta is $h\nu$, where h is Planck's constant and ν is the frequency of light. Thus, the energy of each absorbing electron is increased by $h\nu$. If the energy of a near-surface electron is increased sufficiently, it can be ejected from the irradiated surface. This is the work for which Einstein received the Nobel Prize; it is also a cornerstone of quantum mechanics.

Associating Hallwach's effect, or the photoelectric effect, with the daguerreotype was one way to account for the recurring supposition that "electricity" must have something to do with image formation in daguerreotypes and the experimental observation that current is produced when a daguerreotype plate is irradiated. The photoelectric effect was studied in two ways. First, it could be observed by placing the irradiated plate in opposition and close to a second plate that was connected to a galvanometer. Second, it could best be observed by placing the irradiated plate in a vacuum—that is, essentially a cathode ray tube, the familiar CRT.

In 1927 Julius Reich published a short note describing how Hallwach's effect is related to the daguerreotype.[37] It was then known that both positive and negative ions could serve as condensation nuclei or as sites for chemical reactions. It was also known that ultraviolet light could cause negative discharge from irradiated metal surfaces and that afterwards the discharged metal surface appeared to be positively charged. Reich wrote that this is exactly what happened in the daguerreotype system. Upon exposure, silver iodide is dissociated to form silver ions, and these ions serve as sites for mercury condensation. According to Reich, since Lüppo-Cramer had shown that mercury would also condense on an exposed silver bromide surface, the conclusion could be drawn that all photographic processes are dependent upon the formation of silver ions during image formation. To those versed in modern photographic theory, describing the production of silver ions in latent image formation in such a tentative way might seem peculiar. However, the application of quantum mechanics to latent image theory was only in its infancy at the time of Reich's note. In fact, not until the following decade were models relating the electronic properties of crystals and photoelectron absorption successfully applied to the photographic latent image in a way that matched experimental observation.[38]

The general concern with producing an electronic charge on the surface of a daguerreotype as a result of light exposure and the role of this charge in image formation has been a recurrent theme during this century. In another context this mechanism is the basis of electrophotography as defined by Chester Carlson.[39] In 1928 the French physicist Georges Simon published a paper describing a method to produce daguerreotypes by cathode pulverization, or what is now referred to as sputtering.[40] In this case, the daguerreotype image could be produced in a vacuum by sputtering a suitable metal vapor onto the exposed daguerreotype plate target. A patent was granted to Walter Limberger and Rudolf Wendt in 1968 for a photomechanical process loosely based on the daguerreotype process that used a silver drum as an "endless surface" to produce reproductions of originals.[41] More recently, the applicability of the daguerreotype to x-ray lithography was reported by

R. Prohaska and A. Fisher.[42] Finally, a widely cited paper in *Nature* by Ivor Brodie and Malcom Thackray echoed the same idea by pointing out that exposed silver iodide films on daguerreotype plates are photocharged and that an image may be made on such plates by using positively charged colloidal particles such as titanium oxide. In this way daguerreotypes could be made without having to resort to the use of noxious mercury vapor.[43] All these papers contributed to the observations made about daguerreotypes without providing substantially more to what was known about the process of image formation. The recurrent and persistent interest in the daguerreotype shows the fascination these images have held for the scientist as an unsolved puzzle.

The Conservation of Daguerreotypes

An important area in which scientists have been called upon to contribute to the understanding of the daguerreotype is their care and preservation. In March 1964 the Royal Photographic Society sponsored a day-long symposium in England, "The Recognition of Early Photographic Processes: Their Care and Conservation." This symposium marked the beginning of modern interest in the care and preservation of photographic materials. The interest in conservation as a whole was further awakened a few years later, in 1966, when the river Arno overflowed its banks and Florence was flooded. The international effort mounted to save the contents of the great Florentine libraries and the numbers of volunteers who flocked to Florence to join the rescue efforts marked the beginning of the modern conservation movement. Traditional methods for the restoration of all types of artistic and historic artifacts began to be reconsidered, and new efforts were

made to train conservators in an academic setting rather than through the apprenticeship system that had previously been the norm. Photography benefitted from these events. Each year more articles on the care of photographs began to appear, and the cleaning of daguerreotypes was always mentioned in these early articles (see chapter 11).

A few notes appeared that addressed points related to the conservation of daguerreotypes at a more fundamental level. The first scanning electron micrographs showing the image structure of a daguerreotype were published by Brian Coe in *The Photographic Journal* in 1972 (fig. 7.7)[44]; these were presented with no interpretation. Another scanning electron micrograph of a daguerreotype, published in 1974 by Leon Jacobson and W. E. Leyshon, showed a "daguerreian measle," a defect associated with the use of thiourea cleaning solutions.[45] While both of these papers have interesting micrographs, they did little

Fig. 7.7. First published scanning electron micrographs showing the image structure of daguerreotypes. Source: Coe, "The Daguerreotype Image," p. 119. Courtesy of the Royal Photographic Society.

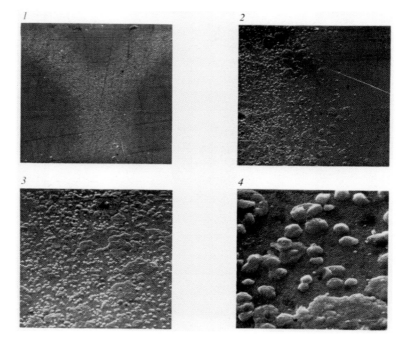

Fig. 7.8. Secondary electron micrograph showing a cross section of a daguerreotype image particle: the image particle (*A*), the daguerreotype plate's silver layer (*B*), and the plate's underlying copper layer (*C*). Source: Alice Swan, C. E. Fiori, and K. F. J. Heinrich, "Daguerreotypes: A Study of the Plates and the Process," fig. 8.

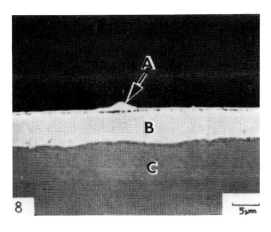

Their study showed that modern materials characterization based on what a daguerreotype actually is and not on what it is assumed to be is the way to make progress in determining ways to care for daguerreotypes.

An extensive scientific effort to understand the daguerreotype in order to determine better ways to preserve these images was begun at the Materials Research Laboratory of The Pennsylvania State University in 1979. The active phase of this study lasted from 1979 until 1984, and the results of that study will be discussed in the following chapters.

more than demonstrate what daguerreotype surfaces look like in a scanning electron microscope.

The ground-breaking scientific work during the 1970s was done by Alice Swan, who in 1975 became conservator at the International Museum of Photography at George Eastman House in Rochester, New York. During her tenure there, Swan published a series of papers on the care of photographs.[46] The most important of these papers, "Daguerreotypes: A Study of the Plates and the Process," described the first extensive materials characterization work that had been done on the daguerreotype.[47] Swan and her co-workers, Charles Fiori and Kurt Heinrich of the National Bureau of Standards, examined the daguerreotype using the scanning electron microscope and an energy-dispersive x-ray spectrometer. They described the method of making daguerreotypes and the results of the examination of both deteriorated and undeteriorated daguerreotype surfaces and provided a catalogue with micrographs of different types of corrosion products commonly found on daguerreotypes. While the Swan, Fiori, and Heinrich micrographs of daguerreotype surfaces were not the first of their kind, their work was the most extensive and detailed to that date (fig. 7.8).

THE SCIENTIFIC INVESTIGATION of the daguerreotype must begin with the most simple of questions: Exactly what is a daguerreotype? The most casual observer can understand that the appearance of the daguerreotype image is related to differences in the optical properties of the various image areas on the daguerreotype plate. Obviously, something must have been deposited onto or etched into the polished silver plate during the processes of sensitization, exposure, and development. A precise explanation of what occurs on the image surface is more difficult. An analysis of the daguerreotype image can be conveniently subdivided into a series of questions:

1. What are the physical characteristics of the daguerreotype surface, and how do these characteristics vary across the tonal areas present on the plate? In the language of contemporary materials science, what is the microstructure of the surface?

2. What is the chemical composition of the surface layer? How does it vary laterally across the tonal areas, and how does it vary in depth between the outside surface and the bulk silver in the plate beneath?

3. How does the surface microstructure permit the sharp image resolution so prized by daguerreotypists?

4. What causes low contrast or what is the relationship between image structure and contrast? How can one recognize "good" structure and distinguish it from "poor" structure?

5. Can the interaction between the microstructure and reflected light explain the visual appearance of the daguerreotype, particularly the odd property of image reversal from positive to negative depending on angle of viewing?

One's preconceptions direct one's experiments, an unfortunate circumstance when the preconceptions turn out to be wrong. At the beginning of these investigations, the

8

The Daguerreotype Image Structure

For beautiful image quality, the best of old daguerreotypes have never been equaled. The positives were made directly on the metal base and the small pictures were esteemed for their exquisite rendering of fine detail.

Edward Weston, "Art Photography"[1]

authors were convinced that the image was made up of surface layers, and our early proposals bristle with high-tech hardware—secondary ion mass spectroscopy, sputter-induced photon spectroscopy, Auger electron spectroscopy—that would reveal the chemistry of the layers. As it happens, the dimension of depth has little to do with the daguerreotype image.

The Image Structure

The image is encoded on the array of particles on the plate's surface. The positional variation in number, size, and organization of these particles is almost the whole story. Understanding the interaction of the particle-studded surface with reflected light allows us to account for image appearance, quality, and the loss of image when daguerreotypes become tarnished or corroded. Understanding the nucleation and growth of image particles is the key to understanding daguerreotype exposure and development. The critical experimental tool is the scanning electron microscope (SEM), and it is the reason that the portraits of historical figures that illustrate the early chapters of this book now give way to portraits of image particles in both formal and rather shabby attire.

Geometry is not quite everything; there are some chemical variations. The image particles are formed as silver-mercury amalgams, and their silver-mercury ratio changes as the particles age. Understanding the role of mercury is critical to understanding daguerreotype development. Gold appears in gilded daguerreotypes and plays an important role in image stability. However, our work shows that a daguerreotype is defined by its microstructure and not primarily by its chemical properties.

Since the nineteenth century it has been understood that the appearance of the daguerreian image is related to the difference in the optical properties of the various image areas on the plate. The chemical and physical composition of the image areas has been studied over the years, leading to various theories of image structure. These efforts were doomed to be inconclusive because the important features of the image are smaller than the resolving power of light microscopes. Only with the invention of the scanning electron microscope was the microstructure of discrete image particles positively identified. The 1979 study of daguerreotype plates and the process by Alice Swan, Charles Fiori, and Kurt Heinrich was a milestone in these investigations.[2] Using primarily scanning electron microscopy and x-ray analysis, they confirmed that the daguerreian image is made up of silver amalgam particles on a polished silver surface. They were not the first to identify the particulate nature of the image using scanning electron microscopy, but they did make the first modern, in-depth investigation of image structure.

The first step in our investigation was to assemble a collection of daguerreotypes for examination.[3] These included nineteenth-century daguerreotypes and also daguerreotype step tablets made in the laboratory. The physical state of these daguerreotypes ranged from very good condition to advanced deterioration. The step tablets allowed for a comparison of daguerreotypes processed in different ways. The combination of nineteenth-century samples and step tablets also allowed for the properties inherent in the daguerreotype system to be separated from artifacts associated with daguerreotype deterioration. In addition, the study of step tablets provided a rigorous way to investigate the link between image appearance and microstructure.

The morphological features of image particles from a gilded daguerreotype step tablet are shown as pairs of scanning electron micrographs (fig. 8.1).[4] These show the typical image morphology, and each pair shows one step of the step tablet at both

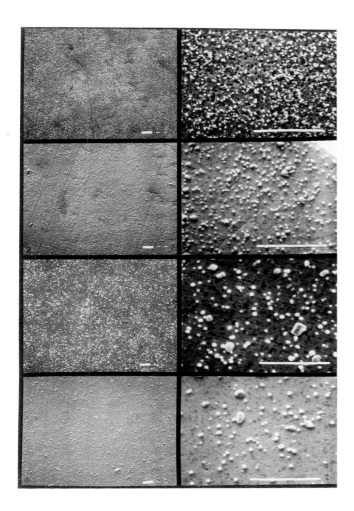 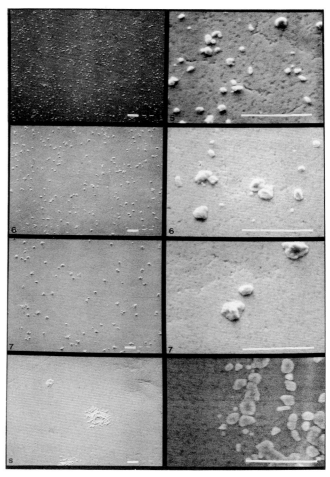

high and low magnification. There are two large classes of image particles: (1) those that are nearly spherical, and (2) those that are oddly shaped, like those found in the shadow regions. The latter are referred to as shadow particle agglomerates; all other particles are simply called image particles.[5] As can be readily seen in the micrographs, the extreme highlight step (step 1, fig. 8.1a) has a very large number of densely packed image particles, while the extreme shadow step (step S, fig. 8.1b) has a very small number of relatively large shadow particle agglomerates. It can also be seen that as the image particle density increases, there is a point, usually in the midtones (about step

4, fig. 8.1a), where particle size becomes fairly uniform and only the particle spacing, distribution, and number are changed as the extreme highlight step is approached. A better idea of the comparison between highlight areas, midtone areas, and shadow areas can be seen in figure 8.2, which was obtained from a daguerreotype rather than a step tablet.

The chemical composition of the image particles was determined by using the energy-dispersive x-ray detector on the scanning electron microscope. This technique is not highly accurate, but it does permit analysis of individual particles and comparison of the composition of the particles with the

Fig. 8.1. Typical microstructure of a gilded, mercury-developed daguerreotype step tablet. Each pair of scanning electron micrographs shows one step; step 1 is at the top. The steps are numbered 1–4 in the first plate and 5–7 on the second plate, with the final image that of step S, or the shadow area of the plate. Both high and low magnifications are given; in each image the scale bar is equal to 10 micrometers.

Fig. 8.2. Comparison of highlight, midtone, and shadow image areas from a gilded, mercury-developed daguerreotype. The image on the right is a blow-up of the small rectangle on the image on the left. The scale bar is equal to 50 micrometers and applies to the image on the left.

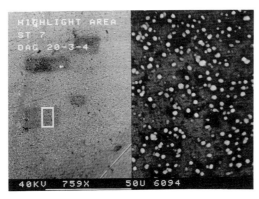

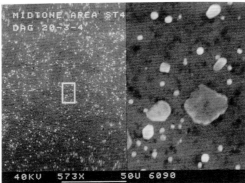

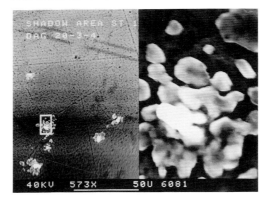

composition of the substrate. Because mercury is used in development, it has been generally assumed that image particles would be some kind of silver-mercury alloy—an amalgam. Swan and her colleagues postulated a specific amalgam, the body-centered cubic gamma alloy Ag_2Hg_3. However, our chemical analysis did not reveal any single compound present in daguerreotype image particles. Chemical composition varies not only from daguerre-

otype to daguerreotype but also from one image area to the next. Comparison of the secondary x-ray peaks of the spectra (fig. 8.3) demonstrates this quite clearly. These analyses are for different types of image particles on daguerreotype step tablets that had been processed in different ways.

The uppermost horizontal row of spectra in figure 8.3 are energy-dispersive x-ray (EDX) spectroscopic analyses of shadow particle agglomerates (step S); the center row spectra are midtone image particles (step 4); and the lower horizontal row spectra are highlight area image particles (step 1). The left column shows analyses of image particles from a freshly made, ungilded daguerreotype; the center column shows analyses of image particles from a two-year-old ungilded daguerreotype; and the right column shows analyses of a gilded daguerreotype. It can be seen that the mercury concentration of mercury-developed daguerreotypes varies according to image area, image particle size, and, in the case of ungilded daguerreotypes, age. Gold concentration in gilded daguerreotypes rises and mercury concentration falls as the image particle size decreases. In general, shadow particle agglomerates have relatively large amounts of mercury with a mercury concentration at the center of the agglomerate.

The scanning electron microscope can also be used in a quantitative way to obtain statistical data on the particle size distributions.[6] Control of the scanning electron microscope is turned over to a computer that manipulates the scanning of the electron beam. The energy-dispersive x-ray detector is set to record the mercury signal, information also fed to the computer. Because the image particles have a detectable mercury concentration and the background silver plate does not, it is possible for the apparatus to differentiate between image particles and the underlying plate despite the fact that both are composed predomi-

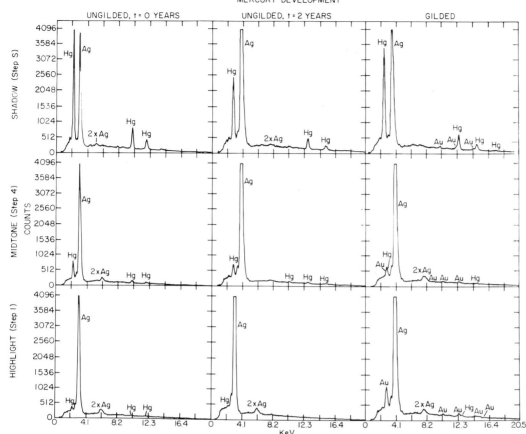

X-RAY ANALYSES OF DAGUERREOTYPE IMAGE PARTICLES
MERCURY DEVELOPMENT

Fig. 8.3. X-ray spectra of daguerreotype image particles from a daguerreotype with mercury development. The peaks represent the characteristic secondary x-ray lines for the elements marked.

nantly of silver. Figure 8.4 shows how the mercury-containing particles are detected and displayed on the microscope's screen. The image data recorded in the computer give the total particle count and a statistical analysis of average particle shape, average particle size (diameter), particle distribution, and surface-volume ratio (roughly how much of the surface is covered with particles).

Typical measurements for image particles in the highlights, midtones, and shadows of a gilded, mercury-developed daguerreotype are given in table 8.1. Average spacing and particle heights were obtained from surface profilometer measurements.[7] The minimum-maximum diameter ratio gives a general measure of particle shape such that for spherical particles the ratio approaches 1 and for oblong particles the ratio approaches 0.

Accurate data were difficult to obtain for

Fig. 8.4. Top micrograph: back-scattered electron image of a midtone image area. Bottom micrograph: sweeping electron beam showing characteristic mercury x-ray emission over the same area. The computer records only mercury-rich areas, i.e., image particles, and not the background. The scale bar is equal to 10 micrometers.

Table 8.1.   Average Image Particle Dimensions of Gilded Daguerreotypes

	Highlight	Midtone	Shadow
Image particle density (particles/mm²)	~2 × 10⁵	~10³	<100 agglomerates
Particle diameter (micrometers)	0.1–1	0.25–2.5	Agglomerate diameter: 10–50
Minimum-maximum diameter ratio	0.85–1.0	0.6–0.8	0.8–0.5
Spacing (micrometers)	0.1–0.6	0.5–1	30
Particle height (micrometers)	0.1–0.3	0.1–0.3	0.1–0.8

the highlight areas on gilded daguerreotypes because these areas did not have sufficient mercury present in individual particles to distinguish particles from the underlying daguerreotype plate, even though the particles are clearly discernible by visual inspection of the micrographs. Likewise, accurate data could not be gathered for ungilded daguerreotypes that had aged for any length of time. For these daguerreotypes the process of image formation continues until the image particles reach some sort of chemical equilibrium with the daguerreotype plate. Again, these particles are clearly discernible by visual inspection. The image particles in Becquerel-developed daguerreotypes are silver only, and these particles are not suited for image analysis based on chemical contrasts. In all these cases particle analyses were performed by inspection of micrographs and by use of profilometer data (tables 8.2 and 8.3).

The results of our own and other authors' scanning electron microscope data make clear that the microstructure of the daguerreian image is made up of discrete image particles of varying chemical composition and varying sizes. These particles are dispersed on a polished silver substrate, and their distribution is directly related to the amount of light exposure in the original scene—that is, the different image areas are distinguished by differences in image particle number and spacing and, in some

cases, image particle size. Image particle size is relatively constant except for image areas corresponding to areas of very high or very low intensity exposures. For most image areas, image particle height is roughly equal to image particle diameter.

Having established the image microstructure, the next task is to construct a physical model that will account for the daguerreotype's observed optical behavior in terms of the microstructure. From this model a set of scattering curves (i.e., diffuse reflectance curves) can be derived to describe daguerreotype behavior.[8] These must account for the physical appearance of the daguerreotype, including the odd positive-negative image reversal that depends on the angle of viewing.

The Optical Properties of Daguerreotypes

Some research has been directed toward understanding the optical properties of daguerreotypes. The major studies concerned with the optical properties of unexposed daguerreotype plates were conducted by John William Draper and Irving Pobboravsky, both of whom were concerned with correlating the color of the unexposed silver iodide layer on the plate to photographic speed and efficiency.[9] Baden Powell mentioned daguerreotypes in a paper on the reflectance characteristics of metals in polar-

Table 8.2. Average Image Particle Dimensions of Ungilded Daguerreotypes

	Highlight	Midtone	Shadow
Image particle density (particles/mm²)	~2 × 10⁵	~10³	<100 agglomerates
Particle diameter (micrometers)	0.1–1	0.1–2	Agglomerate diameter: 10–50
Minimum-maximum diameter ratio	0.85–1.0	0.85–1.0	0.5–0.8
Spacing (micrometers)	0.1–0.6	0.5–1	30
Particle height (micrometers)	0.1–0.3	0.1–0.3	0.1–0.8

Table 8.3. Average Image Particle Dimensions of Becquerel-Developed Daguerreotypes

	Highlight	Midtone	Shadow
Image particle density (particles/mm²)			No agglomerates
Particle diameter (micrometers)	0.1–0.2	0.1–0.2	
Minimum-maximum diameter ratio	near spherical	near spherical	
Spacing (micrometers)	0.1–0.6	0.5–1	
Particle height (micrometers)	0.1–0.2	0.1–0.2	

ized light.[10] However, daguerreotypes do not polarize the light reflected from the image surface; thus, his discussion of them was cursory. Late in the nineteenth century, Otto Wiener published an extensive paper on color photography in which he considered the appearance color in Becquerel-developed daguerreotypes.[11] Finally, J. L. Tupper studied the change in the visual appearance of a fixed daguerreotype as a function of observer position when the daguerreotype was illuminated normal to its surface by a fixed light source.[12] This latter study comes closest to the present problem, but in the absence of knowledge of the surface microstructure, interpretation of the optics of the daguerreotype image is only speculative.

Color—Whiteness, Blackness

The microstructures responsible for the optical properties of materials have been studied extensively.[13] A generalized scheme showing the spacing requirements needed to produce a white scattering material whose primary component is silver has been devised. This scheme (fig. 8.5) defines a white material as white because of its overall uniform reflectance through some broad portion of the spectrum. In this case, the portion of the spectrum of interest is the 380-to-700-nanometer span of the visible spectrum. It also assumes that there is no

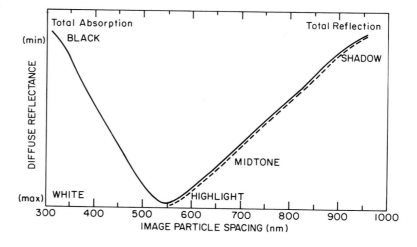

Fig. 8.5. Image particle spacing for a theoretical "white" material.

direct reflected radiation from the material to the viewer; that is, the angle of view does not equal the angle of incident illumination. Last, this scheme assumes that the sizes of the microstructural features will be distributed over a size range approximating the wavelength of visible light. Particle spacing is shown increasing from about 320 to 1000 nanometers. Maximum uniform diffuse reflectance (i.e., maximum scattering) is *white*; no diffuse reflectance is defined as *black*.

The curve shows the relation of microstructural spacing to the appearance of the material. If the spacings are much greater than 700 nanometers, the material approaches black because of total reflectance in the visible region of the spectrum. This black occurs only if the daguerreotype is illuminated by a small source that is reflected away from the viewer. If, on the other hand, the spacings of the microstructural features become very small, approaching 320 nanometers, the material will begin to appear black because of absorbance. The blacks at the two ends of the curve arise from quite different optical mechanisms. For the material to appear white, it should have a microstructural distribution with spacings from 350 to 700 nanometers with a peak in the spacing distribution at about 500 to 550 nanometers. This configuration would give rise to broad-band diffuse reflectance over the visible region of the spectrum with a peak at the top of the luminous efficacy curve for human visual response. If the daguerreotype fits this theoretical concept, then it would be expected to have a microstructure with a diffuse reflectance curve corresponding to the theoretical curve for spacings greater than 550 nanometers. This portion is marked with a dashed line just below the theoretical curve in figure 8.5. The approximate locations for highlight, midtone, and shadow image areas are also indicated.

The predicted scattering curves based on microstructure are shown in figure 8.6. In these curves the broad envelopes of the scattering peaks are related to the size distributions for the hypothetical image areas. The peak heights are dependent on the relative measured amounts of absorbance and specular reflectance for each image area. In addition, peak heights predict the maximum degree of saturation of tone (i.e., blackness) for the various image areas. Saturation is also related to the number of image particles for any given area. In the blue and near-ultraviolet regions of the spectrum, the plasma edge of silver (320 nanometers) dominates and controls daguerreotype optical behavior, as seen by the cut-off of the spectra in the blue regions.[14] For spectral regions of increasing wavelength, image particle shape, size, and height are the critical and controlling factors in image appearance. Scattering peak locations for these curves are dependent upon image particle spacing as indicated in figure 8.6. Curve 1 ($s \gg \lambda\ vis$) corresponds to the scattering from a featureless surface—e.g., an ideal, unused daguerreotype plate. The predicted curves for shadow areas are shown in curve 2 ($s > \lambda\ vis$), curve 3 ($s \simeq vis\ \lambda(a)$), and curve 4 ($s \simeq \lambda\ vis(b)$) and are indicative, respectively, of the behavior of a neutral white highlight and a midtone. A darkened appearance for a daguerreotype is predicted by curve 5 ($s < \lambda\ vis$) and curve 6 ($s \ll \lambda\ vis$) and is due to an increase in scattering in spectral regions below the visible region and the proximity of the silver plasma edge.

Blackness is both a physical and a perceptual phenomenon. Blackness can be a materials property described quantitatively by the percentage of the incident light that is reflected from the surface or by the value scale of the Munsell color system. Blackness is a perceptual phenomenon that depends strongly on the surround because the

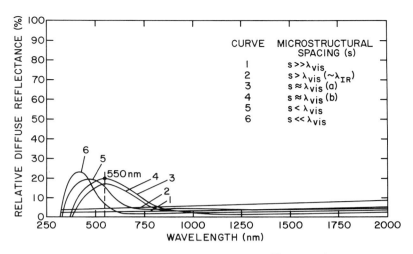

Fig. 8.6. Theoretical scattering curves predicted by daguerreotype microstructure.

human visual system operates by making comparisons. A material with a gray surface will appear darker when viewed against a white background and lighter when viewed against a black background. Proper control of blackness is a problem for all imaging systems, including photography.[15]

A material can appear black in two ways. One is that it is totally absorbing. If all light striking the surface is absorbed, then the absence of any back reflection is seen by the viewer as black. In this case the black appearance is independent of viewing angle. The second way for a material to appear black is for it to have a very shiny, smooth, highly reflective surface viewed in such a way that it reflects no light to the viewer. If all the light striking the surface is specularly reflected away from the viewer, the surface will appear black, especially when viewed next to a surface with a relatively large diffuse reflectance due to scattering. In both cases the appearance is strongly determined by the contrast with the surroundings.

The bulk microstructural requirements for a black absorbing layer have been studied extensively. If a layer is a composite of dispersed particles of size much less than the wavelength of light in some support material of differing refractive index, the layer will absorb light and appear dark in direct proportion to the amount of light scattering from the particles, the number of particles, and the refractive index and thickness of the support material.[16] It is easy to see that this description is applicable to conventional black-and-white photographic materials.

For most photographic systems variations in optical density are produced by variations in the absorbance of light. Optical density is defined as the logarithm of opacity where opacity is the inverse of transmittance. The necessary range of developed grain size and distribution needed to obtain saturated blacks in areas of maximum density has been given a great deal

of thought and analysis.[17] Areas of maximum optical density have a very high density of developed grains with an optimal grain size on the order of 25 to 70 nanometers (i.e., much smaller than the wavelength of light). Further, in conventional systems the developed grains are dispersed in a colloidal support, such as gelatin, which also increases the absorption of light because of its stepgrading of the refractive index between air and the filimentary metal grains (in the case of chemical development) or the metal particles (in the case of physical development). In these systems a reduction in optical density comes about through a reduction in the packing of developed grains. Thus, optical density has a direct relation to the number and packing of developed grains and the total thickness of the image-bearing layer. Figure 8.7 shows a

Fig. 8.7. Sketch of idealized image structure for a conventional black-and-white photographic system.

IDEALIZED IMAGE STRUCTURE FOR CONVENTIONAL AgX SYSTEM

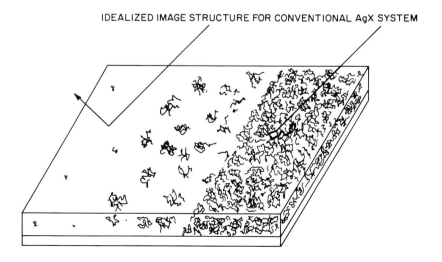

model of the microstructural features that contribute to optical density in conventional photographic systems.

In contrast to conventional photographs, the appearance of black in a daguerreotype is an example of the second way in which materials appear as black. Blackness in a daguerreotype is dependent on a highly polished plate and the proper viewing geometry. Since density is defined for conventional photographic systems in terms of opacity or absorbance, it is immediately apparent that the daguerreotype does not conform to this definition. For that reason, the term *apparent density* is used to refer to the appearance of blackness in daguerreotypes.[18]

As indicated by the particle counts in tables 8.1, 8.2, and 8.3, daguerreotype image structure is the inverse of image structure for conventional photographic systems. Areas of maximum apparent density have the smallest number of image particles, and image particle number and image particle density increase as the highlights become more white. Figure 8.8 indicates how this inversion comes about.

The polished silver plate allows the shadow areas in a daguerreotype to appear as a neutral black surface when viewed such that only scattered light is observed. Silver is a very good material for this purpose because its reflectance exceeds 95 percent and is independent of wavelength through the visible spectrum. Silver produces excellent

shadow areas because it is easily polished to a very smooth, specular finish. As seen in figure 8.6, curve 1 ($s \gg \lambda_{vis}$) shows the scattering behavior predicted by a pristine daguerreotype plate, and curve 2 ($s > \lambda_{vis}$) gives the predicted reflectance for shadow regions based on shadow particle agglomerate spacing.

The appearance of white or gray tones in daguerreotypes is due to the diffuse scattering of visible wavelengths by multiple reflection from image particles. The average particle size for highlight image particles ranges from 100 to 1000 nanometers. The particle size distribution usually peaks at sizes on the order of visible wavelengths in the very white highlights. The particle spacing is also on the order of visible wavelengths since the particles are fully packed. This combination of microstructural features gives the appearance of white because of relatively equal scattering of light throughout the visible spectrum. In the case of daguerreotypes, the visual effect of brightness is considerably heightened because particle size distributions usually peak around 550 nanometers, wavelengths close to the peak of the human visual response in bright light. These effects are predicted by curve 3 ($s \simeq \lambda_{vis}(a)$) and curve 4 ($s \simeq \lambda_{vis}(b)$) in figure 8.6. These effects are also observed in the white opalescent color of aluminum thin films and in the white color of milk. Intermediate tones in daguerreotypes occur primarily because of a reduction of image particle density, an increase in particle spacing, and, to a lesser extent, changes in particle shape or size. The total diffuse scattering is reduced accordingly.

Reflectance data for daguerreotypes support these predictions of microstructural behavior. Figure 8.9 (a and b) shows typical reflectance data for both total (i.e., specular plus diffuse) reflectance and diffuse reflectance alone for different steps of a daguerreotype step tablet.[19] The diffuse reflectance should be, and is for the step tablet, proportional to the spacings, sizes, and distri-

Fig. 8.8. Sketch of idealized image structure of a daguerreotype.

IDEALIZED IMAGE STRUCTURE FOR DAGUERREOTYPE

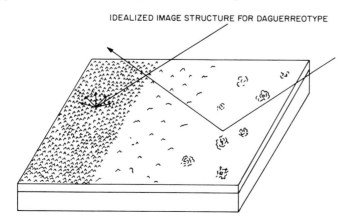

bution of the image particles. The broad peak across the visible spectrum is indicative of a white, diffusely reflecting surface. Likewise, as the height of the peak increases, the brightness of the material increases and remains neutral as long as the peak remains broad over the complete visible region. Diffuse reflectance measurements provide a quick way to obtain an average microstructural picture of a daguerreotype, since the reflectance spectra directly relate to the actual features of the microstructure and to the experience of viewing.

Positive and Negative Images

The appearance of a positive or negative image on a daguerreotype is also related to apparent density and viewing geometry. As the viewing geometry changes, so does the apparent saturation (i.e., blackness) of the shadow tones. The highlight areas have nearly the same luminance regardless of viewing geometry. When the angle of incident illumination equals the angle of view, the image tones appear to reverse, and the image appears as a negative. This reversal is due to the large increase in the luminance of the shadow areas compared with a relatively small change in the luminance of the highlight areas.

The contrast in a daguerreotype is simply the difference between the darkest shadow area and the whitest highlight. Since the saturation of the darkest shadow is dependent on viewing geometry, contrast range is also dependent on viewing geometry. Contrast range grows larger or collapses depending upon viewing conditions. When the daguerreotype is held so that the diffuse reflectance from the highlights exactly balances the specular reflectance from the shadow areas, the contrast is a minimum, and the image is perceived to vanish.

Daguerreotype Sensitometry

The familiar characteristic curve of photographic response has also been used for

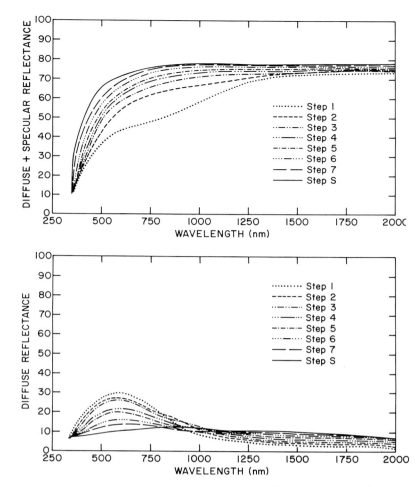

Fig. 8.9. Typical reflectance spectra for a daguerreotype step tablet: specular plus diffuse (i.e., total) reflectance (*A*) and diffuse reflectance only (*B*).

daguerreotypes. From the preceding discussion, it should be clear that the usual characteristic curve, obtained by plotting density against the logarithm of the exposure, must be relabeled in terms of apparent density. More important, the characteristic curve for a daguerreotype made using a reflection densitometer deviates somewhat from the characteristic curves obtained from conventional black-and-white photographic materials. The characteristic curve shown in figure 8.10 illustrates a common deviation, that of a partially "solarized" daguerreotype.

Figure 8.10 plots the measured density of a step tablet against the apparent density of a daguerreotype made by contact printing the step tablet. The point in the characteristic curve that begins to rise in the highlight region is traditionally called the solarization point.[20] This is seen by the viewer as a reduction in apparent density or a reversal of tones coupled with a blue tinge. This has been assumed to be an image reversal due

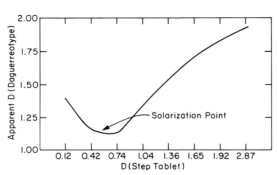

Fig. 8.10. Characteristic curve of a daguerreotype.

to solarization or latent image fading due to overexposure. In daguerreotypes this reduction in the apparent density is caused by a slight increase in particle number with an attendant decrease in particle spacing, shifting the highlight scattering curve toward the blue region of the spectrum, as indicated by curve 5 in figure 8.6. In this case the highlights no longer appear as white because the peak of the scattering curve has been shifted toward the blue region of the visible spectrum. In addition, the image particle spacing has not been changed sufficiently to alter the neutral cast. However, when a daguerreotype is completely solarized, a radical change in color cast occurs because of the large decrease in diffuse reflectance in the green and red spectral regions.

The scattering curves serve to characterize the color response of the daguerreotype. Most of the scattering peaks are broad, spanning the entire visible spectrum, which accounts for the overall whiteness of the response. The height of the peaks describes the neutrality of the tone. A low peak in the blue region of the spectrum will produce a blue color, for example, but it will be a much more subdued blue than if the scattering peak is high and sharp.

The shadow step in a daguerreotype is equivalent to base plus fog for conventional photographic materials. Usually in conventional materials, whatever fog grains are developed are well below the limits of visual

detection, but in a daguerreotype the shadow particle agglomerates are the most mutable of all categories of image particles.[21] Development and gilding conditions determine the size and overall morphology of these agglomerates. Thus, the shadow particle agglomerates moderate the maximum apparent density of the shadow regions by diffuse reflectance because of the complexity of their structure. This reaction is predicted in curve 1 of figure 8.6.

While the actual number of these image particle agglomerates does not seem to increase with development temperature, their size and structural complexity do increase with higher development temperature and in the gilding process. Certain development conditions, such as cold mercury development or Becquerel development, produce very small or no shadow particle agglomerates. It would seem, then, that to get the deepest shadows development conditions that produce no shadow particle agglomerates would be preferred. There is a trade-off, however; development that produces no shadow particle agglomerates also produces image particles of fairly uniform size. Since in this case the number of midtones is then controlled only by particle spacing and particle density, the possible image particle combinations that give rise to midtones are reduced. Thus, the tonal range is collapsed because of microstructural restraints, in addition to the restraints of the viewing geometry already attendant to the daguerreotype system.

Changes in Cast

Under certain conditions daguerreotypes take on a definite cast or color tone. The methods for obtaining various casts by altering exposure or development conditions were well understood and described in nineteenth-century daguerreotype manuals.[22] Changes in cast are caused by changes in the daguerreian microstructure, particularly changes in the image particles in the highlight areas.

For instance, a blue cast (called a blue-front daguerreotype) may be obtained by extreme solarization or by development for a long period of time over cold mercury. In the first case image particle density is increased, and the average image particle size and spacing are similarly decreased, especially in the highlight regions. The height of the scattering peak is increased in highlight regions. Likewise, the scattering peak is shifted toward the blue region because of a decrease in particle size and spacing. The dashed line in figure 8.11 shows reflectance data typical of this type of situation. In the case of cold mercury development the density of particles does not increase significantly; however, the average particle size has been considerably decreased because of the slow development. The small particle size results in a shift of the reflectance curve toward the blue region but no increase or a slight decrease in the amplitude of the peak, as shown by the dotted line in figure 8.11. The solid line shows the typical diffuse reflectance for a white highlight. The model also predicts these behaviors.

Daguerreotypes with a blue cast can also be caused by moisture in the sensitizing boxes when the iodization step is performed or by underdevelopment. Both situations produce conditions similar to cold development—no significant increase in the number of particles but smaller average particle size. Finally, blue-front daguerreotypes can be caused by overgilding, which produces conditions similar to solarization.

The number of image particles seems to be increased, but the particles are smaller, possibly because the gold in the gilding solution nucleates and grows some particles of its own.[23]

Similarly, other changes in cast may be predicted or explained by changes in the microstructure. A narrowing of the diffuse reflectance peak will be seen by the viewer as an increase in apparent color saturation. A shift in the scattering peak, which is seen as a change in cast, indicates changes in particle size or spacing.

It might seem that if the image particle density was greatly increased with an attendant decrease in particle size and spacing the daguerreotype would have a *super highlight*. This is the case to a certain extent, but, as indicated by the colored appearance of blue-front daguerreotypes, the super highlight microstructure causes the highlights to approach black, partly because of the plasma edge of silver and partly because of microstructural changes. Surface microstructural requirements for a perfect surface absorber, as established by Richard B. Stephens and George D. Cody, are that the spacing of surface particles must be less than the visible wavelengths and that the height of the grating made by the particles must be greater than one-fourth of the wavelength of visible light.[24] When the surface texture of a material meets these requirements, it is a perfect absorber. As image particle spacing decreases, the daguerreotype microstructure approaches that of a perfect surface absorber, and the color of the daguerreotype approaches black. Fortuitously, the material restraints of the daguerreotype process produce a characteristic microstructure with image particles that fall within the correct range to produce white diffuse reflectors under a wide range of conditions.

To summarize, the authors' physical model for the daguerreotype derives ultimately from the characteristic daguerreian micro-

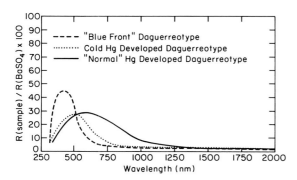

structure. The daguerreotype's appearance depends on the presence of image particles in a specific range of sizes, distributions, and densities. The various light-scattering characteristics of the image particles give rise to the image seen by the viewer. The model accounts for the daguerreotype's unusual optical characteristics, such as the positive-negative reversal and the appearance of casts such as that seen in blue-front daguerreotypes. It also provides a link between the daguerreotype and conventional photographic materials by providing a daguerreotype analogy to the familiar characteristic curve for conventional photographic materials.

The physical model suffers from the usual ailment of scientific theories: it is based on an idealized behavior of materials. Real daguerreotypes are scratched and corroded, and their image particle arrangement is damaged to varying extents. The real test of the model is to determine to what extent it can be applied to actual daguerreotypes as they now exist in galleries, archives, and private collections. To do this, the authors made use of the experimentally determined reflectance characteristics of the daguerreotypes.

Fig. 8.12. Diagram of goniophotometer.

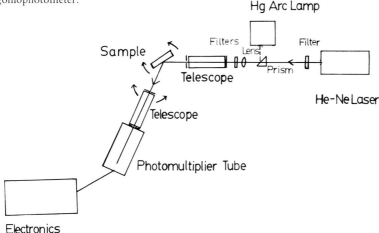

Specular Reflectance Properties of Daguerreotypes

Fifteen nineteenth-century daguerreotypes were used as samples for this analysis. These daguerreotypes were chosen because they were representative of a variety of physical and chemical conditions commonly found among daguerreotypes. The image areas examined were defined as follows: A highlight was the whitest uniform area greater than one millimeter. A midtone was a uniform portion of the background. A shadow was the darkest uniform area greater than one millimeter.

Specular reflectance measurements were made using a goniophotometer as shown schematically in figure 8.12. The reflectance of each image area was measured using three different wavelengths of light: red (632.8 nanometers), green (546.0 nanometers), and blue (436.5 nanometers). A helium-neon laser was used for the red source. The beam from a mercury arc-lamp was diverted into the optical path through an adjustable prism and used in combination with narrow-band filter packs for the green and blue sources.

While measurements were made using the three different light sources, it turned out that all interpretation could be made using the data from only the red illuminant. The reflectance measurements on daguerreotypes using the green and red illuminants gave values that differed only slightly for the same angle and area. The same measurements taken for the blue illuminant typically gave the same curve shape but with 5 to 10 percent lower reflectance than the red and green.

The goniophotometer was aligned using a simple high-precision alignment procedure devised for ellipsometers.[25] Each daguerreotype was positioned on the optical axis of the goniophotometer so that the sample area being measured was at the vertex of the angle formed by the incoming and reflected beams as shown in figure 8.13. Both the light detector and the daguerreotype were rotated about the vertex to maintain $\theta = \theta'$. Once a sample was aligned, its specular reflectance was measured as a function of angle θ' at 2.5-degree intervals for angles 18.5 to 87.5 degrees. The daguerreotype plate was rotated so that θ was always equal to θ'.

The comparison of the specular reflectance from the daguerreotype (R) to the total light intensity (I_o) gave the percentage of reflectance. R is the specular reflectance minus most of the diffuse reflectance of the area sampled. The size of the sampled area is sufficiently large so that a small portion of the diffuse reflectance contributes to the measured reflectance. I_o is the total flux of the incoming beam measured when the telescopes were separated by 180 degrees and there was no sample in the optical path. The reflectance for each sample was measured at all angles and for all light sources before going on to the next sample area.

Figure 8.14 shows schematically the ideal reflectance characteristics of a daguerreotype. The highlight areas are dominated by the diffuse reflectance of the array of image particles. As a result, the extreme highlight area would have a uniform reflectance regardless of the angle θ except at $\theta = \theta'$, where the specular component would have an additive effect. Ideally, the perfect shadow area would behave as a specular reflector and would have a measurable reflectance only at $\theta = \theta'$. The separation between the highlight curve (i.e., diffuse reflectance) and the shadow curve (i.e., specular reflectance) is a measure of the daguerreotype's contrast.

The specular reflectance curves made using the goniophotometer chart the percentage of specular reflectance from a highlight, midtone, or shadow portion of a daguerreotype under the condition when θ is equal to θ' for values of between 18.5 and 87.5 degrees. This is not the same as the

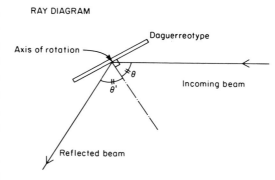

RAY DIAGRAM

Fig. 8.13. Ray diagram showing the optical path of the goniophotometer.

graph in the lower portion of figure 8.14, where the condition of $\theta = \theta'$ is fulfilled only in the exact center of the graph. Measuring the specular reflectance in this way gives an assessment of a daguerreotype's contrast range and the effect of tarnish films, scratches, and the like on a daguerreotype's appearance. The maximum separation between the shadow and highlight curves indicates the maximum contrast range that a particular daguerreotype may

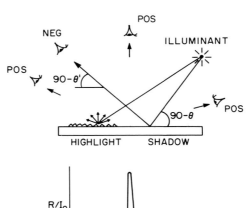

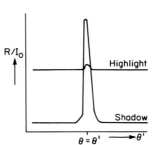

Fig. 8.14. The ideal reflectance curves for daguerreotypes. The upper curve represents a highlight region, the lower curve a shadow area.

have. However, these curves do not indicate the overall visual appearance of a daguerreotype. These measurements are made by examining one image area after another, rather than by examining and integrating all image areas at once, as would be done by an observer.

It appears that the curves for the midtones and shadows are also related to the presence of thin films on the surface of the daguerreotypes. These films may be silver sulfide or silver oxide, as well as silver cyanide, silver thiourea complexes, and various silica materials from the cover glass. Some films of this type form on silver regardless of storage conditions. Others are formed as a consequence of procedures for cleaning daguerreotypes. The presence of maxima and minima in the experimental specular reflectance curves indicates transitions between antireflection and enhanced reflection due to the presence of the films. The specular reflectance data generally show a rising trend in overall reflectance for larger values of θ, θ'. This trend is related to increasing sensitivity to differences in the refractive index compared to the extinction coefficient for the combined reflectance from the daguerreotype plus the thin film.[26] Even though the presence of the microstructure will dampen the overall specular

reflectance from a daguerreotype, the microstructure is not related to the oscillations seen in the specular reflectance curves for the shadows and midtones.

Figure 8.15 shows the specular reflectance data for a nineteenth-century daguerreotype that was removed from an unbroken nineteenth-century seal shortly before the reflectance measurements were made. This daguerreotype was in good condition, as can be seen in the figure; however, scanning electron microscopic examination indicated that the daguerreotype had been cleaned at some point before it was sealed. The highlight region has a relatively flat curve with a maximum reflectance of about 18 percent. The shadow and midtone regions show peaks around 25 degrees, in addition to peaks around 45 and 70 degrees. The maximum specular reflectance for the shadow region is about 90 percent. The contrast range is wide, as indicated by the difference between the maximum specular reflectance of the highlight and shadow regions and by the overall separation of the specular reflectance curves for all the image areas. The wide contrast range can also be seen by examining the photograph of the daguerreotype. The undulations in the midtone and shadow curves show the presence of thin films on the daguerreotype. As can be seen in the photograph, these films are not thick enough to mar or obscure the image.

Mechanical abrasion such as scratching and scraping of the daguerreotype surface alters its reflectance characteristics by interfering with the smooth polished substrate of the plate. The quality of polish determines the blackness of the shadow regions. Scratches may be either diffusely reflecting or specularly reflecting depending upon their size. In either case their presence on a daguerreotype interferes with or, in severe cases, overrides the reflectance characteristics of the image itself, making the image difficult or impossible to see. In addition, abrasion may also remove image

Fig. 8.15. Reflectance data for a nineteenth-century daguerreotype shortly after unsealing.

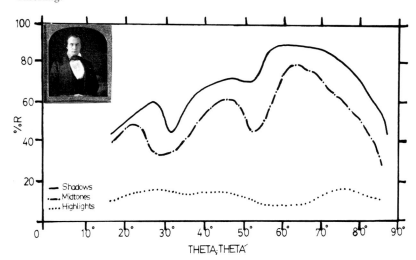

particles, thus altering the reflection characteristics of the highlights. Further, scratches and pitting can act as sites for the initiation of corrosion.

Specular reflectance data for a badly scratched daguerreotype, along with a photograph of the daguerreotype, are shown in figure 8.16. In this case the image, even under the best of conditions, is very difficult to see and is covered with a veil of scratches. The specular reflectance data show that the highlight regions remain diffusely reflecting but that the reflectance varies over a range of 20 percent with the highlight maximum reflectance at about 30 percent at the high angles (θ, $\theta' > 60$). The specular reflectance data for the shadows and midtones show a general reduction of the overall specular reflectance values when compared with daguerreotypes with no surface damage. For some angles the midtones show lower specular reflectance values than the highlights. The maximum reflectance for the shadow areas is about 62 percent. The reduction of specular reflectance characteristic of the shadow regions has resulted in an overall collapse of the maximum contrast range. The overlapping reflectance curves indicate the presence of uneven films on the surface of the daguerreotype, as well as a general blurring of the image due to the lack of clear distinctions between the various image areas.

Daguerreotypes may also be covered with thicker tarnish films that are highly absorbing, thus making the image very hard to see. Daguerreotypes in this condition show none of the distinctive specular reflection characteristics of daguerreotypes in otherwise good condition. Badly tarnished daguerreotypes show reflectance values of less than 10 percent, with little or no difference between the reflectance of the highlight, midtone, and shadow image areas. Daguerreotypes in this condition may be mistaken for other types of cased photographs, such as ambrotypes or tintypes.

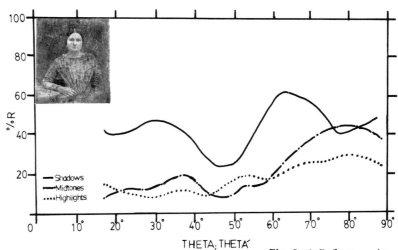

Fig. 8.16. Reflectance data for a scratched nineteenth-century daguerreotype.

Since the daguerreian era tarnish removal has most often been effected by the use of silver solvents. Cleaning solutions tend to cause irreparable damage to daguerreotypes by etching the metal of the silver substrate and by totally or partially removing the image particles. Specular reflectance data for an overcleaned daguerreotype are shown in figure 8.17. This daguerreotype is difficult to see because the image is quite faint, although it appears rather enhanced in this conventional photographic copy. The data indicate that the highlight regions have become highly specular and show behavior more like that expected of the

Fig. 8.17. Reflectance data for an overcleaned nineteenth-century daguerreotype.

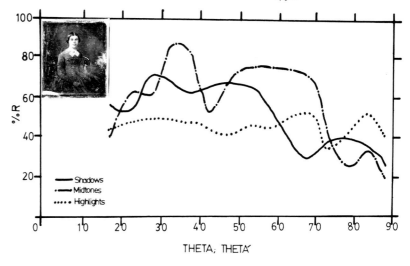

shadow regions of the plate. At some angles the highlight areas show higher reflectance values than either the shadow or the midtone values. The possible contrast range has been collapsed by the increase in specular reflectivity of the highlight regions and the decrease in specular reflectivity of the shadow regions. The loss of contrast range is due to the alteration of the microstructure, as well as etching of the plate surface. The reflectance curves have taken on unusual patterns suggesting the presence of uneven coatings of films or residues left from the cleaning procedure. The combined effect of microstructure damage and cleaning residues is an image that has little or no contrast and thus appears faint.

The model for the visual appearance of daguerreotypes described in this chapter seems to give both a qualitative and at least a semiquantitative interpretation of most of the observed features. The optical properties of daguerreotypes turn out to be moderately complicated although the physics is straightforward. It was a happy twist of fate that the combination of materials and processing used by Niépce and Daguerre happened to produce a characteristic microstructure with the right properties to produce a visible image.

At the primary level the appearance of the daguerreotype is determined by the size, density, spacing, and distribution of image particles, which are the essential feature of the daguerreotype process. The image itself is then seen by a viewer as a superimposition of contributions from specular and diffuse reflectance in different regions of the plate. The image appearance is very sensitively dependent on illumination, whether diffuse or a point source, and on viewing angle. The most dramatic aspects of the optical physics is the positive-negative reversal depending on viewing angle, for which this model nicely accounts. Finally, however, surface damage and the ubiquitous corrosion films further modify the optical characteristics of the surface. Changes in optical characteristics resulting from damage also fit the model.

THESE SENTIMENTS OF James Waterhouse, from the second Traill Taylor Memorial Lecture delivered to the London
Photographic Society in 1899, express the
underlying motivation for most investigations of the daguerreotype's fundamental
basis after the process was no longer practiced. It is hard to resist the thought that
the daguerreian process offers a unique way
to study the most profound elements of silver halide photographic systems because
the daguerreotype is apparently the most
simple of all photographic processes. The
unexposed daguerreotype plate is composed of an unadulterated silver halide layer
with no organic binder present to alter or
influence the photographic reaction. However, as seen in chapter 7, the findings of
most investigations into the scientific basis
of the daguerreotype process were not sufficient either to further silver halide theory
or to relate nineteenth-century reports
about the process's practical working to
later scientific findings. The secrets of the
process have remained a perpetual frustra-
tion to scientists who have explored its fundamental nature since the time it was first
introduced.

There are several very important reasons
for this state of affairs. Early on, as we have
seen, researchers lacked both a coherent
theoretical basis and the analytical tools that
would have allowed them to characterize
the daguerreian image and examine the process of image formation. In addition, the
theoretical basis of the present-day photographic process is still a subject of some debate and is yet to be fully described. In order
to examine the interaction of light with silver halide crystals and the creation of a latent image, visible images must be formed,
and when these images are formed part of
the physical evidence of the latent image is
altered. It is impossible to separate these
two phenomena, and therefore the researcher is always looking at both the physics and the physical chemistry of the pho-

9

Image Formation

Although Daguerreotype has long been abandoned, it
is capable of teaching us a good deal about the action
of light on the silver haloid compounds which form
the basis of our modern dry plates.

James Waterhouse, "Teachings of Daguerreotype,"[1]

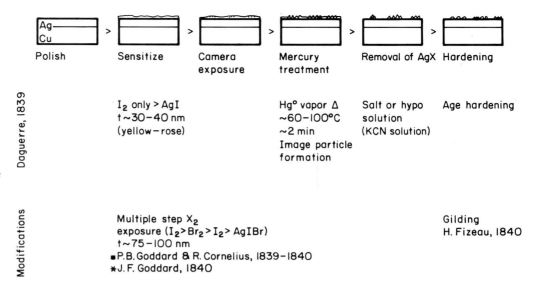

Fig. 9.1. The processing steps for the daguerreotype are shown schematically. The middle row is the process as introduced by Daguerre, and the bottom row shows the modifications to the process and the dates when they were introduced.

	Polish	Sensitize	Camera exposure	Mercury treatment	Removal of AgX	Hardening
Daguerre, 1839	Ag—Cu	I_2 only > AgI $t \sim 30-40$ nm (yellow–rose)		Hg° vapor Δ ~60–100°C ~2 min Image particle formation	Salt or hypo solution (KCN solution)	Age hardening
Modifications		Multiple step X_2 exposure (I_2>Br_2>I_2> AgIBr) $t \sim 75-100$ nm ■P.B. Goddard & R. Cornelius, 1839–1840 ∗J.F. Goddard, 1840				Gilding H. Fizeau, 1840

tographic system. Until the advent of modern characterization instrumentation and techniques such as computer modeling, the best tool available to the photographic scientist for solving practical problems was to run an experiment and see what happened—trial and error, the best of shirt-sleeves chemistry.

To review, the steps of the daguerreotype process are shown schematically in the first row of figure 9.1. The steps as introduced by Daguerre in 1839 are given in the middle row, and the modifications to the process and the dates they were introduced are given in the third row. Image formation in the daguerreotype system raises two intriguing questions: (1) After a latent image is formed by camera exposure, what is the role of mercury in forming the characteristic daguerreian image microstructure and, thus, in forming a visible image? (2) How is camera exposure shortened by the additions of other halogens to the iodine of the original daguerreotype process? Even though shortening the camera exposure of the light-sensitive medium is a preliminary step to forming a visible image, it is necessary to look at how a visible image is produced, regardless of exposure times, before analyzing how the process is accelerated.

In its initial stages of exposure and latent image formation, the daguerreotype should be and is no different from any other photographic process. In its unexposed state the sensitized daguerreotype plate is much like some model systems, such as evaporated sheets of silver bromide and silver halide single crystals, now used to examine latent image formation. For the purposes of this discussion, the commonly held Gurney-Mott mechanism of latent image formation is assumed to hold for the daguerreotype as much as for other silver halide processes. That is, exposure of a silver halide crystal to light creates a photoelectron that is trapped at a sensitivity center or defect in the crystal lattice. Subsequently, an interstitial silver ion from the interior of the silver halide crystal is attracted to and neutralizes the negatively charged site formed by the trapped photoelectron, creating an atom of silver metal. This process is repeated until a sufficient number of silver atoms have aggregated at one site to form a latent image or development center. There are many theories about the most favorable sites for trapping photoelectrons and how silver atoms aggregate; however, these are not pertinent to this discussion.[2]

The most prevalent technical assumption about the daguerreotype is that mercury is a developer that acts in the same way as other developers in more conventional photographic systems. In the broadest sense, this assumption is true. Mercury does play some role in the formation of the image particles that make up the daguerreian microstructure. However, to a photographic scientist the term *development* has a very specific meaning, and it is necessary to discover whether image formation in the daguerreian system is covered by the more narrow convention. Photographic development is, by definition, a reduction-oxidation reaction in which the developer acts as a reducing agent for converting light-exposed silver halide crystals to silver image particles. In this process the developer becomes oxidized. The latent image site acts as the catalytic site, or the electrode, where this reduction-oxidation reaction occurs. The type of developer used determines the way in which image particles are formed and, consequently, their microstructure.

There are two broad classes of photographic development—*chemical development* and *physical development*. These names are carryovers from the nineteenth-century debate about whether the photographic image is the result of the physical (i.e., mechanical) action of light on a light-sensitive material or the result of the chemical action of light on such materials. The difference between these two types of developers is now defined as a function of the source of silver ions needed to form silver image particles. When the source of silver ions for image formation is primarily from outside of individual silver halide grains in the emulsion, the developer is called a physical developer. In this case silver ions for the production of image particles may be provided from an excess of silver nitrate present in the developer or in the colloid (as in the wet collodion process), or silver ions may be provided by general solvent action of the developer on silver halide present in the colloid layer. The latent image is the catalytic site for the reduction of silver ions from the developer solution, and image particles formed in this way have characteristically compact and shotlike microstructures.

Conversely, if the source of silver ions is from within individual silver halide grains of the emulsion, then the developer is called a chemical developer. Here, the latent image is the site for reduction of silver ions from the silver halide crystal itself. The particles formed in this way have a filamentary structure and are like small (10 to 30 nanometers) clumps of steel wool. In both

physical and chemical development the reduction-oxidation reaction is the same, but the mechanisms are different, as manifested in the different image structures. Figure 9.2 shows schematically the difference between physical and chemical development and the image structures that result from their use in conventional photographic materials. This discussion is a broad description of photographic development; there is an entire range of nuance covering developers that have the characteristics of both chemical and physical developers, but these are not pertinent to the subject at hand.[3]

It has been suggested that the mercury in the daguerreotype system is merely a physical developer, both because it is a metal and is therefore assumed to be a source of material for the image formation and also because the characteristic image structure associated with the daguerreotype is small, compact particles similar to those produced by conventional physical developers. This comparison seems valid in a rough way if no empirical evidence is brought to bear. The striking differences between physical

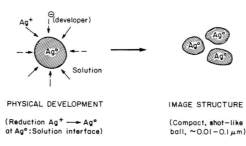

PHYSICAL DEVELOPMENT

(Reduction $Ag^+ \longrightarrow Ag^o$
at Ag^o:Solution interface)

IMAGE STRUCTURE

(Compact, shot−like
ball, ~0.01−0.1 μm)

CHEMICAL DEVELOPMENT

(Reduction $Ag^+ \longrightarrow Ag^o$
at AgX: Ag^o interface)

IMAGE STRUCTURE

(Filamentary bundles,
~1−2 μm)

Fig. 9.2. Schematic representation of physical and chemical photographic development mechanisms and their resulting microstructures. After Metz, "On the Mechanism of Photographic Development," fig. 3.

developers and the daguerreian system are that (1) conventional physical developers usually are based on an excess of silver and not some other metal and (2) daguerreotype image particles are, on average, at least one order of magnitude larger than those produced in conventional photographic materials. One way to test the hypothesis that mercury is merely a physical developer is to substitute a conventional physical developer for mercury and see if it produces the characteristic daguerreotype microstructure. James Waterhouse conducted a set of experiments along these lines during the 1890s and observed that the images he produced using conventional developers were different from other daguerreotypes.[4] Irving Pobboravsky also performed a set of experiments using conventional developers on daguerreotype plates but without much success.[5]

An experiment comparable to those of Waterhouse and Pobboravsky was carried out using dry-plate developers similar to those used by Waterhouse, Metz high- and low-potential physical developers,[6] and the

common developer, D-76, which is a combination chemical-physical developer.[7] Just as Waterhouse observed, some of these developers produced recognizable images that, although on daguerreotype plates, did not look like other daguerreotype images. Some developers produced no discernible visible images. However, when examined using a scanning electron microscope, all the plates were found to have image particles regardless of whether an image could be seen by a viewer. In all cases the image particles produced had very odd microstructures unlike either those found on daguerreotypes or those that would have been produced had the same developers been used to develop conventional films. Some of these microstructures are shown in figures 9.3, 9.4, 9.5, and 9.6. In the cases where no image was produced, image particles were dispersed over the entire plate surface, and only one micrograph of the microstructure is shown. When visible images were formed, micrographs of both highlight and shadow areas are shown.

In general, the image microstructures produced using conventional developers in place of mercury are more like those that would result from using chemical developers on conventional films—filamentary tangles of particles. Solid, shotlike image particles were produced on a few plates; however, these plates did not have visible images, and in addition the entire underlying plate surface was covered with a granular coating, as seen in figures 9.3 and 9.4. Most important, these image microstructures are not like those that define the daguerreotype. Images produced in these experiments were visible because of light absorption, which accounts for their atypical appearance, rather than the light-scattering phenomenon that further defines the daguerreian image. These experiments demonstrate that daguerreotype processing is not simply another example of physical development.

Fig. 9.3. Scanning electron micrograph of the image microstructure that results when a daguerreotype plate is treated with acid-iron dry-plate developer. The scale bar is equal to 10 micrometers.

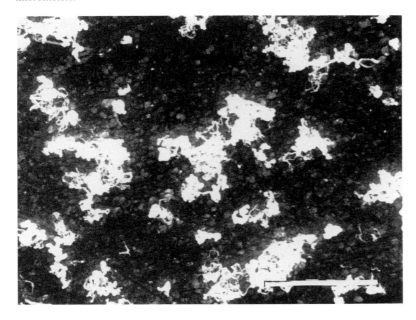

The classic way to examine the results of image formation is to partially develop an image, stop the development action by fixing the image, and then examine the results. As the process is repeated and development times are gradually increased, the entire sequence of image formation can be seen in small steps approximating the separate frames of an animated cartoon. The daguerreotype process was examined this way during mercury treatments to determine the steps of image formation. The initial experiments on image formation were conducted using daguerreotype plates that had been multiply sensitized with iodine and bromine vapor, i.e., the plates were sensitized by being placed over iodine vapor, then bromine vapor, and finally, again, iodine vapor. Even though multiple sensitization is not the original processing sequence introduced by Daguerre, it was the most widely practiced method of daguerreotype processing, and it was studied first. Once image formation on multiply sensitized daguerreotypes has been examined here, the process and function of gilding will be discussed. Then image formation in singly sensitized daguerreotype plates, in daguerreotypes treated with cold mercury, and in Becquerel-developed daguerreotypes will be considered. Finally, the overall relation of halogen sensitization to image formation and photographic speed in the daguerreian system will be described.

Multiply Sensitized Daguerreotypes

Small daguerreotype step tablets (1 by 2 inches) were made as samples for the image formation experiments. After each stage of processing, the progress of image formation was checked using the scanning electron microscope. A procedure was followed to ensure that areas of similar density would always be viewed in the microscope, regardless of the numbers of different plates

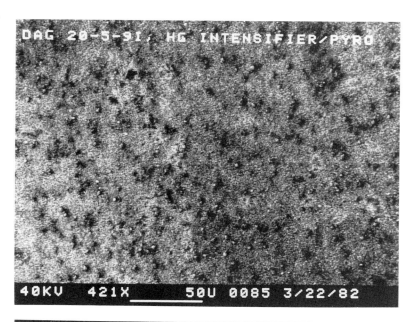

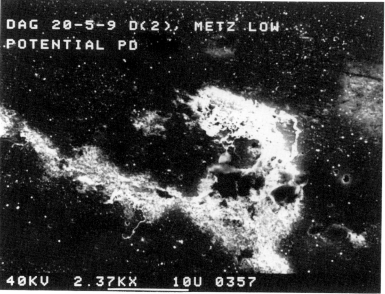

Fig. 9.4. Scanning electron micrograph of the image microstructure that results when a daguerreotype plate is treated with a mercury intensifier followed by an acid-pyro dry-plate developer. The scale bar is equal to 50 micrometers.

Fig. 9.5. Scanning electron micrograph of the image microstructure that results when a daguerreotype plate is treated with a Metz low-potential physical developer. The scale bar is equal to 10 micrometers.

Fig. 9.6. Scanning electron micrograph of the image structure that results when a daguerreotype plate is treated with a Metz high-potential physical developer. The upper micrograph is a highlight area and the lower micrograph a shadow area. The scale bar is equal to 5 micrometers.

coating steps were done in subdued light; the final step was done in complete darkness. The plate was then placed in a contact-printing frame under a step tablet and given a uniform exposure of light from a 3200 Kelvin tungsten photoflood lamp. Mercury treatments over hot mercury vapor (80 degrees Centigrade) were carried out for times ranging from zero to 1.5 minutes. After the mercury treatment unexposed silver halide was removed using a dilute solution of sodium thiosulfate (i.e., hypo). In the first experiments the hypo solution appeared to cause some dissolution of the image particles, and electrolytic fixing was used for the remaining experiments. Electrolytic fixing is done by placing the plate in a saturated solution of common salt and then touching a corner with an aluminum rod. This creates a slight electric current, causing the silver halide layer to be removed from the plate in a wave. After fixing, each plate was rinsed with water and dried with a jet of nitrogen gas before it was examined in the scanning electron microscope.

The evolution of the daguerreian microstructure was studied step by step starting with samples that had been exposed and fixed with no mercury treatment. The micrograph in figure 9.7 shows the microstructure of the initial stage after camera exposure and before mercury treatment. In these samples the entire surface of the sample plates was uniformly covered with small silver particles, all about 10 nanometers in diameter. The left half of the micrograph is oriented so that there is a highlight area on the lower part of the field and a shadow in the upper part, as described earlier. The right half of the micrograph shows the boxed area on the left at ten times higher magnification. If the exposed plate is placed over mercury for less than 10 or 15 seconds, these small image particles are enlarged somewhat, but the random dispersion of particles over the entire surface of the plate persists, and there is no dis-

used in the experiments. First, one corner of the sample plates was clipped to mark the orientation of the step tablet used for exposure; second, a special holder was used to mount the plates in the microscope stage. The holder was positioned so that all samples were oriented with a highlight area in the lower portion of the SEM screen and a shadow area in the upper portion of the screen with the shadow-highlight interface more or less bisecting the screen. The micrographs shown in the figures all have this orientation unless otherwise noted.

For each experiment the plate was polished with jeweller's rouge (i.e., iron oxide) until it had a mirrorlike surface. The plate was placed over iodine vapor until it was a yellow-rose color, then over bromine vapor until it was steel blue, and finally over iodine once again for half the time required to produce the yellow-rose color of the initial iodine coating. The first two halide-

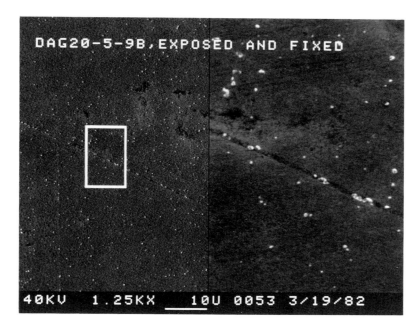

Fig. 9.7. Scanning electron micrograph of the microstructure of small particles that are present on a daguerreotype that has been sensitized and exposed and the silver halide removed before the plate received any exposure to mercury. The field on the left side of the micrograph is oriented so that the portion of the plate shown in the lower half of the micrograph is a highlight image area. The field on the right shows the boxed area of the left side at 10 times the base magnification. The scale bar is equal to 10 micrometers (for the left field).

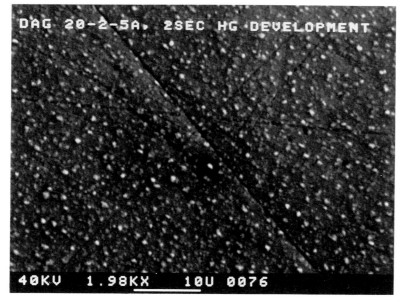

Fig. 9.8. Scanning electron micrograph of a daguerreotype surface after receiving a 2-second exposure to mercury vapor. The daguerreotype is oriented so that the lower half of the field is a highlight image area and the upper portion is a shadow area. The scale bar is equal to 10 micrometers.

cernible arrangement of image particles that would indicate the presence of an image (fig. 9.8).

As mercury treatment is continued for more than 15 or 20 seconds, the evolving microstructure becomes rearranged according to the pattern of light exposure. At this stage the resulting image is not as bright as a finished image, but it can be discerned by a viewer. The microscope examination shows that the random dispersion of small image particles described above has been transformed into an odd microstructure composed of both small, discrete image particles and also image particles perched on pads or platelets. This very odd image

Fig. 9.9. Scanning electron micrograph of the platelet transition stage in image particle growth on a daguerreotype. The right field shows platelets at 10 times the magnification as the left field. The scale bar is equal to 50 micrometers for the left field and 5 micrometers for the right field.

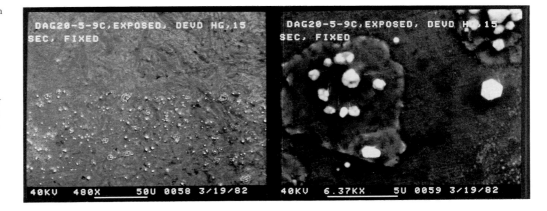

Fig. 9.10. Scanning electron micrograph of the surface of a newly made, multiply sensitized daguerreotype plate. The scale bar is equal to 50 micrometers.

structure is shown on the left side of figure 9.9. Even though the area in the bottom portion of the bottom right micrograph is a highlight area, the number of particles is fewer than would be expected. The right-hand micrograph in figure 9.9 shows some of the platelets with image particles at higher magnification. The wisps that appear to be emerging from some of the white particles on the platelets are deposits, probably mercury volatilized when these freshly made platelets were bombarded by the electron beam of the microscope. When the samples were allowed to rest for a few hours after being made and before being put into the microscope, volatilization did not occur.

The presence of these particle-covered platelets was puzzling in that this type of microstructure had not been observed in the highlight regions of the modern daguerreotype step tablets used to determine the characteristic daguerreian image structure nor on any of the nineteenth-century daguerreotypes in the study collection. These structures occurred in the highlights, and while on casual observation they may resemble shadow particle agglomerates, on closer examination they are not the same at all. Moreover, when the mercury experiment is continued for times longer than a minute the platelets are not then observed. The micrograph in figure 9.10 shows a shadow-highlight interface after 1.5 minutes of mercury treatment. Micrographs of a highlight, midtone, and shadow area of the same daguerreotype are shown in figure 9.11. The newly made image particles are highly crystalline with a more or less hexagonal structure and sharply defined crystal faces. The remnants of some platelets can be seen in the shadow area as the faint outlines around the shadow particle agglomerates.

From this set of experiments it appears that the daguerreotype goes through a complex series of changes during the process of image particle formation. At first, many minute silver particles are dispersed over the surface of the exposed daguerreotype plate. Hot mercury vapor causes the enlargement of these particles until a transitional stage is reached where particle growth is more strongly favored in areas of the plate that were exposed to light. In this platelet transition stage, platelets are formed in the highlight regions of the plate, and, as image particle formation continues, these platelets appear to break up into the small particles typical of daguerreotype highlight regions. The stages of image particle growth are shown schematically in figure 9.12.

What is the chemical composition of image particles, and does their composition change as a result of processing? Typical energy-dispersive x-ray spectra for daguerreotype image particles in shadow, midtone, and highlight areas are shown in the first column of spectra in figure 8.3. Note that the mercury content of individual image particles differs according to their size—that is, the small image particles typically found in daguerreotype highlight regions, even when first made, have almost no mercury, whereas shadow particle agglomerates are large and have comparatively large amounts of mercury when first made. Further, the underlying daguerreo-

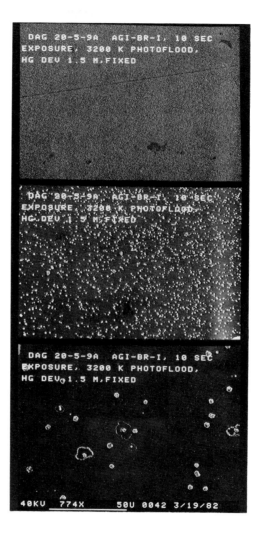

Fig. 9.11. Scanning electron micrograph of the microstructure of a highlight, midtone, and shadow area of a newly made, multiply sensitized daguerreotype. The scale bar is equal to 50 micrometers.

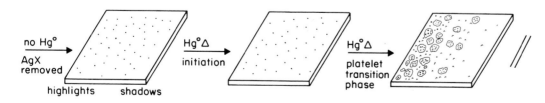

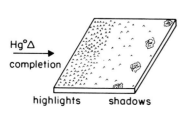

Fig. 9.12. Diagram showing the stages of image particle growth.

Highlights
· IP composition Ag(Hg)
· size ~ λ_{vis}
· ~2 x 10^5 mm^2
Shadows
· shadow particle agglomerates Hg Ag
· size ~50–100 μm
· ~100/mm^2

type plate is not covered with mercury and is merely silver. During the platelet transition stage of image formation, the platelets themselves have some mercury, and the particles on the platelets are mercury enriched. Energy-dispersive x-ray spectra are sufficient to give a rough idea of the elemental composition of the components in these microstructures but not to determine their exact composition. The precise chemical composition of daguerreotype image particles has been discussed and predicted by many experimenters, and the overriding belief has always been that image particles must all have the same composition and that composition must be a mercury-rich amalgam.[8] This belief is derived from reports that the daguerreian image is so fragile that it can be wiped off the surface of a plate with a feather.

Even though the fact that the chemistry of image particles varies is shown in the energy-dispersive x-ray spectra, their relative chemistry can also be inferred from the crystal structure of individual particles. The silver-mercury phase diagram is shown in figure 9.13. A phase diagram is essentially a map that shows the existence and coexistence of crystal and liquid phases as a function of temperature and bulk composition when the system is at equilibrium. One caveat about their use for low-temperature phases (i.e., in the typical temperature

range of mercury treatment) is that equilibrium is not always reached. Further, the presence of dotted lines for the phase boundaries indicates that the exact location of those boundaries is only inferred. The silver-mercury system is not well characterized in the temperature region of interest here.

Alice Swan and her colleagues predicted from energy-dispersive x-ray data that all daguerreotype image particles are composed primarily of the mercury-rich body-centered cubic (bcc) gamma-phase (γ) amalgam Ag_3Hg_4.[9] However, most of the image particles in newly made daguerreotypes (see figures 9.9, 9.10, and 9.11) have a distinctively hexagonal shape and clearly are not cubic. The usual particle shape corresponds to the hexagonal, closest packed (hcp) crystal structure of the silver-rich epsilon-phase (ϵ) amalgam $Ag_{11}Hg_9$, shown in the silver-mercury phase diagram.[10] The presence of the epsilon-phase amalgam and the lack of mercury peaks for highlight image particles, as the energy-dispersive x-ray spectra of figure 8.3 indicate, suggest that most daguerreotype image particles are some silver-mercury solid solution.

Another reason for the ill-defined phase boundaries in the lower temperature range of the silver-mercury system is that amalgams exist as solids that are very near their melting point. These alloys tend to be somewhat disordered at room temperature. Amalgams undergo a process called hardening, which is caused by mercury loss or the more complete amalgamation (i.e., combination) of the alloy over time. In both cases hardening stops when the amalgam has reached a chemical equilibrium. The hardening process of amalgams is of commercial interest because amalgams are used to fill carious teeth. However, in the nineteenth century the simple silver-mercury amalgam of the daguerreotype process was determined to be unsuitable for dental work because this particular amalgam hardens

Fig. 9.13. The silver-mercury phase diagram.

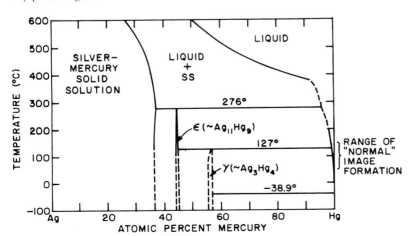

very slowly and also expands excessively during the hardening process, causing teeth to break.[11]

Certainly, the fact that the newly made daguerreian image is fragile and can be wiped away attests to the poor mechanical strength of amalgams. However, since amalgams undergo hardening, the daguerreotype should also undergo some alteration over time if left alone and not gilded. This is indeed the case. Image particles continue forming until they reach some equilibrium state with the underlying daguerreotype plate, at which point the image microstructure has developed significant mechanical strength. The second column of energy-dispersive x-ray spectra in figure 8.3 shows that mercury is lost over time. Ungilded image particles themselves tend to be structurally altered over time. The smaller image particles from the highlight and midtone regions of the daguerreotype gradually collapse, lose their crystal faces, and become rounded, as is shown in the week-old image particles in figure 9.14. Shadow particle agglomerates change very slowly because they have more mercury per particle than any other image particles on the daguerreotype surface. As they harden, they become flattened and porous. The micrographs on the right side of figure 9.15 show ungilded shadow particle agglomerates after two years of aging, compared with gilded shadow particle agglomerates on the left side of the figure.

Gilding

What does gilding do? When he introduced the process of gilding, Hippolyte Fizeau claimed that it did two things: (1) It "fixed" the image, and (2) it made the highlight areas of the image whiter and the shadows darker, thus improving the daguerreotype's overall appearance and contrast range. Sometime during this century a third function was attributed to gilding—namely,

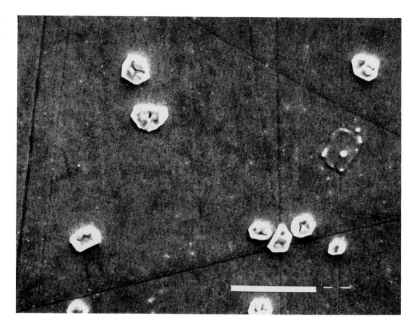

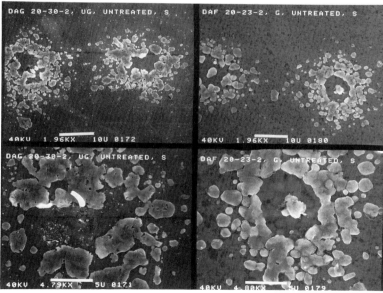

Fig. 9.14. Scanning electron micrograph of image particles of an ungilded daguerreotype after one week of aging. The scale bar is equal to 10 micrometers.

Fig. 9.15. Left: Scanning electron micrograph of ungilded shadow particle agglomerates that have aged two years. Right: Gilded shadow particle agglomerates. The scale bar is 10 micrometers in the upper micrographs and 5 micrometers in the lower micrographs.

that it provided some sort of protection from corrosion, an assumption made because of the relation of gilding to the present-day process of gold toning. Gilding is the forerunner of gold toning for conventional photographic materials; nevertheless, these two processes are not the same. When a daguerreotype is gilded, it undergoes rapid hardening because mercury in the image particles is replaced by gold. This change stops the continuous formation of the image particles that would occur during age hardening, thereby "fixing" the image particles.[12] In conventional gold toning, gold replaces image silver and provides some sort of protective coating for the image.[13] Some silver may be replaced by gold during gilding, but this is doubtful. Daguerreotypes do not become coated with gold during gilding, and no protective function is derived from gilding.[14] However, the rapid hardening caused by gilding does produce a remarkable improvement in the mechanical stability of the image particles.

The rate of gilding depends on the size, number, and mercury content of the image particles, the local surface area, and the amount of mercury per unit area. An image's highlight areas are greatly affected by gilding because they have a large number of particles and, thus, a large surface area and because there is a high concentration of mercury per unit area even though individual highlight image particles have a relatively low concentration of mercury. Shadows are less affected because these areas have fewer particles and tend to be less reactive. The energy-dispersive x-ray spectra for gilded daguerreotype image particles are shown in the third column of figure 8.3. Even though the location of mercury peaks is indicated, probably no mercury is left in either the midtone or the highlight image areas.

In addition to changing the daguerreotype's mechanical properties, gilding also slightly enlarges the size of image particles and, thus, changes its optical properties. The exchange of gold for mercury during gilding is not even: a little more than one atom of gold replaces each atom of mercury. Even though the atomic radius of mercury is slightly larger than that of gold, the uneven exchange causes the size of the image particle to increase slightly. Gilded image particles look plumped up and rounded. The change in size is sufficiently large to alter the light scattering of the images from the bluer regions of the spectrum and move it more to the center of the visible region of the electromagnetic spectrum. A gilded daguerreotype, therefore, appears to have whiter highlights. The shadow regions of the daguerreotype are not made darker, although by comparison with highlight areas they may appear blacker. It is not possible to detect by simple visual observation whether or not a daguerreotype has been gilded. The reports that gilded daguerreotypes appear gold colored are part of the myth of the process.

To complete this morphological discussion of image formation in the daguerreian system, it is necessary to look at the variants of daguerreotype processing to see if the typical microstructure persists regardless of processing. The same type of experiments conducted for multiply sensitized daguerreotypes were also conducted for singly sensitized, mercury-treated daguerreotypes, daguerreotypes treated with cold mercury vapor, and Becquerel-developed daguerreotypes.

Singly Sensitized Daguerreotypes

What microstructure is created when a daguerreotype is processed in the way first described by Daguerre—that is, if only iodine sensitization is used? At the commencement of this subset of experiments, a plate was sensitized in subdued light to determine the amount of time required to produce a yellow-rose iodide layer on the plate. Once this time had been determined, the plates used in these experiments were sensitized in the dark. The initial stages of image particle formation are the same for singly sensitized plates as for multiply sensitized plates except that singly sensitized plates produce significantly fewer image particles during image formation. Just as before, at the outset of image formation image particles were dispersed over the entire plate until some threshold was reached and the platelet microstructure emerged. The platelets found on singly sensitized plates were not as robust as those seen in the multiply sensitized plates. The remains of platelets are visible in the surround of many of the image particles seen in the micrographs of figure 9.16. These micrographs show the characteristic microstructure of the singly sensitized daguerreotypes. Typically, there are very few widely spaced image particles, which are all on the order of 0.1 to 0.5 micrometers in diameter. Also, the individual image particle shapes show little variety. Daguerreotypes made in this way should be—and are—noticeably weaker in appearance than daguerreotypes made using multiply sensitized plates. The chemistry of the individual image particles from these plates is no different from the chemistry of those formed on multiply sensitized daguerreotypes.

Cold Mercury Treatment

One last possibility for forming a daguerreotype image is to use cold instead of hot mercury vapor. In this case a daguerreotype plate is either singly or multiply sensitized, exposed to light in a camera, and then allowed to sit over cold mercury until an image appears. This process can take a few hours or as long as several days, depending on the ambient temperature. The microstructure of an image formed in this way

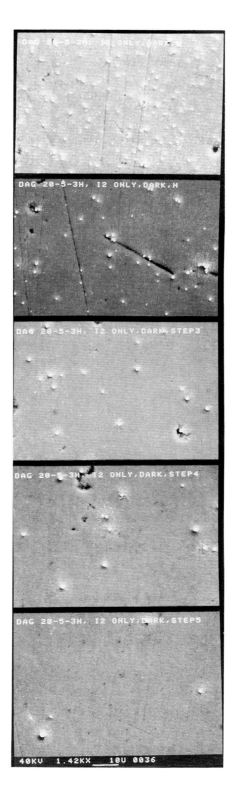

Fig. 9.16. Scanning electron micrograph of different image areas of a singly sensitized daguerreotype after mercury treatment. The top micrograph is the whitest highlight and the bottom micrograph is the darkest shadow. The scale bar is equal to 10 micrometers.

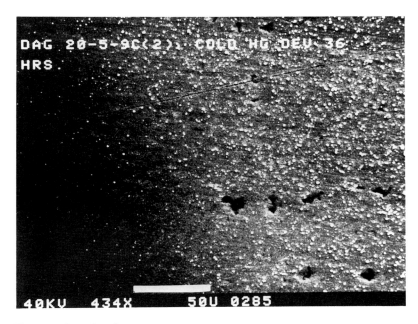

Fig. 9.17. Scanning electron micrograph showing the microstructure of a multiply sensitized daguerreotype after 36 hours over cold mercury. The scale bar is equal to 50 micrometers.

looks like a snowy deposit or the kind of paint surface formed when paint is applied by splattering it through a screen onto the support material, as shown in figure 9.17. All image areas are covered with very small (i.e., a few nanometers in diameter), ill-defined particles confined to areas that were exposed to light. Daguerreotypes formed in this way often have a bluish cast.

Becquerel-Developed Daguerreotypes

The last method of making a daguerreotype does not use mercury at all. Becquerel plates are made by sensitizing a polished daguerreotype plate with iodine vapor only, after which the plate is given a camera exposure. The exposed plate is removed from the camera and given an overall exposure to red light until a print-out image appears.[15] This phenomenon was first observed by Edmond Becquerel in 1840. Essentially, the Becquerel effect is a demonstration of the fact that once a silver

halide crystal has been exposed to light, the latent image can act as a weak spectral sensitizer. Thus, the sensitivity of the silver halide crystal is extended to longer wavelengths than was possible before the image-forming exposure. The exact mechanism of the Becquerel effect is still not understood. The microstructure formed in this way is shown in figure 9.18. All the image particles are of uniform size (about 0.1 micrometers in diameter). The only factors that give rise to variations in density are the differences in the particle numbers and their spacings. The extreme highlight areas of a Becquerel plate can have many more than several hundred thousand image particles per square millimeter. Not surprisingly, image particles formed in this way are composed only of silver, as can be seen in the spectra of figure 9.19.

Halogen Sensitization and Photographic Speed

It should be evident that the traditional processing variations used for daguerreotypes all produce image microstructures that fall within the physical definition of the daguerreotype. Moreover, these image morphologies are not comparable to those produced using conventional photographic developers. However, the initial questions posed in this chapter still remain unanswered. How does multiple sensitization increase the daguerreotype's photographic speed? Does the daguerreotype come under the definition of photographic development? If it does not, does it fit into another category of materials processing?

A series of experiments was devised to see how photographic speed in the daguerreian system is increased by multiple sensitization and also to see whether some contemporary ways of increasing the light sensitivity of conventional photographic materials could be applied to the daguerreian system. Sensitometry, or the mea-

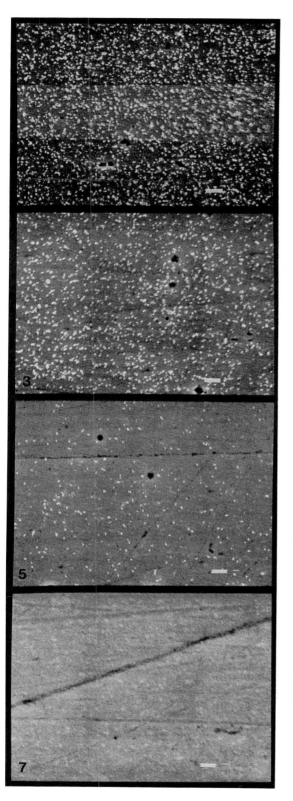

surement of the light sensitivity of photographic materials, is a central issue in photographic science. Standard methods to determine photographic speed have been devised so that comparisons may be made between different photographic materials or between processing methods for the same material. Most of these standard speed criteria are not readily applicable to the daguerreian system. For the purposes of this study, a rough method for comparing the speed of daguerreotypes was devised. Sensitized plates were exposed for 1.5 minutes with a 3200 Kelvin tungsten photoflood lamp at 1.5 feet using a step tablet. Following light exposure, the step tablet was treated with mercury at 80 degrees Centigrade for 1.5 minutes, and then unexposed silver halide was removed from the plate surface. These exposure and mercury conditions were considered the standard conditions for all speed calculations. After image formation the number of steps that appeared on the sample was divided by 1.5, the time of light exposure. The speed number determined for a plate singly sensitized in the dark was set to be X, the basic speed of the daguerreotype system, and all other results were compared to time X.[16]

The sensitivity of conventional photographic materials is affected by a variety of factors related both to emulsion production

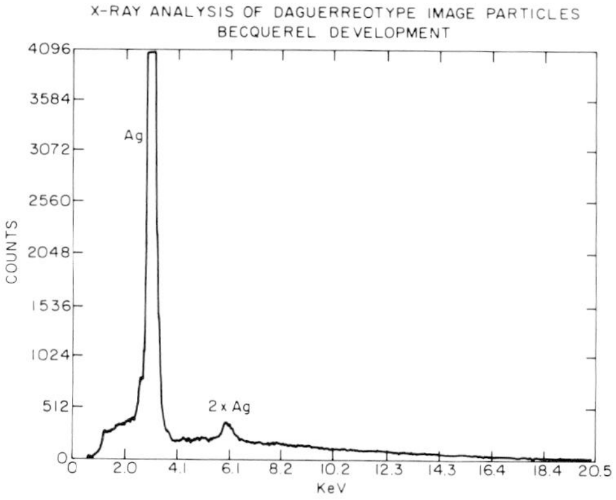

Fig. 9.18. Scanning electron micrograph of different image areas of a Becquerel-developed daguerreotype. The top micrograph is the whitest highlight, and the bottom micrograph is the darkest shadow. The scale bar is equal to 10 micrometers.

X-RAY ANALYSIS OF DAGUERREOTYPE IMAGE PARTICLES
BECQUEREL DEVELOPMENT

Fig. 9.19. Energy-dispersive x-ray spectra of image particles on a Becquerel-developed daguerreotype.

through control of the size distribution and chemistry of the silver halide crystals in a particular emulsion and, later in the camera stage, to control of exposure conditions. For most applications the desire is always to increase sensitivity in order to decrease exposure times. One of the simplest ways to enhance light sensitivity in photographic systems is to use mixed silver halides. Another common method is to expand the sensitivity of light-sensitive salts by adding materials such as certain dyes and semiconductors to an emulsion. This approach works because these additives selectively increase the responsiveness of light-sensitive salts to radiation that otherwise is not absorbed. Additives may also increase the photoconductivity of the light-sensitive material so that electron transfer within a crystal is enhanced and the possibility of forming latent images is thereby improved.

In the daguerreotype process the decrease in exposure times brought about by the addition of bromine and other halogens has been presumed to be simply the result of an increase in the fundamental light sensitivity of the plate caused by the use of mixed silver halides. However, there are several important differences between the silver halide present in the daguerreian system and that present in conventional materials. First, the daguerreotype system is a thin film of silver halide, rather than discrete silver halide crystals suspended in an emulsion. Second, the mixed bromine-iodine silver halide salt most commonly used in conventional photographic materials is not the same as that found in daguerreotypes.

When a conventional photographic emulsion is made, silver halide crystals are precipitated from solution, and if a silver bromide–silver iodide combination is to be used, the emulsion maker starts with silver bromide and adds silver iodide to make silver iodobromide.[17] There is a very good reason for this. Silver bromide can dissolve approximately 30 mole percent silver iodide in its crystal lattice. This solid solution produces a mix that indeed has a broader spectral sensitivity than either silver salt would have if used alone. The daguerreotypist, on the other hand, must start all processing by forming a silver iodide layer on a silver plate. The process does not work unless iodine is used for the first sensitization step. If the daguerreotype is to be multiply sensitized, the daguerreotypist adds bromine after the iodine to form a silver bromoiodide layer.[18] Silver bromoiodide is not the same as silver iodobromide. When the starting material for a mixed iodine and bromine silver salt is silver iodide, it is not possible to add more than a few mole percent of silver bromide. The two silver salts have different crystal habits—silver iodide is hexagonal, and silver bromide is cubic; thus, they cannot form a continuous solid solution. The solid solution is much more extensive on the bromine-rich side than on the iodine-rich side.

The solid solution of silver bromoiodide does not have a particularly different spectral sensitivity from that of silver iodide alone. However, its electronic properties are significantly different from pure silver iodide, pure silver bromide, or the solid solution of silver iodobromide. Silver bromoiodide is known as a fast ion conductor, and its ionic conductivity is approximately one thousand times larger than that of either single halide salt from which it is made. The exact reason for this huge increase has yet to be determined.[19] Bromine appears to be a homovalent dopant for silver iodide. In effect, the hexagonal silver iodide crystal lattice is slightly altered or perturbed by the presence of silver bromide, which causes an increase in the ionic conductivity. Inevitably, this helps increase the sensitivity of multiply sensitized daguerreotype plates because the transport efficiency for the movement of silver ions during latent image formation is improved.[20] The x-ray diffraction patterns in figure 9.20 confirm the perturbed state of the silver halide layer on the daguerreotype plates that develops

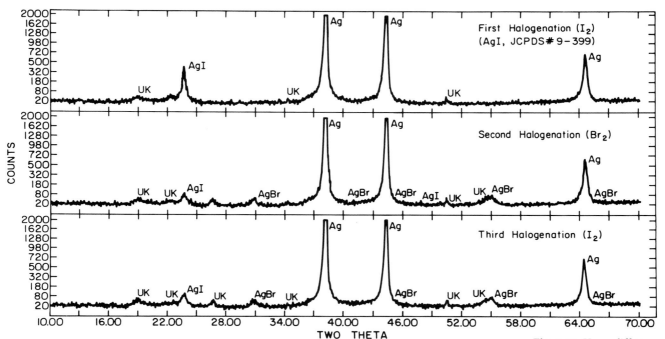

during the process of multiple sensitization. The first row indicates merely the patterns for silver iodide. The second and third rows taken after the bromine and the final iodine treatments of the plate show that the results of these additional treatments is to produce a perturbed silver bromoiodide crystal lattice.

How does the relative speed of daguerreotypes processed using single sensitization compare with that of daguerreotypes processed using multiple sensitization? When the photographic speed of daguerreotypes was tested in the laboratory, there was a concern about fogging the plates during the standard exposures. To avoid this, daguerreotypes were sensitized in total darkness and in subdued light, as would have been done during the nineteenth century. If the speed for a singly sensitized daguerreotype is determined as described above and if that speed is X, then a multiply sensitized daguerreotype sensitized in the dark has a speed at least double that—i.e., $2X$. When multiple sensitization was carried out in subdued light, the speed was at least 2.5 times faster than a daguerreotype sensitized the same way in total darkness—i.e., 2.5 $(2X)$.[21] These are modest increases in speed; however, they represent a considerable increase in the comfort of a sitter.

Those who work with conventional photographic materials find the relatively slow speed of daguerreotypes puzzling and often

feel that it should be possible to enhance their speed by the same techniques used to extend the sensitivity of modern films. Several of these techniques were tried, not to improve the process but rather to see if the mechanism of multiple sensitization could be inferred from other materials. A gamut of sensitizing dyes was provided for this study to see if any could be used to enhance the daguerreotype sensitivity.[22] These dyes were applied in solution as washes for sensitized plates and also as dusting media. None of the sensitizing dyes helped increase the sensitivity of daguerreotype plates: in fact, just the opposite occurred—plates were made totally insensitive. In addition to sensitizing dyes, the plates were exposed to sulfur gas to see if sulfur sensitization could be achieved. This also acted as a strong desensitizer. Further, Pobboravsky attempted to apply gold sensitization to the daguerreian system without success.[23] These experiments did not assist in understanding the daguerreian process. It is probable that the daguerreotype process will never achieve the speeds possible for conventional photographic materials.

One finding about multiple sensitization was particularly striking. The sensitivity of multiply sensitized daguerreotype plates produced in subdued light was greater than the sensitivity of those produced in total darkness; in fact, the increase was greater overall than the simple addition of bromine

Fig. 9.20. X-ray diffraction patterns of the silver halide layer of a daguerreotype plate as the sensitizer layer was being treated with iodine and bromine during multiple sensitization.

would suggest. Something else had to be a factor in sensitivity in addition to the improved electronic conductivity of the silver bromoiodide layer. Since increases in sensitivity are detected by the presence of a stronger image with improved image contrast, the microstructures of daguerreotypes prepared using single and multiple sensitization followed by mercury treatment were reexamined to see how the microstructures varied because of processing changes. Because the daguerreian image is visible as a result of light scattering, a small change in image microstructure would result in a large visible change.

As noted previously, singly sensitized daguerreotypes have relatively few image particles that are all about the same size (about 1 micrometer in diameter), and the image particle morphology varies very little from image area to image area. Multiply sensitized daguerreotypes that were sensitized in total darkness have many image particles with a greater variety in both the size and morphology of the particles. When daguerreotypes are multiply sensitized in subdued light, as would have been the practice, there are many more image particles and even more variety in image particle size and morphology. In particular, image particles in the highlight regions of plates sensitized in the dark are fewer and more widely spaced than those on plates sensitized in subdued light. If all other factors are the same, as they were in this experiment, then the major difference between sensitizing in the dark versus sensitizing in the light is that daguerreotypes sensitized in the light have been partially fogged during the sensitizing process. For conventional materials, this approach is unacceptable; however, the opposite appears to be the case for the daguerreotype.

Chemical Vapor Deposition Processes
Now it is useful to step back and take another look at the process of how image particles are formed during mercury treat-

ments. The authors' goal is to determine the process that leads to the characteristic image structure and to relate this to the foregoing discussion. Let us take an inventory of the components present during image formation: the silver of the daguerreotype plate, the silver bromoiodide sensitive coating, the photolytic silver present because of the fogging of the plate during sensitizing in subdued light, latent image silver from the camera exposure, and hot mercury vapor. All these components are confined in the mercury bath or the apparatus in which the mercury is heated. This physical setup has all the components of chemical vapor deposition (CVD) processing, a method of producing materials by the chemical reaction of gaseous precursors usually by deposition at a heated surface.

Chemical vapor deposition processes have been known for a long time; however, only recently has their application become commercially important and, thus, their technological significance increased. (One common example of this process is the deposition of soot on the interior surface of a chimney.) A technologically important application is the production of silicon carbide whiskers used as reinforcement in composite materials. Chemical vapor deposition processing can be used to produce materials with structures varying from epitaxial and amorphous films to structures such as platelets, whiskers, and single crystals. This method of processing is also of interest because it can be used to produce certain types of materials at lower temperatures or with lower concentrations of starting materials than would be expected using other methods of production.

In general, the chemical vapor deposition process involves the adsorption of mobile atoms called monomers on a substrate. These monomers migrate to form embryonic precursors that, after sufficient growth, become stable nuclei. As with many other processes, the nuclei tend to settle on disturbances and defects in the sub-

Table 9.1. Structure-Growth Relationships for Condensed Species Produced by Chemical Vapor Deposition (CVD)[a]

Effect of increased supersaturation		Effect of increased temperature
	*Epitaxial growth	
	*Platelets	
	*Whiskers	
	*Dendrites	
	*Polycrystals	
	*Fine-grained polycrystals	
	*Amorphous deposits	
	*Gas-phase nucleated "snow"	

[a] This table is taken from John M. Blocher, Jr., "Structure/Property/Process Relationships in Chemical Vapor Deposition (CVD)," *Journal of Vacuum Science and Technology* 11 (1974): 680. The table as published in the Blocher paper was for the structure-growth relations for physical vapor deposition processing, in which supersaturation decreases with increasing temperature. This table has been corrected according to Blocher's observation that in chemical vapor deposition processing supersaturation increases with increasing temperature. The table has been published elsewhere without this correction. Asterisks indicate different classes of structures.

strate surface. Once stable nuclei are formed, their growth kinetics are controlled primarily by temperature, which determines the mobility of the reacting species, and by supersaturation or the concentration of the adsorbed monomer on the substrate surface.[24] Changes in temperature and supersaturation cause predictable changes in the structure of the condensed (i.e., solid) materials formed in the chemical vapor deposition process. Table 9.1 shows the schematic relation between supersaturation and temperature and the structure of materials produced by this process. Several of the morphologies characteristic of daguerreotype image structures can be found in table 9.1.

If daguerreotype image particle growth is an example of chemical vapor deposition, then mercury vapor is the monomer species of the system and the exposed daguerreotype plate is the receiving substrate. Reviewing the micrographs presented earlier in this chapter, it is possible to see that daguerreotypes treated in mercury do indeed go through a stage of nucleation (see figs. 9.7 and 9.8), which is followed by particle growth at favored sites. It is also possible to see that the general morphologies of image particles are in keeping with what would be predicted by the conditions listed in table 9.1. For instance, a daguerreotype

produced over cold mercury, which would also be a low supersaturation condition, does indeed have an image structure that resembles "snow" (see fig. 9.17). In addition, some conditions produce fine-grained polycrystals and platelets. Whiskers, dendrites, and epitaxial films are not observed. Morphological evidence suggests that chemical vapor deposition is a plausible mechanism to explain daguerreotype image formation. However, additional evidence is necessary before that mechanism can be confirmed as the method of image formation. It is essential to determine the roles of the individual components in the daguerreian system and find a way to corroborate experimental findings using evidence from nineteenth-century daguerreotypes.

Historically, the extraction of silver from its ores was usually accomplished through the use of solvent metals (called mineralizers) until the beginning of the twentieth century, when lixiviation or alkaline solvent leaching and electrolytic methods for silver extraction were brought to the point of commercial viability. The most common mineralizer for silver extraction was mercury. Lead also was used, but its use is not of interest here. Several major silver extraction processes were based on the fact that mercury, when mixed with finely ground silver ores, will collect and concentrate silver metal by amalgamation. Referring to the silver-mercury phase diagram in figure 9.13, it is easy to see why mercury was used for this process: there are few silver-mercury crystalline phases present at ambient temperatures; rather, the predominant phases are silver-mercury solid solutions and liquid mercury. In reality, mercury acts as a liquid transport agent for solid silver. No reduction-oxidation or other chemical reaction occurs. Following amalgamation, silver can easily be separated from the mercury first by squeezing the amalgam mass and then by distilling it in an iron retort.[25] Mercury is convenient to use because it is liquid at room temperature, and the entire extraction process can be carried out without heating. Also, it is economical because essentially all the mercury can be collected for reuse.

In chemical vapor deposition systems, sometimes a gas-phase component is added as a mineralizer to dissolve solid components or enhance their surface diffusion. Mineralizers may be used alone or in conjunction with other materials to improve the thermodynamics of a particular reaction over another. The advantage of using a gas-phase mineralizer is that the amount of mineralizer needed for a reaction may be infinitely small in relation to the amount of solid material present. Therefore, often it is not necessary to remove the mineralizer from crystals or other products formed during this type of reaction.[26] This is not the case when a liquid mineralizer is used. Very often these gas-phase mineralizers are used to convert amorphous or metastable substances into more stable crystalline species, especially if the substance being transformed cannot be treated at an elevated temperature because it is thermally unstable or volatile.

While mercury is most often thought of as a liquid, it can also exist as a crystalline solid even at ambient temperatures. These crystalline species of mercury are grown from the gas phase and are formed when the mercury is adsorbed on a suitable surface such as glass. The most prevalent habits for these crystals are whiskers and platelets. Mercury crystals and their growth mechanisms were well studied during the 1950s by Gerald W. Sears, whose interest was to elucidate the growth mechanisms of crystals from the vapor phase.[27] Of particular interest in this discussion is the growth mechanism of mercury platelets. A gas phase may be adsorbed onto a solid surface where it can migrate to and concentrate at an energetic site, such as a dislocation or other crystal defect in the substrate surface. The concentration site may develop into an active nucleus from which a crystal may grow. Any parent nucleus exerts a major influence over the type of crystal that is allowed to grow at that site. Other factors, such as temperature and pressure, also exert strong influences on the thermodynamics of crystal growth. Sears found that both mercury whiskers and platelets formed at screw dislocations located at crystalline defects in the glass substrate surface. He postulated that when all other conditions were the same, the growth of whiskers over platelets was determined only by the direction of the screw axis of the dislocations. In the case of platelets, the growth conditions are such that the developing crystal grows only in a two-dimensional fashion and must reach a certain critical size before it can thicken.

Returning to the daguerreotype, what is the condition of the daguerreotype plate during and after exposure to halogens during sensitization? In other words, what kind of substrate for crystal growth is the daguerreotype plate? Halogen vapor essentially corrodes the surface of the daguerreotype plate during halogenation. When a silver plate is examined in the scanning electron microscope after it has been polished but before it has been exposed to iodine, the plate is essentially featureless except for a random scratch or piece of dust. After the first iodine exposure, the surface of the same plate will appear covered with a fairly smooth coating.[28] As the halogen layer is built up during multiple sensitization, if the plate has been sensitized in subdued light, the plate surface becomes significantly rougher after each subsequent sensitizing step. After the final sensitizing step, which is carried out in darkness, the plate is somewhat smoothed, but it is still rough compared to a bare silver plate. The roughened surface of a multiply sensitized plate is shown in figure 9.21. Following camera exposure, it is possible to detect more surface roughness in what will become the highlight regions of a particular plate, i.e., those areas that received the most light during exposure.

The roughness in the light-sensitive layer of the daguerreotype is due to the formation of photolytic silver within the halide layer. Two types of photolytic silver are formed on the daguerreotype plate: (1) random photolytic silver (Ag°_{phot}), which is the result of sensitizing in the light, and (2) latent image photolytic silver (Ag°_{li}), which is distributed according to the light intensity of the scene recorded on the plate. Of

![Scanning electron micrograph]

DAG 20-5-9A AGI-BR-I, 10 SEC EXPOSURE, 3200 K PHOTOFLOOD

05KV 09.4KX 1U 0025 3/19/82

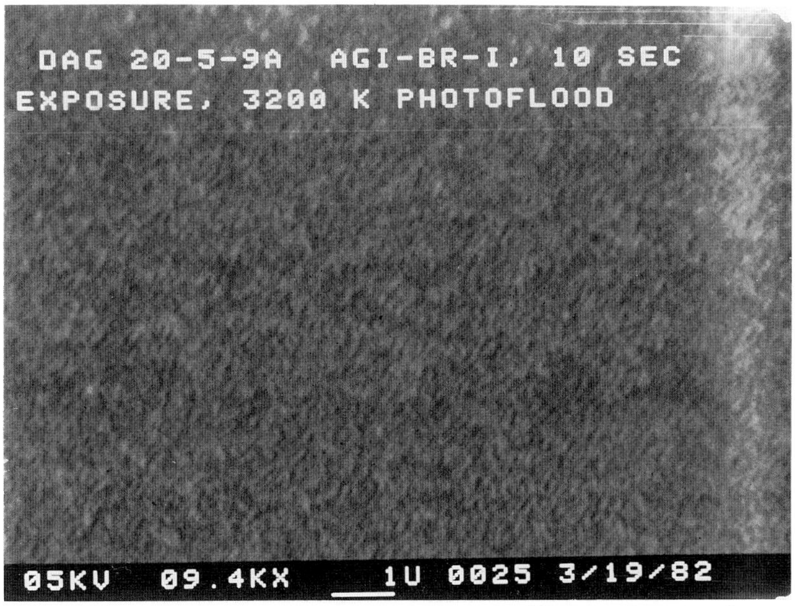

Fig. 9.21. Scanning electron micrograph of the surface of a daguerreotype plate after light exposure and before mercury treatments. The scale bar is equal to 1 micrometer.

course, these are the same material, but the location of the photolytic silver in or on the silver halide layer is important for image formation. Photolytic silver is metallic silver that has been formed by the light reduction of silver halide to silver. It is a metastable form of silver because it usually exists in a finely divided state and, as such, is more reactive than the strongly ordered

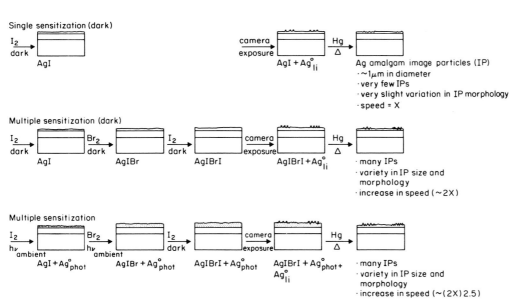

Fig. 9.22. Schematic diagram of the relationship of different daguerreotype sensitization methods and image particle growth.

bulk silver that makes up a silver mass such as that in the daguerreotype plate surface. Thus, three distinctly different forms of silver are present on a daguerreotype plate when it is placed over mercury: (1) random photolytic silver, (2) latent image photolytic silver, and (3) bulk silver.

Morphological evidence along with the elements of the previous discussion can now be put together to describe the process of image particle formation. A schematic representation of image particle formation in relation to the sensitizing of the daguerreotype plate is shown in figure 9.22. During halogenation the surface area of the silver halide layer increases because of the random formation of photolytic silver. This formation of photolytic silver creates defects and other favorable sites for reaction during the image formation process. It also creates a more stable interface for image particles to adhere to by making available reactive silver for image particle growth. During the last sensitizing step in the dark, the outermost portion of the halide layer is converted back to pure silver halide. A latent image is formed during camera expo-

sure, and it essentially "breaks" the silver halide layer open in proportion to light exposure.

At this point the exposed plate is put over hot mercury vapor. During the initial stages of image particle growth, particles are formed everywhere on the plate; however, the preferred nucleation centers for image particle formation (and those that become most stable) are the latent image sites formed during camera exposure. The mercury acts as a vapor-phase mineralizer collecting both latent image silver and the random photolytic silver and causing the metastable photolytic silver to recrystallize into stable silver image particles. Mercury plays the role of a solvent for crystal growth; it is not a chemical reactant in this process. Multiply sensitized plates have a greater amount of photolytic silver available for image particle growth, which is why they should—and do—have more image particles and a greater variety of image particles than singly sensitized plates. The limiting factors for image particle growth are the availability of reactive photolytic silver and the temperature and concentration of mercury at the plate surface during image formation. If the mercury temperature (and, therefore, concentration) is too high, image particles cannot form; if it is too low, the daguerreian microstructures most favorable for viewing are not produced.

The composition of individual crystals changes during the mercury treatment as particles change in volume and become more dense. The general trend during the mineralization step is for the image particles to change from high-energy, reactive photolytic silver to the lowest energy silver (or silver-mercury solid solution) crystalline state. The image particle growth sequence appears to be the drive to transform image particles and platelets from the mercury-rich, gamma-phase, body-centered cubic amalgam to silver-rich, epsilon-phase, hex-

Fig. 9.23. Scanning electron micrograph of platelets formed during mercury treatment of a multiply sensitized daguerreotype. The scale bar is equal to 10 micrometers.

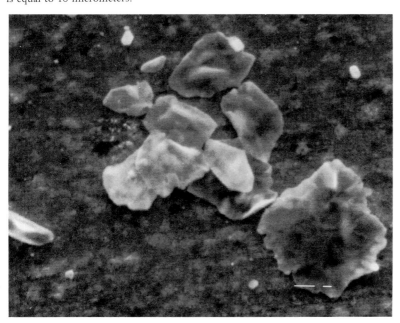

agonal closest-packed amalgam to a silver-rich solid solution and finally to a pure silver, face-centered cubic phase. In effect, during the process of image formation the compositions move from the right side toward the left side of the phase diagram (fig. 9.22).

The production of platelets is a transitional stage in image particle growth. If particle growth proceeds to completion, there should be no evidence of platelets on the plate surface. Platelets would be expected to persist under certain conditions—for example, poor control of temperature of the mercury (and, therefore, the saturation of mercury at the plate surface), the type of defects present in the silver halide layer, and the amount of active silver available for image particle production. The appearance of platelets is most likely tied up with the degree of control over crystal growth conditions. It would be expected, then, that platelets would be more widespread on plates made by novices than on plates made by those who have gained some degree of skill with the process. To test this premise, the microstructure of the first daguerreotypes made in the laboratory were examined in the scanning electron microscope for the occurrence of platelets. Many of these early experimental daguerreotype plates did have platelets in the highlight areas, as shown in figure 9.23.

Unexposed silver halide in the shadow regions of the plate acts as a barrier to most nucleation and particle growth during the image formation process. However, some particle growth does occur, which is manifest in the shadow particle agglomerates. These agglomerates tend to be mercury rich because they occur in areas of the plate that do not have latent image silver available for initiating the mineralizing amalgamation process that occurs in image areas receiving more exposure.

The conditions that favor optimal image formation include careful plate preparation so that the presence of defects (i.e., scratches and dirt) on the plate surface are minimized; careful sensitization and camera exposures so that there is the proper balance between random photolytic silver and latent image photolytic silver; and, finally, carefully controlled mercury conditions so that the temperature and supersaturation of mercury vapor within the mercury bath favors crystal growth at latent image sites rather than everywhere on the plate surface. These factors are exactly those emphasized by daguerreotypists during the nineteenth century. The amount of control needed in this process is rigorous. As is stated over and over again in daguerreian manuals, the daguerreotypist must be persistent while getting the process to work and then must be diligent in managing the process. Thus, the control of the process requires an intangible element—a degree of skill.

This description of the process of image formation is based on experimental evidence derived from laboratory experiments with the process. In 1983 the authors were given a unique opportunity to verify this empirical evidence by comparing these experimental results with daguerreotypes produced by Joseph Saxton, Robert Cornelius, and other early daguerreotypists working in Philadelphia. In preparing the exhibition "Robert Cornelius: Portraits from the Dawn of Photography," William F. Stapp of the National Portrait Gallery asked to have as many of the Cornelius daguerreotypes examined in the Materials Research Laboratory as could be arranged. The owners of about half of the then known Cornelius daguerreotypes gave permission for their daguerreotypes to be analyzed.

The results of these analyses are described extensively in the catalogue that accompanied the exhibit.[29] The most remarkable aspect of this opportunity is that these daguerreotypes reflected Cornelius's progress as a daguerreotypist throughout his career—from when he started working on the

Fig. 9.24. Portrait of Pierre Étienne Du Ponceau made by Robert Cornelius during the spring of 1840. This is the earliest daguerreotype identified as having been made using multiple sensitization. Courtesy of the American Philosophical Society, Philadelphia.

process in 1839, to his addition of bromine and multiple sensitization, to the addition of gilding, and through to the end of his career. Thus, it was possible to monitor how his ability to make daguerreotypes changed and improved over the years. Especially because of the analytical work on the Cornelius daguerreotypes and the Saxton daguerreotype of the Central High School in Philadelphia (fig. 4.3), it was possible to verify our experimental work by actual comparison with this seminal nineteenth-century set of daguerreotypes. In many cases it was possible to match micrographs from experimental plates to micrographs from early plates on the basis of image particle size, shape, and distribution.

Some two hundred micrographs were made of the various daguerreotypes examined in the Cornelius set. Many of those micrographs are included in the exhibition catalogue, and two are shown here. These were obtained from the Cornelius portrait of Pierre Étienne Du Ponceau (fig. 9.24). This daguerreotype was made as a demonstration plate and given to the American Philosophical Society in Philadelphia as a gift by Paul Beck Goddard on May 15, 1840. It is the earliest plate that has been identified as having been made using multiple sensitization. The micrograph in figure 9.25 shows a highlight area of the plate with some platelets and poorly formed image particles. This kind of microstructure indicates that the process is being handled by a novice, which indeed Cornelius was at the time. Figure 9.26 shows an ungilded shadow particle agglomerate and is a good example of the structural changes associated with the flattening and expansion of the agglomerates during age hardening.

Coming full circle, the daguerreotype process is a unique method of forming images by chemical vapor deposition. In the image formation process mercury acts as a mineralizer to concentrate and assist in the

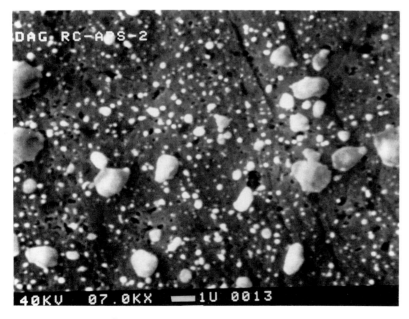

Fig. 9.25. Scanning electron micrograph of a highlight area of the Du Ponceau portrait by Robert Cornelius. The scale bar is equal to 1 micrometer.

recrystallization of silver image particles in a pattern corresponding to the light intensities of the original scene recorded by the camera. This method of image processing is an example of photographic development only in its broadest sense; however, in the strictest sense the term *photographic development* is inappropriate when used in the context of the daguerreotype process. No reduction-oxidation reaction occurs between mercury and the other components in the daguerreian system. Further, because image particle formation is a crystal growth process, processing the daguerreian image involves a certain degree of skill, which is a wild card. Image particle formation is not static once the mercury treatment has been completed and the unexposed silver halide removed from the plate surface. Rather, because image particles are amalgams, they continue forming until an end point is reached—either age hardening occurs, or the plate is gilded. Gilding is a process of rapid hardening in which mercury is replaced by gold, causing the image particles to become slightly larger and thus improving the optical properties of the daguerreotype and also considerably improving the mechanical properties of the image.

This method of image formation is not intuitively obvious, and the fact that Daguerre devised this process is even more remarkable in light of its true technological details. This unusual method of image particle growth is an example of a materials processing method that has become commercially important only within the last twenty years.

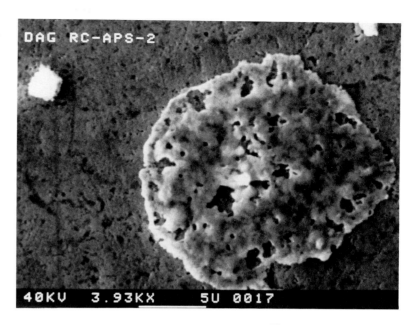

Fig. 9.26. Scanning electron micrograph of an ungilded, aged shadow particle agglomerate in a shadow area from the Du Ponceau portrait by Robert Cornelius. The scale bar is equal to 5 micrometers.

10

Image Deterioration

Only a few years ago a lady came to me with a half-sized picture, and you could not see anything at all upon it. She wanted to know if I could clean it. . . . In about five minutes I brought it to her. . . . She fainted dead away. . . . It was her husband who had been dead 20 years, and she had not seen the picture in 15 years. It was so completely covered with a film that there was nothing to be seen and I brought it up as good as it was originally. As I say, the lady fainted immediately. It was just as if her husband had been brought back from the grave for her to see.

Abraham Bogardus, "Trials and Tribulations of a Photographer"[1]

DAGUERREOTYPES AGE AND CORRODE over time; like other silver objects, they become tarnished. Tarnish layers are usually opaque and may build up to such thicknesses that the image is completely obscured. The corrosion of daguerreotypes is of interest to those wishing to devise better ways to preserve historic and artistic objects. If the corrosion process can be understood and the corrosion products identified, ways of avoiding corrosion and perhaps repairing corrosion that has already occurred can be prescribed.

Extensive investigations of many daguerreotypes from various sources and stored in a variety of environments show that daguerreotype damage can be categorized as follows: (a) physical damage caused by scratching or bending the soft daguerreotype plates; (b) corrosion caused by chemical reaction of gases in the atmosphere with the daguerreotype plate; (c) corrosion caused by previous attempts to clean the plates; and (d) debris and reaction products caused by corrosion and spalling of the cover glass used in the daguerreotype package.

Details of daguerreotype deterioration are controlled by image structure and composition, packaging methods, and the long-term effects of 150 years of aging and handling. There are some interesting trade-offs between corrosive attack on the silver plate, which is greatly inhibited by the daguerreotype package, and some corrosion activity brought on by the package itself. Nineteenth-century daguerreotypists, as it turns out, were wise in their choice of packaging system although some improvements can be made with modern materials.

Physical Damage

A common form of physical deterioration in daguerreotypes is surface abrasion, bending, or peeling away of the silver layer from its copper substrate. These are primarily mechanical problems related to the physical properties of silver metal and, to some extent, the manufacturing methods used for a given daguerreotype plate.

The hardness of a material is an expression of its resistance to penetration by some other—usually harder—object or its susceptibility to abrasion by another material. Silver is a relatively soft and ductile metal. It can be scratched without much difficulty. However, since the preparation of a material has some bearing on its ultimate hardness, some daguerreotype plates are inherently more resistant to abrasion. The silver layer of plates made by cold-roll cladding followed by annealing is made up of comparatively large silver crystal grains, while the silver layer of electroplated plates is made up of small grains. Materials of large grain size tend to be softer because single crystal grains are less resistant to deformation than are fine-grained polycrystalline materials. In practice, this means that cold-roll clad plates are initially more difficult to polish because they are softer, and over time they are more susceptible to abrasion. On the other hand, because electroplated plates are harder they are easier to polish to the mirror precision required to obtain the deepest shadows—thus, the nineteenth-century observation that these plates produce daguerreotypes of greater contrast. These plates are not as prone to surface abrasion as roll-clad plates.

American process daguerreotype plates, in which the daguerreotypist electroplates an additional layer of silver onto a cold-roll clad plate, has the advantage of a harder silver layer, which can be more easily polished. Thus, this additional step does provide daguerreotypes with a greater contrast range. Moreover, the plate may not be as easily scratched as a roll-clad plate, especially if the scratch does not penetrate the electroplated layer. However, the advantage of electroplating during plate preparation has more bearing on the quality of the image produced than on its long-term abrasion resistance.

The thickness of the daguerreotype plate and the ductility of both silver and copper causes the plates to bend and dent easily. This pliableness was desirable for attaching the plate to polishing blocks and other accessories used during the processing of daguerreotypes. The traditional daguerreotype case usually guards against this sort of damage.

From time to time the silver layer of a daguerreotype peels away from its copper substrate. This peeling is most often observed in cold-roll clad plates and indicates incomplete or inadequate bonding between the silver and copper layers during the process of making the plate. Flaws introduced in manufacture may lie hidden and not produce detectable plate failure through peeling until years after the daguerreotype has been made. The appearance of this phenomenon may have many causes. It usually occurs when some internal flaw in the plate gives way under stresses produced by mechanical, thermal, or chemical conditions or a combination of these. Often the peeling in clad plates follows the direction of rolling.

The electroplated layer in American process plates may also peel, again because of the silver layer's poor adhesion to its clad silver substrate. Poor adhesion may be due to inadequate cleaning before electrodeposition or the use of contaminated plating baths. This type of peeling most often occurs in image areas with a high image particle density—i.e., highlight areas, areas at the plate edges, or areas covered with dense tarnish, all high-stress areas for the relatively thin electroplated layer. When examined under a light microscope, the underlying silver of American process plates appears etched because the simple silver cyanide plating baths used to make these plates cause both plating and etching to occur on the plate surface during deposition.

The most prominent and disturbing forms of daguerreotype deterioration are

Fig. 10.1. Daguerreotype with advancing corrosion fronts. The fronts of the plate and mat edges can be seen.

corrosion films, the highly colored tarnish films that form on the plate surface and can completely obscure the image. Corrosion simply means the reaction of the metal with its nonmetallic environment, resulting in the continuing destruction of the metal. The extent of corrosion is determined by thermodynamic considerations, kinetics, and the availability of materials for reaction. The key to understanding this form of daguerreotype deterioration is to examine the daguerreotype's particular nonmetallic environment.

In almost any discussion of tarnish on daguerreotypes, it is assumed that because daguerreotypes are made on silver plate they must behave like silverware. The major corrosion product is assumed to be silver sulfide. Silver objects, especially those used around food, do indeed become blackened with silver sulfide because ambient hydrogen sulfide (H_2S) vigorously attacks and corrodes silver, and there are many sources of hydrogen sulfide and other sulfur-bearing volatile species in the ambient environ-

ment. Daguerreotypes that share the same ambient environment that pervades a kitchen do indeed become blackened in a similar way. However, only uncased daguerreotypes share the common environment with most other silver objects. The vast majority of daguerreotypes are cased, and their primary environment is confined to the daguerreotype package. This fact leads to some surprising findings.

If a daguerreotype is removed from its package, two corrosion fronts or places from which tarnish seems to spread are usually evident. The first is along the plate edge, as shown in figure 10.1, and the second along the mat edge. Direct analysis of these corrosion fronts shows that they are made of a variety of compounds and that they differ in composition not only from place to place on one daguerreotype but also from daguerreotype to daguerreotype. For the time being, only the simplest and major components of these two corrosion fronts will be considered. The primary constituent of the plate edge corrosion front is silver sulfide. The mat edge corrosion front is primarily silver oxide. Thin films of silver oxide and silver sulfide are very similar under visual inspection. Both first appear as colored films on silver surfaces, going through a progression of interference colors in accordance with the thickness of the growing film.

The tarnish layers have received the most attention and are responsible for the long history of efforts to clean daguerreotypes. The next concern, then, is to identify the tarnish layers and show how they accumulate.

The accumulation of silver sulfide on a daguerreotype surface is easy to demonstrate. A polished, blank daguerreotype plate was cased and sealed with a paper tape and kept in a hydrogen sulfide environment averaging 15 parts of hydrogen sulfide per billion parts of air (fluctuating over a range of 10 to 50 parts per billion) at room temperature for one and a half years. Half of the daguerreotype was loosely covered with a card to protect it from light without restricting the access of hydrogen sulfide. A strip of another polished, blank daguerreotype plate was placed beside the sealed plate to act as a control. Both these experimental plates are shown in figure 10.2. The unsealed control was heavily corroded in a matter of days. The cased daguerreotype showed no visible signs of tarnish even after 1.5 years of exposure. The hydrogen sulfide exposure over the duration of the experiment was the equivalent of 30 to 40 years of exposure in a polluted atmosphere.

The daguerreotype package is very effective in protecting the plate from atmospheric contaminants. Demonstrating the effects of package failure is also possible. A polished, blank daguerreotype plate was exposed to a hydrogen sulfide atmosphere as before. However, this plate was glazed with several glass laboratory slides butted next to each other to simulate a broken cover glass. Half of this sample was loosely covered as in the first experiment. After one month in the hydrogen sulfide atmosphere (equivalent to approximately two years of exposure under normal conditions), the plate areas under the glass joints were heavily corroded where the plate had been exposed to light. The covered portion was just beginning to show signs of tarnish at the end of the experiment, as seen in figure 10.3.

From these experiments it appears that if hydrogen sulfide can freely react with a daguerreotype, the rate of corrosion is similar to that of other silver objects. However, the conventional daguerreotype seal provides an effective barrier to the ambient environment. Thus, the corrosive reaction is inhibited by the prevention of gas transport through the seal. Light accelerates the kinetics of the hydrogen sulfide reaction as seen by the experiment with the simulated broken cover glass. It may be more likely

Fig. 10.2. Cased blank daguerreotype plate after exposure to 15 parts per billion hydrogen sulfide in ambient atmosphere for 1.5 years. The left side of the plate was exposed to light during this period; the right side was masked. The control strip shows the effect of direct exposure of the plate to the same atmosphere for the same period of time.

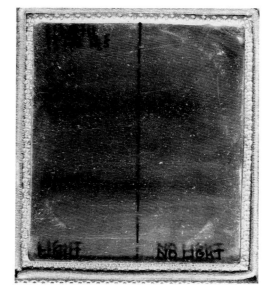

Fig. 10.3. Blank daguerreotype plate exposed for one month to an ambient atmosphere containing 15 parts per billion hydrogen sulfide. The plate was protected with three microscope cover glasses simulating a broken cover glass. The position of the "cracks" is shown by the two horizontal strips where corrosion is much more intense. The left side of the plate was exposed to light while the right side of the plate was masked. Corrosion beneath the cracks was more intense on the lighted side of the plate.

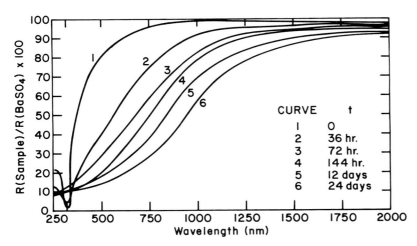

CURVE	t
1	0
2	36 hr.
3	72 hr.
4	144 hr.
5	12 days
6	24 days

Fig. 10.4. Combined diffuse plus specular reflectance of a daguerreotype plate exposed for various lengths of time to hydrogen sulfide. Silver sulfide film growth is shown by the decreasing reflectivity of the plate. Labels on the curves show the time of exposure.

that light plays a role in the conversion of silver oxide to silver sulfide rather than in the conversion of silver metal to silver sulfide.

The darkening of the daguerreotype plate with tarnish film buildup can be measured quantitatively by the reflectance spectra of the plate. Figure 10.4 shows the diffuse plus specular reflectance of a blank daguerreotype plate that has been exposed to a hydrogen sulfide atmosphere for the periods of time ranging from no exposure to 24 days' exposure. The reflectance shifts from that of silver metal, which is highly reflecting in the near infrared and visible portions of the spectrum down, to the onset of the plasma edge in the long-wave ultraviolet. The developing silver sulfide films quickly block the blue region of the spectrum, and as the film thickens the absorption edges move across the visible spectrum. At the longest exposure times the film thickness is sufficient to absorb most of the incident light well out into the near infrared, and the visual appearance of the plate is black. The optical properties of silver sulfide tarnish films have been extensively investigated.[2] The optical absorption spectrum can be used to determine film thicknesses for tarnish films on polished silver surfaces.

Damage Caused by the Chemical Reaction of Gases

There seems to be no doubt that silver sulfide films form easily on silver surfaces. However, this fact does not answer the question about the identity of the dark-colored films appearing on daguerreotypes that have not been exposed to sulfur-rich atmospheres. Thermodynamics is of some help. If the silver metal simply oxidizes, the reaction is

$$2\,Ag + \tfrac{1}{2}\,O_2 \rightleftarrows Ag_2O.$$

The equilibrium constant for this reaction is

$$K_O = P_{O_2}^{-1/2}$$

where P_{O_2} is the oxygen partial pressure in the ambient atmosphere. The numerical value of K_O at 25 degrees Centigrade can be calculated from the free energies of formation of the reacting compounds and is found to be 91.66.[3] The corresponding oxygen pressure at which silver would be exactly in equilibrium with silver oxide is 1.19×10^{-4} atmospheres. At any higher oxygen pressure silver would spontaneously oxidize to silver oxide. This threshold is much less than the 0.21 atmospheres pressure of oxygen found in the normal atmosphere. Thus, from the point of view of thermodynamics, silver metal will tend to oxidize spontaneously in the ambient atmosphere at 25 degrees Centigrade. The thermodynamic calculation, of course, says nothing about the *rate* at which oxidation takes place; it says only that the reaction *may* take place. The rate would be determined by catalytic agents present on the silver surface and the availability of oxygen. The experiments with blank daguerreotype plates described earlier suggest that light may be an important catalytic agent. Moisture is likely to be another.

The reaction of silver metal with sulfur or hydrogen sulfide gas is somewhat more complicated and does not lend itself to a simple calculation. The direct reaction with hydrogen sulfide is

$$Ag + H_2S \rightleftarrows Ag_2S + H_2.$$

The hydrogen that is formed would react with oxygen to form water, thus involving two additional gases. Alternatively the reaction can be written

$$4 Ag + S_2 \rightleftarrows 2 Ag_2S.$$

The S_2 can represent sulfur directly present in the ambient atmosphere or obtained from the dissociation of hydrogen sulfide

$$2 H_2S \rightleftarrows 2 H_2 + S_2.$$

Depending on the assumptions made about other gases in the ambient atmosphere, one can obtain various numbers for the hydrogen sulfide partial pressure at which silver metal is in equilibrium with silver sulfide. They have in common that the calculated hydrogen sulfide partial pressure is extremely low, in the range of 10^{-30} atmospheres. Although perfectly acceptable as a thermodynamic measure of hydrogen sulfide activity, these numbers are absurdly small as measures of an actual partial pressure of a gas. What they mean is that any amount of hydrogen sulfide, even in the part per trillion to part per billion range, is sufficient to corrode metallic silver into silver sulfide. Of course, some sulfur must actually be present to form a significant mass of tarnish, but the thermodynamics leaves no question about the direction of the reaction.

The chemistry considered here assumes that silver and its oxide and sulfide are reacting directly with atmospheric gases. If moisture had collected inside the daguerreotype case so that liquid water was present, other possible chemical reactions could be written. The calculations show that both silver oxide and silver sulfide are stable under the conditions present inside the daguerreotype case. Direct laboratory investigation using modern characterization techniques must be used to determine the actual corrosion products.

A comprehensive analysis was made to determine the chemical and structural character of the tarnish films. Because the films are very thin, surface methods of analysis proved to be most effective.[4] These methods were Auger electron spectroscopy, energy-dispersive x-ray spectroscopy, Raman spectroscopy, and Fourier transform infrared spectroscopy. X-ray diffraction, which is usually used for bulk analysis, was used to confirm both the presence of certain corrosion products and the complexity of the tarnish films. In addition, the presence of water-soluble films was determined using atomic emission spectroscopy analysis of the effluent from washing daguerreotypes in deionized water. The characteristic morphology associated with certain corrosive action was studied using scanning electron microscopy. However, this last technique cannot be used to discern most corrosion films themselves because the films do not have sufficient thickness to be seen by scanning electron microscope examination. Summaries of the findings from each analytical technique are given in tables 10.1 and 10.2.

Auger electron spectroscopy measures the chemical composition of a very thin surface layer, typically 2 to 10 nanometers in thickness. It is particularly useful for determining the chemical composition of thin surface films such as the tarnish layers. However, the results are in the form of elemental compositions with no indication of how the elements are arranged into compounds. In addition to a dominant signal from silver, typical Auger spectra of the tarnish layers show various amounts of sulfur, oxygen, chlorine, carbon, silicon, nitrogen,

Table 10.1.   Analysis of Corrosion on Daguerreotypes: Elemental Data

Element	Atomic Emission Spectroscopy	Auger Electron Spectroscopy	Energy-Dispersive X-ray Spectroscopy	Fourier Transform Infrared Spectroscopy	Raman Spectroscopy	X-ray Diffraction	Gandolfi X-ray Diffraction	Scanning Electron Microscopy
					Analysis Technique			
Ag			X				X	D
Al	X		X					D
B	X							
C		X						
Ca	X		X					D
Cl	X	X	X					
Cu			X					
F	X							
K	X		X					
Mg			X					D
N		X						
Na	X		X					E
P		X	X					
S			X	X				
Si	X		X	X				D

E = Etching   D = Debris

and sometimes iodine. This may be taken as evidence for such compounds as silver oxide, silver sulfide, silver cyanide, and perhaps others. Although the iodine that appeared occasionally is likely from the sensitizing step, the minor chloride is more likely an atmospheric contamination product. The Auger spectrum analyzes such a thin layer that it is very sensitive to trace contaminants on the surface.

The energy-dispersive x-ray detector on the scanning electron microscope also gives elemental composition, but because of the focused electron beam it gives elemental composition of individual grains and particles. This makes energy-dispersive x-ray spectroscopy the method of choice for identifying the many small bits of debris scattered over the daguerreotype surface.

It should be noted that some of these techniques give elemental data while others give information on structure either on the atomic scale or on the scale of grains and particles. It is important to keep this in mind because the identification of any particular element by simple elemental analysis is not sufficient to determine the presence of a particular compound formed from that element. For example, the presence of sulfur on a daguerreotype may imply but does not confirm the presence of silver sulfide. Often silver sulfide is present on daguerreotypes in only very small amounts. Other sulfur-containing compounds, such as silver thiourea complexes and sodium sulfate, are routinely found on daguerreotype surfaces and produce the same sulfur peak as silver sulfide in elemental analysis. The as-

Table 10.2. Analysis of Corrosion on Daguerreotypes: Compound Data

Compound	Analysis Technique							
	Atomic Emission Spectroscopy	Auger Electron Spectroscopy	Energy-Dispersive X-ray Spectroscopy	Fourier Transform Infrared Spectroscopy	Raman Spectroscopy	X-ray Diffraction	Gandolfi X-ray Diffraction	Scanning Electron Microscopy
Ag_2O		X		X		X		
Ag_2S				X		X		
AgCN		X		X		X		E
Ag_3PO_4				X		X		E
$Ag-(SC(NH_2)_2)^n$		X (?)		X	X			E
AgCl		X				X		
CuO							X	A
Malachite-like copper carbonate							X	A
NaCOOH							X	D
$Na_3[Cu_2(CO_3)_3(OH)]\cdot 4H_2O$							X	A
NO_3^-	X							
PO_4^{-3}	X							
$PbSO_4$							X	D
SO_4^{-2}	X						X	D
SiO_2 fibrils — red, cream, blue							X	D
SiO_2 (gel?) opal structure							X	D

E = Etching   D = Debris   A = Accretions

sumption that the de facto presence of a sulfur peak indicates silver sulfide has been a strongly held idea, and the authors have shared this view in the past, but it leads to many erroneous and misleading conclusions. Conversely, once the general array of compounds in these corrosion films is known, informed elemental analysis may be sufficient to make qualitative determinations of what corrosion products are present.

One of the most useful nondestructive tools for obtaining structural information from corrosion films on daguerreotypes is diffuse-reflectance Fourier transform infrared spectroscopy. Silver has no infrared spectrum but is strongly reflecting in the infrared region. The infrared spectrum obtained by measurement of the reflectance spectrum of the plate is, therefore, the characteristic absorption spectrum of the corrosion film alone. This technique has been invaluable for gaining an understanding of the overall corrosion cycle of daguerreotypes. A typical spectrum is shown in figure 10.5. The ideal thickness for measurement of an infrared spectrum is about one micrometer. Because the corrosion films are thinner than this, the spectral bands are typically weak, but the spectrum is often quite complicated. The individual absorption bands shown in the spectrum corre-

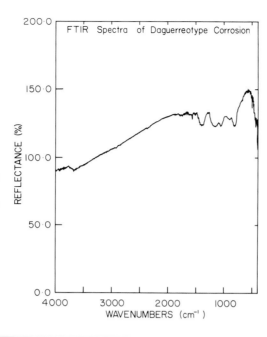

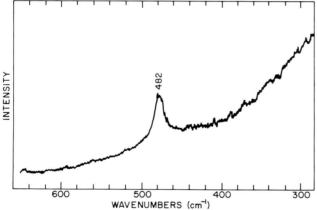

Fig. 10.5. Infrared spectrum of the tarnish layer on a daguerreotype taken with a Fourier transform infrared spectrometer in the reflectance mode. The extension of the curve above 100 percent reflectance is an artifact of the particular arrangement of specimen and reference mirror and has no significance for the interpretation of the spectrum. The abscissa of the spectrum is in units of wavenumbers (cm^{-1}), which is the reciprocal of the infrared wavelength given in units of centimeters.

Fig. 10.6. Raman spectrum of a low image particle density area of a daguerreotype using the 488-nanometer blue line of an argon ion laser as an excitation source.

spond to the modes of vibration of the compounds that make up the tarnish layer. Oxide bands, sulfide bands, cyanide bands, and bands caused by sulfate, phosphate, and thiourea complexes can be recognized. Likewise, the presence of any organic material such as a polymer coating or varnish is immediately evident by a distinctively different infrared spectrum.

Raman spectroscopy is an inelastic light-scattering experiment in which the exciting source is a laser beam that can be focused to a spot size of about one micrometer. The spectrum of the inelastically scattered light, like the infrared spectrum, is a measure of the modes of vibration of the crystals and molecules and should provide a recognizable fingerprint for their structural identification. The authors originally had great hopes that Raman spectroscopy would be useful for obtaining structural information about the particulate corrosion products on daguerreotype plates because of the ability to focus the beam on individual particles. Figure 10.6 shows the Raman spectrum of a low image particle density area. The single band at 482 cm^{-1} is characteristic of silver sulfide. The intensity of the Raman spectrum gradually decreased during the course of measurement. The intense concentration of light at the focal spot of the laser caused photodecomposition of the tarnish layer, an interpretation confirmed by scanning electron microscope examination of the target area after the Raman measurement. The method has potential as a characterization tool for corrosion products, but some substantial improvement in technique is necessary.

The data given in tables 10.1 and 10.2 show the surprising variety of corrosion products typically found on daguerreotypes. These materials are derived primarily from corrosion caused by past cleaning treatments or from the environment formed by the primary daguerreotype package. It is important to keep in mind that thin films of most of the materials listed in the table have relatively high absorption coefficients over the range of the visible spectrum. Highly colored or black films can be formed from very thin layers of the materials, and in visual appearance these are very similar to silver sulfide or silver oxide tarnish.

Another important fact derived from the analysis of corrosion products is their location, distribution, and structural features across the daguerreotype plate. Corrosion tends to be most severe in the more reactive portions of the daguerreotype. Silver grain boundaries found in the silver layer of the daguerreotype plate are active sites for reactions because of the energy tied up in broken bonds and unsatisfied charges found at these locations. Image areas with high image particle density (i.e., highlight areas) have a large surface area, which makes them highly reactive. Further, the spaces between image particles can trap and become filled with corrosion products and debris. The geometry of the daguerreian package causes both the plate edge and the area adjacent to the mat to be more vulnerable to corrosion than other locations on the plate.

The buildup of corrosion at these different active sites may or may not be visible to the unaided eye. For example, the evidence of silver oxide formation at silver grain boundaries can be seen from the effects of preferential etch patterns caused by certain cleaning procedures. This silver tarnish at grain boundaries may not be sufficiently thick to form visible films. Likewise, debris trapped between image particles may be of sufficient size to scatter visible light, and thus appear the same as an uncorroded image. The daguerreotype itself may in some cases act as a catalytic surface for certain reactions. Its composition of silver, mercury, and gold, as well as the particulate structure of the image, may contribute to this effect. It has not been determined if oxide films on the daguerreotype

surface would lessen or stop catalytic reactions from occurring. These analyses suggest that it would be useful to rethink the process of daguerreotype corrosion in order to account for the variety of materials found. The simple silver sulfide–silver oxide model can account for tarnish made of those two compounds. Other corrosion products need more careful consideration.

Damage and Corrosion Caused by Cleaning Attempts

Important sources of plate damage and corrosion products are the reagents used for daguerreotype cleaning. Daguerreotypes traditionally have been cleaned in solutions called "silver-dip" cleaners (to be discussed in chapter 11). Until the early 1950s the active ingredient in these cleaners was potassium cyanide. Thiourea-based cleaners, which are considerably less toxic than cyanide, were introduced for silverware in 1953 and were rapidly adopted for daguerreotype cleaning.[5]

Cyanide Cleaners

Cyanide, a good silver complexing agent, is also a commonly used silver etchant for metallography. It has been used in daguerreotype processing for the removal of unexposed silver halide and was also used for "brightening" freshly made daguerreotypes. When dull white plates were quickly dipped into cyanide, the white areas became whiter. When it is used to remove unexposed silver halide from a freshly made daguerreotype or for "brightening," some minor etching doubtless occurs. However, the exposure to the cyanide solution is brief, and there is not much time for severe etching to occur. Further, the cyanide–silver iodide reaction is fairly vigorous, but when the cyanide reaches the underlying silver plate, the action would be considerably slower. Although it is unlikely that a freshly made daguerreotype would be tarnished by silver oxide during processing, any oxide present would also be vigorously corroded by the cyanide. The nineteenth-century literature does not give a clear idea as to how prevalent the use of cyanide for processing was, but sodium thiosulfate (i.e., hypo) and hot salt solution were also used. It is not possible to determine how silver halide removal was accomplished by experimental examination of presently existing plates.

When cyanide is used as a cleaner, the results are far more drastic and damaging. Thermodynamic data indicate that cyanide will react readily with silver oxide and silver iodide, moderately with silver metal, and not at all with silver sulfide. Thus, when a tarnished daguerreotype is cleaned in cyanide, different areas of the plate surface will be attacked at different rates according to the composition of the corrosion products. The nineteenth-century literature on cleaning daguerreotypes indicates that the cleaners used at that time, all based on cyanide, were unpredictable. Sometimes a daguerreotype was cleaned well, sometimes the cyanide caused visible etching, and sometimes the image disappeared altogether. The random action is probably a reflection of the variable attack by the cyanide on different corrosion products, all of which appeared to nineteenth-century daguerreotypists as a black tarnish.

Surface analysis shows that varying amounts of silver cyanide have been found on daguerreotypes cleaned with cyanide. The highest concentrations correspond to the mat edge corrosion front, which is primarily silver oxide. Further, scanning electron microscope examination reveals a characteristic etching associated with daguerreotypes that have been cleaned in cyanide. The overall patterning of the plate surface shown in the micrograph in figure 10.7 is due to cyanide. The etching pattern is probably associated with the presence of silver oxide in areas of heavy corrosion, as

well as along the silver grain boundaries on the plate. Since grain boundaries are localized high-energy sites, and therefore potential sites for the initiation of corrosion reactions, it is likely that oxides form almost immediately at the grain boundaries. Gradually, certain areas, such as the mat edge and plate edge, build up higher concentrations of silver oxide because other corrosion mechanisms take over. Silver sulfide present on a daguerreotype plate cleaned in cyanide is probably removed by being carried along as other corrosion products are stripped away.

In addition, cyanide cleaners can cause image particles to be etched or broken up and redeposited elsewhere on the plate surface. Individual image particle composition (in particular, mercury content), gilding, and the daguerreotype's age all appear to influence the cyanide vulnerability of individual image particles. Highlight areas show the largest losses. The spaces between image particles in highlight areas can also become full of debris either from the deterioration of the package or from corrosion. In such instances cyanide cleaners appear to eat tunnels through the debris. As a result the altered structure of the debris may act to scatter light in more or less the same manner as the underlying image particles. Thus, the image may be improved in appearance, but corrosion has not been removed.

Cyanide cleaners react at different rates with different corrosion products on the daguerreotype surface, thus causing the unpredictable etching and the disruption of image particles. These two effects cause the image to appear to fade, lose contrast, or, in drastic cases, disappear. It is difficult to assess the detrimental effect of cyanide cleaning by simple visual inspection. Since the observer must adjust each daguerreotype to achieve its best viewing conditions, it is possible to attribute difficulty in seeing the image to factors other than damage from cleaning.

Thiourea Cleaners

Thiourea cleaners work differently from the cyanide cleaners they replaced. The theoretical reactions involved in silver sulfide removal by acidified thiourea cleaners were presented by Howard Brenner in 1953 and formed the basis for all succeeding formulations of these cleaners.[6] He postulated that a minute amount of silver sulfide would dissolve in water and ionize. In the ionized state thiourea would form silver-thiourea complexes with the silver ions, and the sulfide ions would combine with free hydrogen to form hydrogen sulfide. The rate and direction of these reactions are controlled by the pH of the solution. Brenner provided the pertinent equilibrium constants and calculations to determine the best theoretical cleaning action, which he calculated would occur if the pH were kept below 1, and recommended that strong mineral acids such as hydrochloric or sulfuric acids or acid salts such as sodium disulfate be used. These calculations considered only silver sulfide removal and did not include silver oxide or other corrosion products.

In the Missouri Historical Society formula, phosphoric acid was substituted for the hydrochloric or sulfuric acids recommended by Brenner. Presumably, phosphoric acid was substituted to avoid the formation of silver sulfate or silver chloride as possible deleterious side reactions and also to prevent etching of the plate surface. Phosphoric acid does not react with silver metal.

Thermodynamic calculations can be used to compare the action of the three mineral acids with silver, silver oxide, silver sulfide, and silver cyanide. In all cases it was assumed that solid silver or silver salt was present on the plate surface and that the solution was maintained at pH = 1. An extremely low free energy of reaction for the reaction of silver oxide with phosphoric acid indicates that the choice of this acid was indeed unfortunate. Phosphoric acid not only acts to adjust the pH for the thiourea

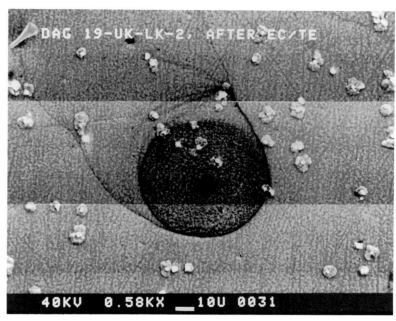

Fig. 10.7. Scanning electron micrographs of damage caused by daguerreotype cleaners. The fine-grained pattern is due to intense etching along silver grain boundaries caused by cyanide cleaning. The circular stain results from corrosive material dropped from the cover glass. The scale bar is equal to 10 micrometers.

complexing action but also reacts with silver oxide, etching the plate and leaving silver phosphate as a corrosion product.

Daguerreotypes treated in thiourea often develop brown spots, commonly called measles, after they have been cleaned. An analysis of daguerreian measles by electron microprobe in 1974 showed the measles to be crystals composed of silver, chlorine, and mercury with traces of sulfur and silicon.[7] The sulfur was attributed to thiourea residue on the daguerreotype. It was concluded that crystal occlusions on the plate trapped a small amount of the cleaner, which then reacted and turned brown some time after treatment. It was recommended that the daguerreotypes be thoroughly washed to remove all traces of thiourea.

Analysis of the measles by energy-dispersive x-ray emission and by Gandolfi x-ray diffraction has not conclusively demonstrated their composition either. The chemistry seems to differ very little from

Fig. 10.8. Portrait of Paul Beck Goddard taken by Robert Cornelius in 1839. This daguerreotype was cleaned in thiourea and has since become retarnished, so that the image is almost obscured. Courtesy of the American Philosophical Society.

Fig. 10.9. Scanning electron micrographs of corrosion blemishes caused by reaction with buffered board mat. Left: A secondary electron image. Right: A back-scattered electron image. The scale bar is equal to 10 micrometers.

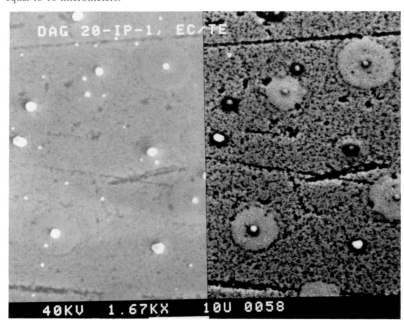

the underlying plate. Measles appear to grow at minute holes in the surface of the daguerreotype plate. Since no copper has been found in the analyses of the measles, it is unlikely that these holes go through to the copper substrate of the plate. It has been suggested that inadequate rinsing of the daguerreotype after cleaning causes measles to grow.

More recent investigations have shown that the complexing action of thiourea on silver is so strong that these complexes cannot be removed by chemical means. Researchers investigating the application of surface-enhanced Raman spectroscopy for the detection of thin films use thiourea films on silver as a calibration tool.[8] Raman spectroscopy has been used to confirm the presence of thiourea films on daguerreotypes cleaned by this method.

Thiourea cleaners do cause some etching, which is difficult to distinguish from cyanide etching. The practice of cyanide cleaning was so widespread that daguerreotypes that do not exhibit its characteristic effects are rare. It is estimated from scanning electron microscope examination of some two hundred nineteenth-century daguerreotypes that only 2 percent of the extant daguerreotypes have never been cleaned. It is difficult, therefore, to find a daguerreotype that has only been cleaned in thiourea.

The major corrosive effects of both cyanide and thiourea cleaners are etching of the plate surface, dissolution and rearrangement of the image microstructure, and deposition of new corrosion products such as silver cyanide, silver-thiourea complexes, and silver phosphate. Various areas of the daguerreotype surface are affected in different ways because of the diversity of reactions with the many corrosion products present. Daguerreotypes treated with these cleaners remain ultimately more vulnerable to further tarnish. Cleaned metal surfaces are more reactive, and the etching increases the surface area of the plates (see fig. 10.8). The new corrosion products themselves are relatively reactive and make up the new colored corrosion films that inevitably begin to appear on the carefully cleaned daguerreotypes.

Damage Caused by the Daguerreotype Package

The daguerreotype package—including the mat, the preserver, and the cover glass itself—does a good job of protecting the daguerreotype from attack by atmospheric pollutants and retarding the oxidation and sulfidation of the plate, but it is itself a source of corrosive attack.

The mat is a corrosion source only when buffered boards are used. The alkaline earth carbonates used to maintain mildly alkaline conditions in the board cause corrosion blemishes on daguerreotype plates (fig. 10.9). This reaction is not limited to recased nineteenth-century daguerreotypes. A modern daguerreotype step tablet mounted with a buffered board was stored for a period of time. Figure 10.10 clearly shows a corrosion ring beginning to advance over the plate from the edge of the mat.

Energy-dispersive x-ray analysis of certain odd-shaped, greenish-blue particles found on daguerreotypes reveals that these particles are copper compounds (fig. 10.11). The most likely source of copper is from the copper in the underlying daguerreotype plate. These growths occur where the silver layer has been penetrated and are usually found on the more central regions of the plate. Similar growths can often be seen on the brass mats of daguerreotypes and in those cases the copper comes from the brass itself. In both cases the appearance of these copper accretions requires some sort of active reagent that, as will be shown, probably arises from the cover glass but is certainly related to the environment within the daguerreian package. Direct analysis of the copper-bearing accretions using Gandolfi x-ray diffraction techniques has shown that they are made of a variety of copper carbonates, hydroxides, and hydrates. No copper sulfides or sulfates have been identified. None of the materials

Fig. 10.10. Twentieth-century step tablet cased using buffered board mat. A corrosion ring caused by the buffer material is spreading from the mat edge.

DAG 19-110

40KV 0.88KX 10U 0014

Fig. 10.11. Scanning electron micrograph of malachite-like copper-containing accretion formed on a daguerreotype surface. The scale bar is equal to 10 micrometers.

so far identified corresponds to natural copper minerals, although the phase shown in figure 10.11 is closely related to malachite ($Cu_2CO_3(OH)_2$). This particular assemblage of copper compounds is unexpected if one sticks to the old idea that sulfur is the major active agent in daguerreotype corrosion. A more reasonable conclusion is that most available reactants are carbon dioxide and water from the air, rather than hydrogen sulfide. Given high pH conditions from glass corrosion and some trapped moisture in the package, the copper corrosion products should be similar to those actually observed.

The most important culprit is the cover glass, which slowly reacts with the atmosphere inside the daguerreotype package. Far from being inert, the glazings used by nineteenth-century daguerreotypists created an amazing variety of degradation products. Sorting out the reaction and corrosion products involving the cover glasses produced a few surprises.

It has been observed that the interior surfaces of many cover glasses dating from the nineteenth century become dusty and dirty with age. The glasses are said to "weep" and become "greasy" over time. The debris shown on the glass in figure 10.3 is typical of the appearance of aged nineteenth-century cover glasses. Other problems commonly observed on these glasses are patchy areas of discoloration and discrete spots or clumps of amber-colored "crystals" and "mold growths." There were also nineteenth-century reports of the presence of mites and their "excrementious matter" on cover glasses.[9]

Nineteenth-century daguerreotypists were selective about the glasses they used to construct daguerreotype packages, but glasses were selected for optical quality rather than chemical resistance. Glasses were put over a picture or other object for protective glazing; they were not chosen because of specific durability or corrosion-resistant characteristics. Certain special fashions, such as domed glazing and bevelling play a role in the selection of casing materials, but these factors have little or no direct bearing on the ultimate chemical durability of the cover glasses.

To get some idea of the range of glass compositions used in the nineteenth century, the authors assembled a collection of some two hundred cover glasses from daguerreotypes. These cover glasses were gathered from a large number of sources and represented a variety of source conditions and ages. Some were removed directly from nineteenth-century daguerreotypes with their original seals intact; some were from daguerreotypes that had been cleaned, resealed, and recased in the more recent past using a new or a cleaned vintage cover glass. Some glasses came from stockpiles of vintage daguerreotype cover glasses that had been removed from daguerreotypes either in a museum setting or by private collectors and stored for reuse at some future date. The geographic locations of the cover glasses were random and wide ranging; there is little likelihood that the samples could have come from the same glass houses.

The glasses were grouped on the basis of their characteristic luminescence colors: yellow-orange, orange-magenta, and blue. A total of ten glasses was chosen for complete chemical analysis on the basis of the percentage of the total glass sample with similar luminescence colors. Thus, the largest number of glasses (five) was primarily soda fluxed and had a common yellow-orange luminescence. The next largest group (three) was made up of soda-lime fluxed glasses with a common orange-magenta luminescence. A small percentage of lead glasses, only about 5 percent of the total, had blue luminescence under the ultraviolet lamp. However, their blue luminescence was so striking that two such samples were analyzed.

The bulk compositions shown in figure 10.12 indicate that the formulae of the glasses within the groups are remarkably similar. They all have unexpectedly high levels of total flux. Water solubility is related to flux content and would be expected to play a role in the degradation of the glasses. Pane glass made with lead is fairly unusual because of the comparatively high cost. For that reason these glasses may have been mirror glass or some other specialty glass, or the lead may have been derived from lead glass cullet used in the glass melt. These results, in general, support the idea that these glasses were not chosen for any reason other than overall clarity and easy availability.

Optical microscopy was used to check for the presence of glass stones (i.e., crystals), seeds (i.e., bubbles), and bits of refractory debris. This information reflects the melt history of the glass and its degree of homogeneity and is indicative of the ultimate stability of a particular glass. In order of decreasing prevalence, the following glass stones were commonly found in nineteenth-century glasses: devitrite, alpha and beta wollastonite, diopside, cordierite, and quartz (i.e., unmelted sand). None of the stones is unusual given the bulk chemical compositions. Unmelted refractory brick and sand and seeds were also present in most of the glasses. Glasses that appeared to be of more recent manufacture had fewer or no discernible glass stones, reflecting improving glass technology over the years.

A wide variety of body colors was found in the sample. These body colors are indicative merely of the presence of very minor constituents; however, the variety suggests the random assortment of the glasses used in the study. The colors range from clear or colorless to various greens and blue-greens and to purple and pink casts.

The degradation of glass surfaces is usually first noted visually by the appearance of a faint haze or clouding of the surface

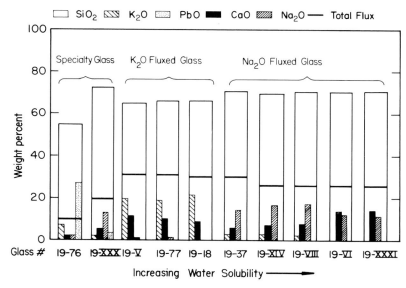

Fig. 10.12. Chemical composition of a selection of nineteenth-century daguerreotype cover glasses shown in bar graph form. For a complete listing of chemical compositions in numerical form, see Barger, Smith, and White, "Characterization of Corrosion Products on Old Protective Glass."

when viewed in transmitted light. This haze is called *weathering*. Weathering phenomena, unlike aqueous corrosion of glass, are the result of atmospheric interaction of water vapor and other airborne gases with the glass surface. There are two types of weathering: Type I, where sufficient moisture collects on a surface so that it periodically runs off and carries away reaction products, and Type II, where moisture evaporates before runoff can occur. Type I weathering is usually characterized by localized areas of total glass breakdown, which leaves a roughened and pitted glass surface. The presence of surface reaction products, such as salts and leached residues, is usually indicative of Type II weathering.[10]

Water or moisture is easily trapped inside a container with a small opening such as the daguerreotype package, which is not completely airtight. Air exchange within the container is dominated by barometric variations or similar mechanisms, and the possibility of dilution from the ambient environment is diminished. Once the glass corrosion cycle begins, corrosion products

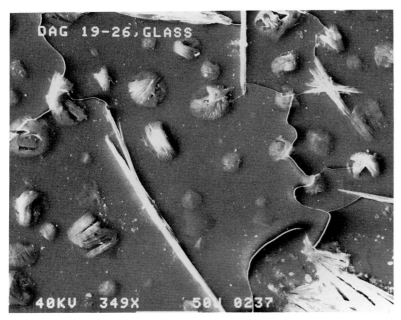

Fig. 10.13. Scanning electron micrograph showing the inside surface of a daguerreotype cover glass with moderate corrosion. The scale bar is equal to 50 micrometers.

The inside surfaces of the cover glasses could usually be identified by examination under the optical microscope because that surface was always more seriously degraded than the outside surface. This difference was also true of cover glasses that had long since been removed from daguerreotype packages. The different surfaces of cover glasses that had been cleaned for reuse could also be identified in petrographic examination, albeit with more difficulty.

The inner surface of cover glasses is often covered with sticky droplets commonly referred to as weeping glass. These droplets are primarily composed of amorphous sodium silicate. The exact composition of these gels varies according to the bulk compositions of the glasses from which they were derived. Qualitative pH measurements of the droplets made by burnishing pH paper over the interior surface of freshly uncased cover glasses show that the pH of the interior surfaces varies from 10 to 14. The interior glass surface from freshly uncased daguerreotypes with wet droplets usually feels greasy or soapy to the touch. The location of these gel droplets appears to be controlled by the geometry of the package or by the chemical inhomogeneity in the glass itself. Typically, the most severe weeping was associated with cover glasses from the soda-fluxed bulk composition group, whose nineteenth-century seals were intact.

Figure 10.13 shows the inside surface of a typical daguerreotype cover glass. The micrograph covers an area of about 3 square millimeters. On simple visual examination the rounded patches are seen as droplets of weeping glass, and the spiky bundles are seen as dust. The cracked material on the surface is the hydrated silica layer formed during the process of sodium leaching. Once this layer is cracked, it can spall off the glass surface onto the underlying daguerreotype. The crystalline features on the

are trapped and contaminate the interior environment and surfaces of the container. Trapped condensate initially allows for selective dissolution of sodium from the glass surface. This reaction leaves a hydrated silica-rich layer on the surface along with sodium hydroxide. Over time the effective pH of the condensate rises as more and more sodium hydroxide is formed. As the pH rises, dissolution of the glass surface increases resulting in pitting and the formation of various dissolution products. Thus, the geometry of the container itself can promote the evolution of a potentially hostile interior environment that actually increases the rate of glass corrosion.[11]

The deterioration products found on these cover glasses fall into four categories: amorphous gels, crystals found in these gels, products formed by gel interaction with the surface of the daguerreotype plate or its brass mat, and other products. Many of the products were not completely identified. They were characterized by x-ray powder diffraction patterns and sometimes by scanning electron microscopy.[12]

interior glass surface grow from the gels formed during the weathering cycle. They typically are needlelike crystals that form in sheaves or spiky balls, laths, dendritic crystals, cubes, weakly crystalline silicate fibrils, and, occasionally, spherical balls. The chemistries of these crystals closely follow the chemistry of the glasses from which they were derived. They typically contain silica, sodium, calcium, potassium, phosphorus, and aluminum as revealed by energy-dispersive x-ray spectra of the individual crystals. Figure 10.14 is a typical response. The crystal debris is shown in figure 10.15. The spectrum drawn as a dashed line is from the large group of crystals on the left side of the micrograph; the spectrum drawn as a solid line is from the darker background area. Only rarely was it possible to obtain good x-ray diffraction patterns from the crystals and thus make a clear-cut structural identification.

Some glasses develop blisterlike crystal growths that appear to have a small, darker colored core (fig. 10.16). Usually these blisterlike crystals are sulfates such as sodium sulfate or barium sulfate. These sulfates are likely derived from the presence of sulfur dioxide in the furnace gases during glass melting or from the intentional use of sulfur dioxide atmospheres during the annealing process. Adsorbed sulfur dioxide and water act as a sink for migrating sodium ions, leaving a dealkalized surface layer on the glass. Carbonates can be formed by the interaction of carbon dioxide and water from the atmosphere with various products of the glass dissolution reactions, particularly the sodium hydroxide. Although these compounds were expected and have been found on weathered glass surfaces of other glasses, none was identified in the study of cover glasses.

The various dissolution products and crystals formed on the cover glasses can dry up and spall off the glass surface as a result of seasonal cycling or some similar change

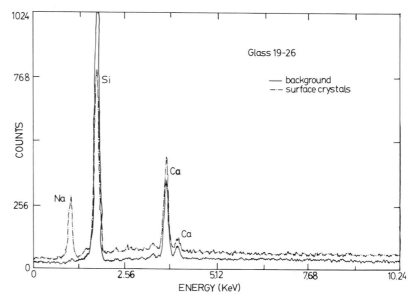

Fig. 10.14. Energy-dispersive x-ray spectrum of the crystalline debris shown in figure 10.17.

Fig. 10.15. Scanning electron micrograph of the interior surface of a daguerreotype cover glass. The scale bar is equal to 50 micrometers.

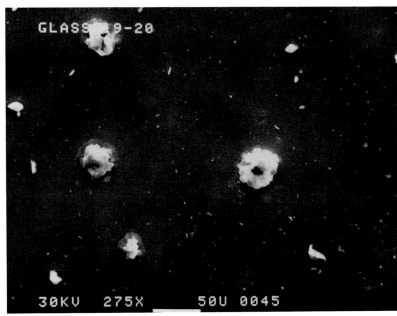

30KV   275X   50U   0045

Fig. 10.16. Scanning electron micrograph of sodium sulfate blisters on the interior surface of a daguerreotype cover glass. The scale bar is 50 micrometers.

Fig. 10.17. Daguerreotype with corrosion spots caused by cover glass degradation products.

in the ambient humidity. This debris can be found littering the daguerreotype surface, as seen in the daguerreotype shown in figure 10.17, and is easily seen in the scanning electron microscope images. In some cases these products act as transfer media to carry metal and metal salts from the daguerreotype or brass mat back to the glass surface.

Silver, silver phosphate, and silver chloride have been identified on glass surfaces and are clearly derived from this kind of interaction. The scanning electron microscope image shown in figure 10.18 shows some unidentified copper-bearing crystals that are most likely derived from copper found in the brass mat; see also figure 10.11. Energy-dispersive x-ray data on the corrosion products often show traces of silver, copper, zinc, and sometimes tin.

The most surprising silicates that appear on both daguerreotype cover glasses and daguerreotypes themselves are moldlike masses such as those shown in the composite scanning electron microscope image of figure 10.19. The underlying surface in the micrograph is the silver surface of a daguerreotype. The moldlike mass is clearly related to the white shard that has fallen from this daguerreotype's cover glass. These masses have been identified as mold by other experimenters on the basis of morphology.[13] The masses resemble the morphology of candida, a very common yeast-like fungus. However, these masses are consistently composed of silicon and sodium with smaller amounts of potassium and calcium. A careful examination revealed no evidence of spores or the presence of DNA or RNA. The authors have concluded that many of the "molds" are actually inorganic materials derived from some crystal growth mechanism. These growths occur only in those daguerreotype packages with the least durable cover glasses—that is, the high-soda glasses, which have yellow-orange fluorescence under ultraviolet excitation. For more detail on the mold question, see "Characterization of Corrosion Products on Old Protective Glass."[14]

The daguerreian package with its closed, moisture-trapping environment, active and perhaps catalytic metal surfaces, and reactive silicate cover glass is a chemical reactor of amazing complexity. Far from the inert

protective container for the delicate daguerreotype plate that is usually visualized,
the daguerreian package provides a reactive
environment for slow but ultimately de-
structive processes of corrosion and crystal
growth, illustrated on two scales in figures
10.20 and 10.21. The daguerreotype in figure 10.20 exhibits an advancing corrosion
front from the upper-left corner with a
complex, branching pattern characteristic
of dendritic crystal growth, but in this case
it is a pattern of dissolution. The micrograph in figure 10.21 shows this intense
plate etching due to glass corrosion as well
as a silicate mass with bead-and-thread
structure.

Some cover glasses have small golden-
yellow patches on their interior surfaces.
These patches usually appear to be small,

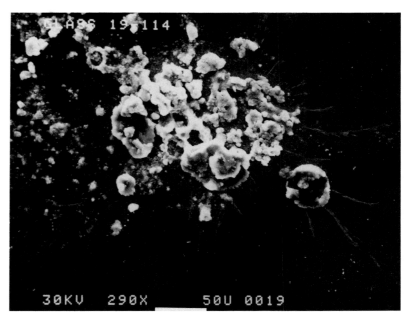

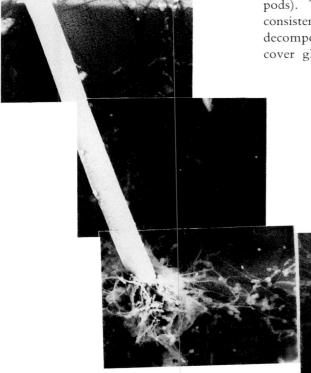

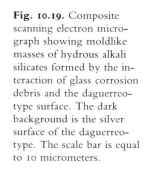

viscous spots, but with the aid of magnification they look like small bugs (arthropods). Two types of arthropods were
consistently found in various states of
decomposition on the interior surface of the
cover glasses. Figure 10.22 shows mites

Fig. 10.18. Scanning electron micrograph of cop-
per-bearing crystal on the
interior surface of a corroded daguerreotype cover
glass. The scale bar is
equal to 50 micrometers.

Fig. 10.19. Composite
scanning electron micro-
graph showing moldlike
masses of hydrous alkali
silicates formed by the interaction of glass corrosion
debris and the daguerreo-
type surface. The dark
background is the silver
surface of the daguerreo-
type. The scale bar is equal
to 10 micrometers.

Fig. 10.20. Daguerreotype showing dendritic corrosion front advancing from the upper left corner.

Fig. 10.21. Scanning electron micrograph of a daguerreotype showing three kinds of plate damage: (1) catastrophic etching of plate due to glass corrosion, (2) silicate mass with bead-and-thread structure, and (3) some cyanide etching seen mainly in the gray background. The scale bar is equal to 50 micrometers.

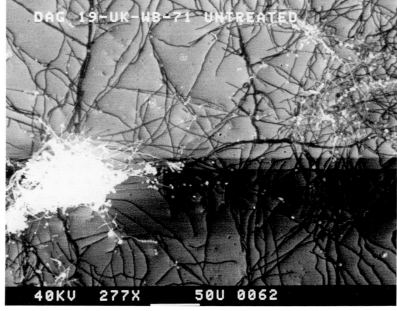

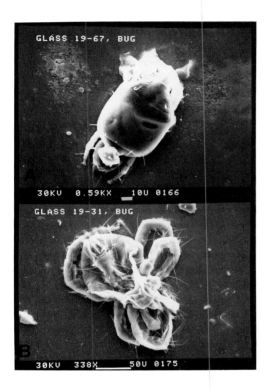

(arachnid, order *Acarina*) and book lice (*Liposcelis divianatorius* and related genera of *Psocoptera*). It is unlikely that these arthropods have anything to do with causing or initiating the glass corrosion cycle; however, they do get caught up in the cycle. Mites make up a large portion of house dust and are pests commonly found in stored materials. Book lice, too, are common pests found among stored organic materials. The animals probably come into the daguerreian package because they are initially attracted to the organic components, such as adhesives, cloth, and wood in the cases and sealing materials. They are further attracted into the package by the elevated relative humidity of the package interior. The silica gels on the inside surface of the cover glasses are very sticky, and the arthropods become entrapped and cannot survive the environment of pH 10 to 14. All the arthropods found on the cover glasses were dead and had become silicified to some degree.

Among the unexpected reaction products found commonly within the daguerreotype package was sodium formate. Formates have been found among the corrosion products on many types of objects encased in glazed vitrines and on deteriorating building stones. Formic acid and formate salts have also been implicated as factors in the corrosion of optical glasses.[15] The widespread instances of formic acid and formate salts on enclosed glass surfaces indicate that these species should be easily formed corrosion by-products occurring during the aging of these materials under common physical conditions. In the case of the daguerreotype cover glasses, the quantities of formic acid and formate salts observed, their locations, and the frequency of their occurrence are greater than can be accounted for by only an organic pathway. One possible source for these formates is the fixation of carbon dioxide and water from the atmosphere, perhaps with methanol as an intermediate step.[16] Although these are purely inorganic routes, they do require reducing environments and probably some sort of catalytic surface. It was suggested that formic acid or formaldehyde can be formed at the silver–silver oxide interface. These products, given off as volatiles, could then attack the inner glass surface, leaving sodium formate by reaction with the sodium hydroxide remaining from glass hydrolysis. The identification of the formate salts is reasonably well established, but the reaction chemistry is mainly speculative at present.

The deterioration of the daguerreotype image turns out to be a quite complicated business. Other than such straightforward matters as mechanical damage to the plate due to bending or scratching and the substantial damage done to plates due to well-intentioned efforts at cleaning them, much of the protection-damage issue is related to the daguerreian package itself. On the pos-

Fig. 10.22. Scanning electron micrographs of arthropods found on the interior surfaces of corroded daguerreotype cover glasses: a mite (A) (scale bar is equal to 10 micrometers) and a book louse (B) (scale bar is equal to 50 micrometers).

itive side, the combination of case, mat, preserver, and cover glass, designed almost immediately after the invention of the daguerreotype process, has held up well over a century and a half of use. It is hard to look back and see how the package could have been designed better. On the negative side, the atmosphere inside the case and the cover glass are more reactive than early daguerreotypists would have ever suspected. Over the long years slow chemical reactions within the package have conspired to produce a rich assortment of corrosion products that must be dealt with when conserving and preserving daguerreotype images.

"CLEANING" DAGUERREOTYPES—the removal of corrosion from daguerreotypes—has been a problem fraught with peril since the daguerreotype was introduced. Daguerreotypes are produced on silver plates, and silver corrodes, as has been discussed. When these jewellike images become covered with dark films, it is difficult to deter their owners from cleaning them. Cleaning a daguerreotype is as rewarding visually as it is damaging to the image. After removing the tarnish that once obscured the image, the viewer can tilt and move the image so that it appears "as it did when it was first made."

From the beginning of the daguerreian era there was concern that the daguerreotype was not stable and that images recorded on daguerreotype plates could fade. The durability of daguerreotypes with respect to light, air, heat, and various pollutants was investigated both formally and informally and compared to other types of photographic materials.[2] Most of these studies determined that daguerreotypes were fairly stable if properly made and sealed. James T. Foard, the successor to the English daguerreotypist Richard Beard, claimed that "there never was a faded daguerreotype, and . . . the image written on a tablet of silver fixed by the alchemical agency of gold and mercury is, in its own nature, eternal."[3] Yet, nineteenth-century literature is riddled with references to fading daguerreotypes. The term usually referred to daguerreotypes that had become covered with tarnish and whose images were obscured by corrosion films. The remedy for this problem was cleaning. Photographers commonly advertised and performed this service in addition to making paper copies of daguerreotypes brought to them by their customers.

Cyanide Cleaning

During the nineteenth century and up until the early 1950s the usual method of cleaning

11

Corrosion Removal

To clean Daguerreotypes perfectly; so many ways have been proposed, so many Daguerreotypes have been spoiled beyond redemption, by so many ways being tried, by so many who did not understand cleaning them. If a photographer, or even the old Daguerreotypist, follows the following instructions closely he will never injure one, and clean it perfectly, so that it will be as brilliant as the day it was taken, if it has not been defaced by rubbing the surfaces of it.
"How to Clean Daguerreotypes"[1]

daguerreotypes was to put them in a solution of potassium or sodium cyanide. Most recommendations for compounding cleaning solutions call for a lump of cyanide the size of a pea or a bean. More exact formularies were published from time to time, but more often it was stated that the amount of cyanide in water made no difference because the quality of commercially available cyanide varied widely and cleaning solutions had to be adjusted according to the quality available. Cleaning instructions usually cautioned that cleaning solutions and wash waters should be distilled water. Some mention that it is best to use hot water. The cleaning regimes always followed more or less the same steps: The daguerreotype was uncased and gently brushed with a soft camel's hair brush. It was then washed with water and possibly alcohol to remove surface grime and oils. Next the daguerreotype was placed in or washed with the cyanide solution and rocked gently until stains disappeared. Finally,the plate was washed well in running tap water with a final distilled water rinse. The daguerreotype was then dried over a spirit lamp to prevent staining from the final wash water.[4]

Cyanide was considered the best choice for cleaning daguerreotypes because it was believed to be a solvent for silver sulfide and "not for the mercury forming the image."[5] It was also thought that cyanide could be completely removed from the plate surface by simple washing. Some researchers felt that simple cyanide solutions were too aggressive for daguerreotypes and published formulas that included other ingredients, such as potassium carbonate, common salt, borax, and various buffering agents. Some, such as M. P. Simons, felt that cyanide should be used only for the most difficult cases or as a last resort: "[W]hen a daguerreotype from bad usage is in the last stages of decay, and the picture would be worthless unless cleaned, then it may be experimented on with cyanide, or anything else,

as powerful medicines are sometimes to be administered to the sick when there is but little hope of recovery."[6] Other cleaning solutions occasionally recommended included full-strength ammonia and simple hot water washes, but these were never as popular as cyanide cleaners.

The frequency with which articles about cleaning daguerreotypes were published suggests that this was a very common practice during the nineteenth century. Most notices said that primary requirements for the successful cleaning of daguerreotypes were care and some modicum of skill. These notes encouraged the tyro to forge ahead and clean daguerreotypes. Occasionally, there were reports of disasters, especially when the daguerreotypes of deceased relatives were lost to the ravages of cyanide baths. In 1875 a suit was brought to the Southwark County (England) Court over the loss of a daguerreotype to cleaning. In *Sandy v. Dean* Sandy sued Dean for five pounds sterling because Sandy had employed Dean to clean and copy a daguerreotype of his deceased mother. After three months the daguerreotype was returned, "but in such obliterated condition as to be barely discernible, and as he [the defendant] would not make any compensation, the plaintiff brought . . . action for the loss he sustained."[7] Dean claimed that he had done nothing to injure the daguerreotype and that it was so faint when given to him none of his munitions were effective. The judge ruled that Sandy be given compensation only for the intrinsic value of the daguerreotype, or ten shillings. While this suit was unusual, notices of the loss of daguerreotypes because of cleaning appeared from time to time.

One particularly tragic loss to cyanide cleaning was the portrait of Dorothy Draper that Dr. John Draper had sent to John Herschel, claiming that it was the first portrait taken from life in America. This portrait has obtained a great deal of importance not because it was the first portrait taken but because Draper's repeated claims about having made the first portrait have been taken up by certain historians not familiar with the history of photography. In 1933 the Draper portrait, which had remained in the Herschel family, was given to an "expert" to clean, but the image was lost. About the same time Robert Taft, who was working on the research for his book *Photography and the American Scene,* had written to the Herschel family asking for a copy of the Draper portrait. The family sent the daguerreotype to Taft because he had offered to try to salvage it; unfortunately, there is no record as to whether he actually made any such attempt.[8]

Thiourea Cleaners

The use of cyanide cleaners remained the state-of-the-art until the early 1950s, when thiourea cleaners were introduced. In 1952 a new silver dip cleaner that used an acidic solution of thiourea in place of cyanide was introduced.[9] Thiourea cleaners for cleaning daguerreotypes were first adapted by Charles van Ravensway of the Missouri Historical Society in 1956.[10] His method was published in *Image,* the journal of the International Museum of Photography at the George Eastman House, along with photographs taken before and after cleaning to document the cleaner's efficacy. The editor of the journal gave an enthusiastic introduction and endorsement. Van Ravensway listed the advantages of the process: Stains were completely removed without changing the mirrorlike quality of the plate; daguerreotypes could be left in the solution indefinitely without adverse effects (thus, badly stained daguerreotypes could be cleaned without worry); and plates could be cleaned repeatedly without harm. He did mention that since daguerreotypes previously cleaned in cyanide "lose their original lustre," daguerreotypes that had never been cleaned before produced better results.

This new cleaning process was greeted with a great deal of excitement. Eastman House set up facilities to use the new cleaner. Hirst Milhollen, of the Prints and Photographs Division of the Library of Congress, was first acquainted with the thiourea process while on vacation in 1957. In 1950 the Library of Congress had rediscovered an important collection of more than 350 daguerreotypes by Mathew Brady that had been in storage since given to the Library in the late 1920s. Since their discovery there had been great concern and caution about cleaning these daguerreotypes, and Milhollen thought that the Missouri Historical Society cleaning method might be a way to restore these valuable plates. In a memo to Edgar Breitenbach, acting chief of the Prints and Photographs Division, Milhollen proposed, "As this method had been successfully used by George Eastman House, it might be useful in the restoration of certain Brady daguerreotypes."[11] Breitenbach forwarded the Milhollen memo, along with his own memo noting that the Brady daguerreotypes were "greatly in need of preservation" and recommending that an order could be placed to the photoduplication service "for the application of this new method to our deteriorating daguerreotypes."[12] Such a program was established at the Library shortly afterwards. The Library proceeded with some caution, and when Dorothy Meserve Kunhardt suggested that the Library's portraits of Abraham Lincoln and Mary Todd Lincoln be restored for the sesquicentennial celebration of Lincoln's birth, the National Bureau of Standards was asked to study the safety of the thiourea process.[13] The Bureau evidently gave its approval because the Lincoln portraits were sent to Eastman House to be cleaned in late 1958.[14] In like manner, the thiourea cleaning method spread and was recommended in many venues and in many publications.

As the value of daguerreotypes and other photographs began to increase during the late 1960s and into the 1970s, the alarm began to sound on the safety of thiourea cleaning systems. Collectors and curators began to notice that some daguerreotypes cleaned in thiourea developed brown spots, the daguerreian "measles." The scanning electron microscope examination of these measles by Leon Jacobson and W. E. Leyshon was unable to fully identify their chemistry. Jacobson and Leyshon attributed their occurrence to the incomplete removal of the cleaning solution.[15] This report caused new worries to spread through the photographic community about the safety of cleaning daguerreotypes. Research into better formulations of thiourea daguerreotype cleaners was begun by Alice Swan and others. The general modus operandi was to devise less active formulations of the thiourea silver dip formulas for daguerreotypes and determine better ways to remove cleaning residues from the plate surface.[16]

Problems in Cleaning Daguerreotypes

The removal of corrosion products and the restoration of the daguerreotype image presents a formidable problem. The corrosion products, which are silver compounds, coat the silver image particles and the polished silver substrate between the image particles. The trick is to remove the silver corrosion products without removing image particles, etching, or otherwise damaging the specular reflectance characteristics of the silver substrate. Most silver cleaners work by removing a minute amount of metal in addition to the tarnish. What is perfectly acceptable for dinnerware or silver plate could be a disaster for a daguerreotype. What is needed is a cleaning method that is highly controllable yet forgiving. Unpredictable results from traditional cleaning treatments—including fading and the loss of valuable images—brought about a general recommendation in the 1970s that daguerreotypes should not

be cleaned at all until a safe method had been found.

The traditional silver dip cleaners have certain other problems. The entire object was simply dipped into the cleaning solution for treatment. When the object appeared to be clean, it was removed and rinsed with water. The only controls offered by these methods of cleaning are the concentration of the active ingredients in the cleaning solution and the time of treatment. These methods cannot work on specific areas of an object while leaving other areas untouched.

Silver solvent solutions cannot be used for daguerreotypes that have been colored with pigments. These daguerreotypes were usually colored by grinding pigments with gum arabic or isinglass and then dusting the pigment mixture on the area of the daguerreotype to be colored. Obviously, anything colored in this fashion would probably be quite fragile and not able to withstand any but the most delicate cleaning methods. Needless to say, cleaning by dipping a colored daguerreotype into a solvent solution is dangerous and risky, at best.

The cyanide ion, CN^-, forms strong complexes with many metallic ions. Silver cyanide complexes are highly stable, and thus tarnish compounds on the daguerreotype surface tend to break down and form silver cyanide. Thiourea, $(NH_2)_2C=S$, is also a complexing agent but is applied in an acid solution. Hydrochloric and sulfuric acids were recommended by Howard Brenner in the first paper on thiourea silver dip cleaners.[17] When Brenner's formulas were adapted for use on daguerreotypes, phosphoric acid was substituted.[18] Evidently, it was thought that phosphoric acid would be less harmful to the daguerreotype plate and image. Unfortunately, silver oxide is the most prevalent corrosion product on cased daguerreotypes' surfaces, and phosphoric acid is extremely reactive with the highly tarnished areas of the plate.

The microscopic effect of silver dip cleaners can be seen in the secondary electron image (fig. 11.1). The overall patterning on the plate surface is the characteristic etching associated with cyanide cleaners. Most extant daguerreotypes have been cleaned with cyanide at some point. Of the almost two hundred daguerreotypes examined in the Materials Research Laboratory, only 2 percent showed no signs of previous cleaning treatments. The presence of a nineteenth-century seal is no guarantee that a daguerreotype has never been cleaned or that the seal is original to the daguerreotype. The only sure way to identify uncleaned daguerreotypes is to examine their microstructure using the scanning electron microscope. Examining the optical properties using reflectance spectroscopy is also useful. Thiourea cleaners have some of the same effects as cyanide, but they appear to be somewhat less aggressive. Analysis of the daguerreotype corrosion products indicates that no amount of rinsing will remove thiourea because it is such an active complexing agent for silver. There are very few examples of daguerreotypes that have been cleaned only in thiourea.

When used on daguerreotypes, silver dip cleaners cause damage to the plate surface and image microstructure. This damage appears as image particle alteration and loss, overall etching of the daguerreotype plate, and more severe etching in previously tarnished areas. Image particle alteration or loss and etching change the optical properties of the daguerreotype surface and make the image appear faded or weak. Further, the daguerreotype becomes more vulnerable to recorrosion caused by the cleaning treatment itself because the etching action leaves a fresh surface of greater area that is more reactive than the uncleaned surface.

Electrolytic Cleaning

Silver objects can be cleaned electrolytically. In this method the silver object is

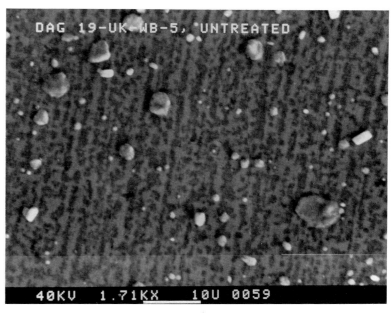

Fig. 11.1. Scanning electron micrograph of a midtone region of a daguerreotype. The darker pattern on the substrate is the result of etching caused by cyanide cleaning treatment. Scale bar is equal to 10 micrometers.

placed into an aluminum vessel filled with a hot, slightly alkaline salt solution.[19] Trisodium phosphate is usually used for commercial applications of the technique, sodium carbonate or bicarbonate for "home" use. The silver object and the aluminum vessel become the electrodes in a galvanic cell containing an alkaline electrolyte. Contact between the silver object and the aluminum vessel short-circuits the cell and allows the electrolytic reaction to proceed. Silver ions from silver sulfide or oxide corrosion film are reduced to silver metal while simultaneously small portions of the aluminum vessel become oxidized. The only role of the electrolyte is to provide a conductive pathway, and the chemical reaction is essentially one of metal-metal exchange.

Electrolytic cleaning is an example of sacrificial corrosion in that the aluminum is corroded, or sacrificed, in the process of removing corrosion from the silver. The aluminum container can be replaced by an aluminum electrode that will continue to provide electrons for the reduction of silver ions until a passivating film of aluminum oxide is formed on its surface or until there are no more silver ions available to the reaction. The same effect can be achieved, but not as efficiently, using magnesium or zinc in place of the aluminum. In principle any metal that has a more negative electrode potential than silver could be used. No external source of current is needed to make this reaction go forward; it is driven by the large free-energy difference between silver oxide and aluminum oxide. In addition, the reaction cannot be reversed. Silver metal, whether in the image particles or in the underlying plate, is not attacked because it is already in its reduced state.

Electrolytic cleaning, like the silver dip or solvent cleaners cannot be controlled except by diluting the electrolyte solution or by changing the treatment time. An object that is heavily corroded may require several treatments. Unlike solvent cleaners, however, this is not a harmful cleaning method because the reaction is self-limited. It cannot continue once all the corrosion products have been removed. In addition, no corrosion by-products form on the silver surface as a result of the process.

Very early in the authors' investigations of daguerreotypes, it was apparent that some improvement in cleaning methodology was needed. Many tarnished daguerreotypes exist that might possibly be restored to something like their initial quality if only a reliable method was at hand. Yet it was abundantly clear that traditional silver dip methods, although they removed tarnish, also did irreversible damage to the distribution of image particles. We examined two techniques: (1) reactive sputtering and (2) electrocleaning.

Sputter Cleaning

Sputter cleaning is a completely new and viable approach to the removal of tarnish from daguerreotype surfaces, although it is less effective than electrocleaning and requires specialized equipment. Sputtering is a highly controllable process using radio

frequency or direct current plasmas for either the deposition of materials on a substrate or the removal of materials from a substrate. The latter process, sputter cleaning, had been investigated in connection with cleaning museum objects. Different approaches were reported first by V. D. Daniels in 1981 and by M. Susan Barger, S. V. Krishnaswamy and R. Messier in 1982. These two groups used different means to reach the same ends—namely, the removal of silver sulfide from daguerreotypes. Further investigations based on both groups' work were carried out by Anker Sjøgren and Mogens S. Koch at the Conservation School of the Royal Danish Art Academy.[20] All these investigators observed that while sputtering removed tarnish and produced a visible improvement in the appearance of cleaned daguerreotypes, faint white films were left behind, especially in areas that had been highly corroded. The presence of these films indicated that there had been some alteration of the optical properties of the daguerreotype surface.

In sputtering the energy of a glow discharge—that is, a low-pressure plasma—is used to generate active species that may be atoms, radicals, or ions from gases such as argon, hydrogen, other gases, and gas mixtures. The generated species are extremely reactive and have high kinetic energies. When they bombard the specimen (called the target), they can be used to physically remove material from a surface (this is called physical sputtering) or to react chemically with material on a surface. Reactive sputtering refers to sputtering conditions in which there is both physical removal of material and chemical interaction at the target surface. In sputter cleaning the object to be cleaned is the target, and any of the three modes—physical sputtering, plasma reduction, or reactive sputtering—can be used. Essentially, the object to be cleaned is placed in a suitable vacuum chamber and attached to the cathode portion of the plasma electrodes. The chamber is then filled with sputtering gas at the proper pressure, and the direct current or radio frequency (rf) field is applied to create a plasma between the electrodes of the sputtering chamber.[21]

The sputter cleaning of silver objects has an intrinsic limitation. The sputtering rate, or the rate at which material is removed under the same conditions, varies in different materials. Silver metal has one of the highest sputtering rates of common materials, while silver sulfide and silver oxide have very low sputtering rates. As a result a sputtering system for the removal of silver sulfide from a silver surface must provide a way to protect the silver itself from being removed while the silver sulfide is being removed; otherwise, pitting would occur.

The primary method for controlling the sputtering process is altering various components—sputtering time, substrate-to-target distance, gas pressure, rf voltage, power, and current. In the experiments described here, the substrate-to-target distance was kept constant. In all but one case, the rf voltage was kept low (about 200 volts) so that the energy of positive ions would be held to about 200 electron volts (eV) and, therefore, the sputtering rate would be very small. Three parameters were changed: the ratio of argon to hydrogen partial pressures, the total gas pressure, and the sputtering time (table 11.1).

Each sample set was made up of a portion of two daguerreotype step tablets, one tarnished and one not, an unused daguerreotype plate as a silver blank, and a piece of tarnished daguerreotype. Some of the daguerreotypes and silver blanks were tarnished artifically by placing them uncased in a room with hydrogen sulfide levels on the order of 5 to 20 parts per billion. These daguerreotypes became totally blackened in a matter of days.

Table 11.1. Sputter Cleaning Conditions

Sample Number	P_{H_2} (mTorr)	P_{Ar} (mTorr)	P_{tot} (mTorr)	Rf (volts)	Power (watts)	Current (amps)	Time (min.)
1	9.0	0.0	9.0	500	100	0.18	5
2	9.25	0.0	9.25	200	25	0.16	4
3[a]	9.25	0.0	9.25	200	25	0.16	1
	16.5	0.0	16.25	200	25	0.16	9
4	9.25	0.75	10.0	200	25	0.16	10
5	9.25	1.75	11.0	200	25	0.16	10
6	7.8	2.2	10.0	200	25	0.16	10
7	4.0	6.0	10.0	200	25	0.15	10
8	7.75	2.25	10.0	200	25	0.16	480
9	7.75	2.25	10.0	200	25	0.16	5
10	7.75	2.25	10.0	200	25	0.16	2
12[b]	16.5	0.0	16.5	200	25	0.16	5

[a] The sputtering process was interrupted and restarted in order to check changes caused by reiterative sputtering.

[b] This was a test on a blank silver plate to test hydrogen passivation of the surface.

The overall results of the sputtering experiments were good. Daguerreotypes representing the entire spectrum of sputtering conditions reported in table 11.1 were cleaned, and no one set of conditions appeared to work better than another. All daguerreotypes, especially the tarnished step tablets, developed a faint white surface in areas from which tarnish had been removed, as was observed earlier.[22] This effect could not be avoided no matter how the sputtering conditions were altered.

Surface analysis using scanning electron microscopy, energy-dispersive x-ray spectroscopy, and Auger electron spectroscopy showed that the white areas are not films but rather the result of microscale etching of the plate surface caused by sputtering. The upper micrograph (*a*) in figure 11.2 shows a shadow area of a daguerreotype before sputtering. As an aside, by careful examination the viewer can see the characteristic etching of the plate surface from some past solvent cleaning treatment. The lower micrograph (*b*) is of a similar shadow area on the same daguerreotype after sputter cleaning. As can be seen, the surface shown in micrograph *b* is textured and the white image particles are slightly more distinct on the plate surface. In previously heavily tarnished areas of a daguerreotype, this microtopography is often more pronounced than elsewhere on the plate surface.

Figure 11.3 shows the surface of one of the heavily corroded silver blanks that was partly covered by a small glass cover slip to protect part of the surface during sputtering. The sputtered portion of the surface is marked *c* and the unsputtered portion is marked *uc*. It is easy to see that area *c* is quite textured and looks something like an aerial view of a dense forest.

Corrosion occurs in an uneven fashion over a metal surface according to which portions of the surface are more favored thermodynamically or chemically. Thus, high-energy spots caused by either mechanical or chemical factors corrode first and usually have the most accumulated corrosion. When an unevenly corroded surface is cleaned with a plasma, all other conditions being equal, the depth and extent of sputtering will be determined by the difference in the sputtering rates of the corrosion products and the base metal, as well as the thickness of the corrosion films. Since silver has such a high sputtering rate compared to its corrosion products, no matter what conditions are used, the sputtering of heavily tarnished areas will always proceed at a slower rate than that of other areas. The resulting microtopography is the physical manifestation of this phenomenon. Surfaces that have been relatively smooth and absorbing because of the presence of tarnish films are changed to comparatively rough surfaces that scatter visible light, seen by the viewer as a white film as illustrated in figure 11.4.

Fig. 11.2. Scanning electron micrographs showing daguerreotype surface before (a) and after (b) sputter cleaning. Scale bars are equal to 10 micrometers.

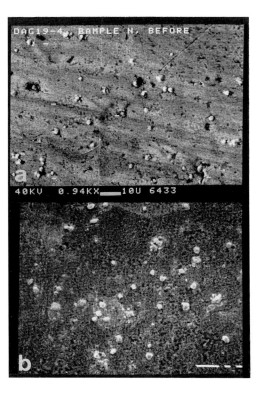

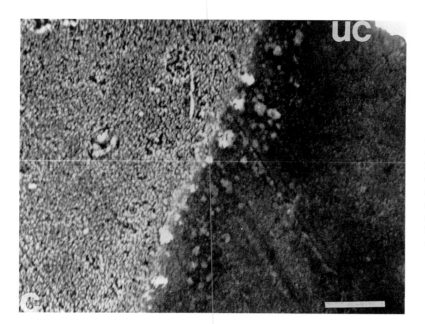

Fig. 11.3. Scanning electron micrograph of the surface of a silver blank tarnished in a hydrogen sulfide atmosphere and sputter cleaned. A small glass cover slip was used to protect one section of the plate from treatment so that both silver loss and the effect of sputtering on highly tarnished areas could be monitored. The cleaned portion is marked *c*, and the uncleaned section is marked *uc*. Scale bar is equal to 10 micrometers.

Fig. 11.4. Daguerreotype with one half cleaned by sputtering and the other half left untreated. Note the faint white films on the side that has been sputter cleaned, especially in the central area of the plate near the mat edge.

Chemical analysis of these whitish areas using Auger electron spectroscopy demonstrates that these areas have slightly more oxygen than surrounding areas without the optical effect. A sputtered surface is most certainly more chemically reactive than the same surface before sputtering. It is likely that a newly sputtered daguerreotype immediately adsorbs oxygen as it is being removed from the sputtering chamber. The amount of oxygen adsorbed will be proportional to the extent of microtopographic development caused by sputtering. In this respect sputter cleaning produces results similar to solvent cleaners. The increased surface area produced by both techniques makes cleaned daguerreotypes more vulnerable to recorrosion. Unlike solvent cleaners, however, sputtering does not leave new corrosion products on the daguerreotype surface as a result of treatment.

The experimental evidence suggests that while sputter cleaning is not as drastic as solvent cleaning, its use should be limited to daguerreotypes that need to be cleaned without wetting, for example, daguerreotypes that have been heavily hand colored. For these sputter cleaning offers a dry way to remove corrosion without harming the coloring materials on the surface. A broad range of sputtering conditions will produce cleaning. The process can be done in an iterative fashion without harm to the daguerreotype. A practical drawback to the technique is the need for sputtering equipment and skilled technicians to run the equipment.

Electrocleaning

In spite of the interesting results from the experiments with sputter cleaning, it was obvious that the problem of daguerreotype cleaning had yet to be solved. Of the various methods that had been examined, electrolytic cleaning had the most advantages. Making the daguerreotype into the anode of a battery was at least a benign procedure so far as attack on the metallic silver was concerned. Most of the other objections to electrolytic cleaning could be removed by a new method that made direct use of the electrochemical reaction between silver and the silver corrosion products. The daguerreotype was made part of an electrochemical cell but not part of a battery.

The basic experiment was to again place the daguerreotype in an electrolyte solution.[23] However, this time the driving energy was supplied by an electrical current from an external power supply. One lead of the power supply was connected directly to the daguerreotype, the other to a movable silver electrode. The movable electrode was used to direct the current and thus allow for concentrated cleaning on specific areas of the silver object. By using a direct current, the polarity of the silver object and that of the movable electrode can be reversed so as to make use of both anodic (i.e., reverse) cleaning in addition to cathodic (i.e., direct) cleaning. This process is called *electrocleaning* to differentiate it from the electrolytic cleaning of the simple galvanic cell. Each form of electrocleaning, cathodic and anodic, has advantages, especially for objects with heavy corrosion: switching back and forth from anodic cleaning to cathodic cleaning can break up corrosion layers, making them easier to remove. The key differences between electrocleaning with a movable electrode and electrolytic cleaning with the galvanic cell is that in electrocleaning the overdriving potential provided by the applied current improves the kinetics of the cleaning reaction, allows control over the type of cleaning that can occur, and allows control over the extent and location of cleaning.

The oxidation of silver to form its various oxides or the reaction of silver with sulfur or hydrogen sulfide to form its various sulfides each have a characteristic electrochemical potential that is a function of

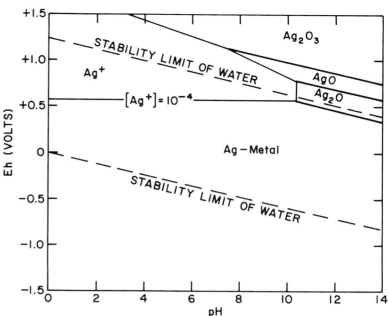

Fig. 11.5. Pourbaix (pH-Eh) diagram for the system silver-oxygen-hydrogen (Ag-O-H). Phase boundaries showing the regions of stability of the solid silver oxides are drawn as heavy lines. The solubility of silver is calculated for a concentration in solution of 10^{-4} molal (10.8 parts per million). These solubility boundaries are shown as light lines. The two parallel dashed lines show the upper and lower stability limits for liquid water at 25 degrees Centigrade. The equations used for calculating this diagram were taken from Pourbaix, *Atlas of Electrochemical Equilibria in Aqueous Solutions,* chap. 4, sec. 14.2.

pH when the metal and oxides are in equilibrium with an aqueous solution. These potential pH relations can be plotted to give a map, called a Pourbaix diagram (fig. 11.5), which shows the stability ranges of the various compounds. Silver resides high on the electromotive series; thus, silver metal occupies much of the stability region in the Pourbaix diagram. In acid solutions silver dissolves directly to form monovalent silver ions in solution. The oxides are the only stable solids in contact with aqueous solutions at high pH. The wide stability field for silver sesquioxide lies above the stability limit for water. The stability boundaries in turn determine whether the metal will be stable, corrode, or form a passivation film on its surface (fig. 11.6).

The stability field for silver oxide (Ag_2O) at pH 12 lies just within the corrosion boundaries. The other silver oxides—Ag_2O_2 and Ag_2O_3—exist in the passivation region of the diagram, but these are all relatively unstable at room temperature and standard pressure. These oxides can be formed at pH 12 by applying a relatively small overpotential. At this point it is useful to design an electrocleaning system that can take advantage of both the instability of the various silver oxides and the low potentials needed to form these oxides at pH 12.

To set up an electrocleaning system it is necessary to find an electrolyte solution that can be maintained near pH 12. However, if the electrolyte was also a solvent for silver oxide, another method to control the cleaning process would be added. Ammonium hydroxide does have these characteristics; its reactions with silver are shown in figure 11.7. The solution used for daguerreotype cleaning is two parts water to one part concentrated ammonium hydroxide (28 percent) by volume. Silver oxide is very soluble in this solution. The solution will act only on the silver oxide present and not on any other corrosion product or the silver of the daguerreotype plate. Ammonium hy-

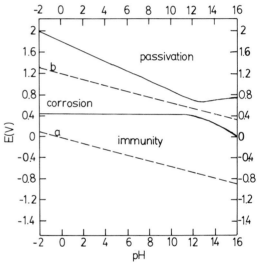

Fig. 11.6. Corrosion diagram for silver based on Pourbaix diagram showing chemical reactions between metallic silver and the aqueous solution.

Fig. 11.7. Some regions of compound formation in the system silver oxide-ammonia-water (Ag_2O-NH_3-H_2O).

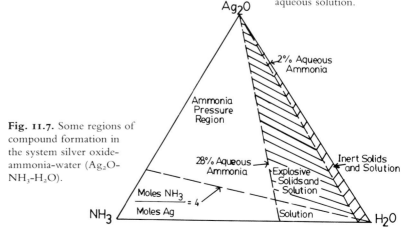

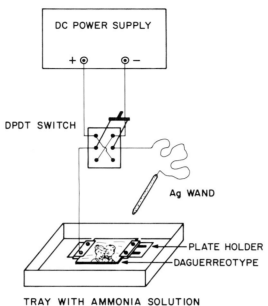

Fig. 11.8. Electrical connections for the electrocleaning experiment.

DC POWER SUPPLY

+ ⊙   ⊙ −

DPDT SWITCH

Ag WAND

PLATE HOLDER
DAGUERREOTYPE

TRAY WITH AMMONIA SOLUTION

droxide will complex available silver ions to form water-soluble ammonium complexes that are easily washed away leaving no surface residues behind. The phase diagram also indicates a region where explosive silver azides and related compounds can form. These present no safety hazard; there is never enough silver oxide present to worry about the formation of these explosive silver compounds.

For the application of electrocleaning to the daguerreotype problem, two electrodes—the daguerreotype itself and a wand electrode—were attached to a dc (direct current) power supply through a double-pole, double-throw switch (fig. 11.8). In the first switch position the daguerreotype is the anode, and the wand electrode is the cathode. In the second switch position the poles are reversed. The dc power supply provided various voltages up to 100 volts. The bath was operated at room temperature.

The daguerreotypes used for all these experiments covered a broad range of quality and ages—from very early daguerreotypes

(made in 1840) to one made in the late 1970s—and came from many sources. The bulk of the samples came from the Materials Research Laboratory working collection of daguerreotypes that had been specifically gathered for experimentation (see chapter 8, note 3). The daguerreotypes were photographed before treatment to document their initial state. However, photographs of daguerreotypes can be misleading because this mode of copying automatically enhances their contrast range. For experimental purposes it is important to maintain a record in the form of cleaned and uncleaned portions of the same daguerreotype in addition to the photographic records. Thus, a few of the experimental daguerreotypes were cut into pieces before treatment so that direct comparisons could be made between treated and untreated portions. Daguerreotypes from several private and public collections were also provided for field testing of this cleaning method. These daguerreotypes were loaned by their owners for this use with the full understanding of the experimental nature of the work. None of the loaned daguerreotypes or others of lasting value was cut or otherwise damaged. Further, there was an additional agreement with the owners that the daguerreotypes would be returned for monitoring and testing to follow the long-term results of the cleaning procedure.

A special holder of nylon and silver (fig. 11.9) was made so that a daguerreotype could be securely held in a flat position and

Fig. 11.9. Plate holder for daguerreotypes in electrocleaning solutions.

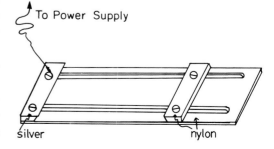

To Power Supply

silver   nylon

connected to the circuit without being scratched. Each daguerreotype cleaned in the laboratory and during the field testing period was examined using scanning electron microscopy and energy-dispersive x-ray analysis before and after treatment. In addition, some daguerreotypes were examined using Fourier transform infrared spectroscopy to check for the presence of thin surface films subsequent to cleaning. Daguerreotypes were photographed following treatment. Overall, some one hundred daguerreotypes were cleaned.

In some of the early experiments the wand electrode was made of aluminum. The aluminum electrode worked erratically with wild fluctuations and spikes in the voltage over a 20-to-60 dc volt range. This erratic behavior continued until the aluminum was seasoned, that is, until it grew a thick porous coating of aluminum oxide. Once the oxide layer had built up, the aluminum wand could be used indefinitely. Often a black tarnish would form on the edge of the daguerreotype plate when the aluminum electrode was used. If the electrode accidentally touched the daguerreotype surface during cleaning, an explosive connection would be made, forming a molten pit composed primarily of silver and aluminum, along with some copper if the silver layer of the plate was pierced. Figure 11.10 shows a micrograph of one of these pits.

More disturbing than the pitting is that the silver layer on a daguerreotype could be entirely separated and peeled away from the copper substrate during cleaning (fig. 11.11). At first it was thought that this peeling was associated with daguerreotypes made on American process plates—that is, cold-roll clad daguerreotype plates that have an electroplated silver layer added by the user. The peeling was attributed to poor adhesion between the silver electroplated layer and the silver-clad layer of the plate. However, it became apparent that if an alu-

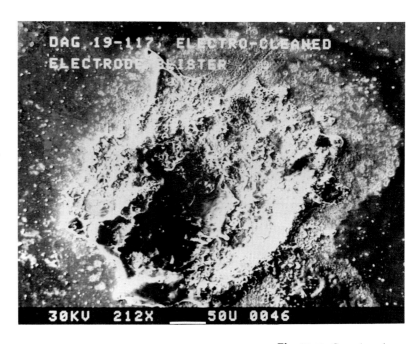

Fig. 11.10. Scanning electron micrograph showing pit resulting from the use of an aluminum wand electrode that touched the daguerreotype surface during cleaning. Scale bar is equal to 50 micrometers.

Fig. 11.11. Daguerreotype showing areas of peeling and delamination of the silver layer of the daguerreotype plate caused by the use of an aluminum electrode during electro-cleaning.

minum electrode was used for cleaning, peeling occurred regardless of the method of plate manufacture. An investigation of this phenomenon using silver blanks showed that peeling occurred at power levels as low as 29 dc volts and 10 to 15 milliamps.

The aluminum electrode is clearly not an appropriate material for use in this application of electrocleaning. The natural aluminum-silver galvanic cell interferes with the driving potential of the applied voltages. Further, the aluminum tends to act like a capacitor and store charge, thus accounting for the wild voltage fluctuations.

To avoid the problems associated with the aluminum electrode, electrodes of more noble metals—copper, brass, platinum, and silver—were tested. The cleaning results and control improved markedly with electrodes of increasing nobility. When the silver electrode is used, the natural galvanic cell is eliminated, but electrolysis from the applied potential is maintained. Thus, when a silver electrode is used, there is no seasoning period, no fluctuations in voltage, no peeling or pitting, and no blackening of the edges, and cleaning is more rapid and complete. In addition, if the silver electrode is used, the voltage remains constant at 2 to 5 dc volts. The current of the cell varied between 8 and 25 milliamps according to the distance from the electrode to the plate surface. The variation in current appears to have no effect on either the rate or the effectiveness of cleaning. In a few instances slightly higher voltage was necessary to remove unusually resistant corrosion. However, lower voltages are recommended. As in all conservation treatments, use only the amount of energy that will accomplish cleaning. Some experimenters have used platinum electrodes for electrocleaning daguerreotypes; however, these offer no real advantage over silver electrodes, and the cost per electrode is considerably higher.

Essentially, electrocleaning uses the applied voltage to form silver oxide thin films several atomic layers thick at the corrosion-plate interface when the daguerreotype is in the cathode position. The silver oxide film is extremely unstable because of its thinness, its solubility in the ammonium hydroxide solution, and the high pH (pH = 12) of the electrolyte solution. The instability of the oxide layer helps loosen and break up the thicker tarnish layer. When the polarity of the cell is reversed, silver ions at the plate surface are reduced to silver, and the tarnish layer is lifted from the plate. The tarnish layers come off in sheets or disappear into solution depending upon the daguerreotype treated and the extent of its corrosion. Cleaning is continued by switching back and forth from anodic to cathodic phases until there is no longer any visible removal of tarnish or improvement in image appearance.

A certain amount of microscopic improvement in the daguerreotype plate surface occurs as a result of slight electropolishing caused by the electrocleaning (fig. 11.12). Electropolishing is related to anodic phase cleaning and is used to produce bright metal surfaces without mechanically working the surface. Brightening observed on daguerreotype plates is due to very slight elimination of surface microroughness and microirregularities on the order of 0.1 to 0.01 micrometers. Anodic film formation on a metal surface helps promote the uniform removal of unwanted material by the electrolyte; thus, the surface is evenly "polished." Electropolished surfaces have unique properties in that they are very smooth and have few voids. As a result, they also are more corrosion resistant than mechanically polished or etched surfaces. The brightening effect is seen visually as an increase in the specular reflectance of the metal surface.[24] This increase in specular reflectance improves the appearance of the electrocleaned daguerreotypes by increasing their contrast range since the maximum

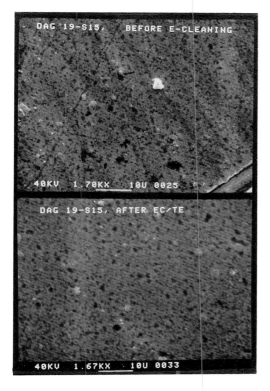

types cleaned by this method show no recognizable alterations in either the image structure or the plate surface at the level of scanning electron microscopic examination. With conventional silver dip cleaners, the effects are quite marked when examinations are carried out in the same way. Unfortunately, electrocleaning can never restore a daguerreotype to its original pristine condition. The etch patterns characteristic of silver dip cleaners can never be removed. Figure 11.13 shows a section of plate that had at some time in the past been heavily etched by cyanide treatment. Electrocleaning brings out the etch pattern in exquisite detail but does not help the deeply corroded grain boundaries heal. At best only a slight healing of the plate surface occurs, and it is not sufficient to alter the overriding optical behavior of the damaged plate and image particle microstructure.

On a more cheerful note, figures 11.14 and 11.15 give comparisons between uncleaned and cleaned daguerreotypes. In fig-

Fig. 11.12. Scanning electron micrograph of daguerreotype surface before and after electrocleaning. Scale bars are equal to 10 micrometers.

Fig. 11.13. Scanning electron micrograph of daguerreotype surface severely cleaned by cyanide and then electrocleaned. The cyanide solutions etched away the silver in grain boundary regions, leaving the grains standing in relief. The grain boundaries were covered with later corrosion products that were duly removed by the electrocleaning procedure, thus revealing the damaged surface in fine detail. The scale bar is equal to 10 micrometers.

amount of blackness observed on daguerreotypes is directly related to the specularity of their surfaces.

Electrocleaning does not affect the image particle microstructure in any way. Indeed, there is some evidence that an electrocleaned daguerreotype may closely resemble what a daguerreotype looked like when it was first made, before it had any corrosion films or had been subjected to any kind of solvent cleaning treatment. The brightening seen visually after electrocleaning is a combination of two effects: primarily, the removal of tarnish films and, secondarily, a slight amount of electropolishing. Since the original microstructure is usually unknown, the relative proportions of these two effects cannot be quantified. However, the electropolishing takes the daguerreotype plate a bit closer to its original freshly polished condition. Modern daguerreo-

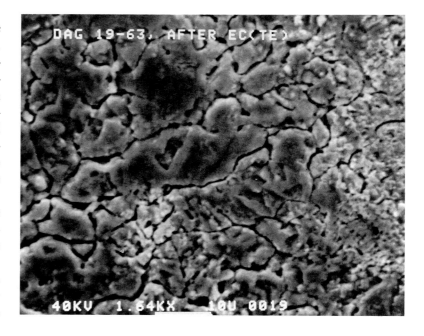

Fig. 11.14. Partially cleaned daguerreotype. The right side has been electrocleaned. The left side is uncleaned.

ure 11.14 only the right half of the daguerreotype was electrocleaned, with immense improvement in detail in the woman's portrait.

Having established that electrocleaning was definitely superior to sputter cleaning and any of the chemical dip methods previously used, there remained the question of whether the authors could effect a technology transfer and recommend electrocleaning for general use. We decided that out-of-laboratory testing by a working conservator was vital to the development of electrocleaning as a useful methodology. Thus, daguerreotypes were given the usual preparatory examinations and sent to a private conservation shop for an application of the electrocleaning treatment.[25] The conservator had seen the method in the laboratory setting and was given written guidelines and the daguerreotype holder for use in field testing. All other equipment necessary for the field testing was put together on site. Close to 40 daguerreotypes were cleaned in the first group of tests. When the work was completed, the daguerreotypes were brought back to the laboratory and given postcleaning analyses. Using the observations and experience gained in field testing, more daguerreotypes were treated, and the process took on the shape in which it has now been presented. The field testing was deemed a success. Of about 120 daguerreotypes cleaned and evaluated, the tarnish could be removed from most with substantial improvement in the appearance of the object.

All the experimenting with cleaning methodology reinforced a long-known fact: daguerreotypes were and are extremely delicate objects. Immersing them in solutions of any kind must be done with care, and certain procedures must be avoided. In the water-washing stage of the electrocleaning process, the daguerreotype should only be soaked and rocked gently from time to time so that fresh water

Fig. 11.15. Comparison of electrocleaned daguerreotype with its pretreatment condition. Cleaning by Tom Edmondson, summer 1988. Courtesy of Paul Katz.

Fig. 11.16. Daguerreotype with filiform delamination across the central portion of the image caused by treatment in an ultrasonic cleaner with water. This daguerreotype was previously heavily corroded, and most of the image was occluded by tarnish. The tarnish was cleaned away using a silver wand electrode and the electrocleaning regime described in the text.

reaches the plate surface to carry away water-soluble tarnish products. Warm water, not hot water, may be used for this step. Under no circumstances should ultrasonic agitation be used for this portion of the cleaning or for any other treatment of daguerreotypes. It was found that daguerreotypes subjected to ultrasonic treatment can develop unpredictable filiform delamination of the plate surface, an example of which is shown in figure 11.16. The filamentary lines moving across the face of the daguerreotype were formed during ultrasonic cleaning in water and are the result of delamination of the silver layers of the plate. No remedy has been found for this problem.

Ungilded daguerreotypes cannot be electrocleaned. Without the strengthening of the image particles in the gilding step, the microstructure will not withstand electrocleaning.

Daguerreotypes with extensive hand-applied color should not be placed in solutions because of the risk of losing the pigments. Daguerreotypes of this type that need cleaning should probably be sputter cleaned, as there is less risk that color will be removed by sputtering. A colored daguerreotype should be checked using an optical microscope for the presence of pigment. Often daguerreotypes that appear slightly colored (e.g., natural skin tones in portraits or blue skies in landscapes) are colored because of optical effects of the image microstructure and not because of applied color.[26] When this is the case, electrocleaning will not harm or remove the color. Inherent coloration of this type is caused by image particle spacing and size and will not be altered in the cleaning process.

WHEN FRANÇOIS ARAGO announced the discovery of the daguerreotype in August 1839, he noted that these images were fragile and needed to be either varnished or encased in order to be preserved. The most common misconception about the daguerreotype image is that it is and remains so fragile that the image can be wiped off the plate surface with the slightest touch. While the daguerreotype image is initially fragile, the images become less fragile over time either because of age hardening of the silver image particles or because of gilding. However, while the image itself may gain mechanical stability, the silver of the daguerreotype plate remains a relatively soft metal vulnerable to scratches and corrosion. A daguerreotype must be protected from dust and damage caused by handling. Thus, from the outset, these images were always encased or enclosed in some sort of package that was protective and yet allowed the daguerreotype to be displayed.

Daguerre mentioned that he had used varnishes on daguerreotypes, and the earliest daguerreotype manuals also suggest various varnishes, such as copal, damar, wax, and India rubber, that could be applied to a daguerreotype surface to protect the image. However, Daguerre also pointed out that all these coatings weakened the image by diminishing the whites and altering the depth of the blacks. Further, these varnishes caused the image to disappear in a short time, presumably because of the aging of the varnish. None of these coatings was widely used, and daguerreotypes that were varnished after they were made are rare. However, one example is the Joseph Saxton daguerreotype of the Central High School in Philadelphia, which was varnished with manilla copal.[2]

Daguerreotypes were usually placed in a protective package that consisted of the daguerreotype, a paper or metal mat (to hold the cover glass away from the daguerreotype surface), and a cover glass, all of which

12

Daguerreotype Preservation and Display

A daguerreotype is permanent; the picture cannot be rubbed off the plate; the plate cannot be broken by accident; the picture will bear microscopic investigation. Other processes have their merits, but this is the triumph of the photographic art, and a boon to science.

Thomas Sutton and George Dawson, *A Dictionary of Photography*[1]

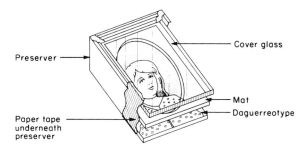

Fig. 12.1. Sketch showing the component parts of the daguerreotype package.

Fig. 12.2. A daguerreotype case.

were bound together with a paper tape (fig. 12.1). This package was then placed in a frame or a small case that could be closed. Various kinds of frames were used for daguerreotypes from plain wooden ones to elegant, engraved passe-partout frames made especially for the daguerreotype trade.

The small cases used for daguerreotypes were the same type of cases that had been used for painted miniatures on ivory.[3] Initially, casemakers who were already making cases for jewelry, instruments, and other small items began making cases for daguerreotypes. Soon after the introduction of photography, some casemakers began to specialize in daguerreotype case manufacture and sold their goods directly to daguerreotypists. Cases of American manufacture are still common in collections and

in the antique trade. By the early 1850s the demand for cases was so great that wholesale photographic supply houses such as Scovill in Connecticut and E. and H. T. Anthony in New York began large-scale manufacture of these items.[4]

These small cases were made of wood or pasteboard boxes sheathed with a decorative covering made of embossed leather, cloth, or paper. The inside lid of the case usually had a pad covered with silk or velvet to protect the cover glass. Often an inner frame of embossed soft brass called a preserver was folded over the paper tape seal on the daguerreotype package and was used to help snug the daguerreotype into the box of the case. The outside of the case usually had some sort of clasp (fig. 12.2). Cases were also made of other materials such as papier-mâché with inlays of mother-of-pearl and stone; some had painted decorations, and some were covered with imitation leathers, imitation tortoise shell, and velvet or plush. Daguerreotype cases were one of the first items to be manufactured from thermoplastic materials. These plastic cases, called union cases, were manufactured by a few American companies after the early 1850s and became a part of the foreign trade in daguerreotype and photographic goods.[5] Daguerreotypes were also set in lockets, broaches, and all types of jewelry, as well as incorporated into pieces of furniture and architecture, even memorial stones on graves (fig. 12.3).

The mats used in the daguerreotype package were sometimes made of embossed paper or thin paper card. More often they were made of stamped brass and came with variously shaped central windows. Initially, brass mats were relatively simple and of heavy brass. The overall surface finish of the mats varied, and the name and address of the daguerreotypist may have been stamped into the metal. As time went by, mats became more and more elaborate. In the later part of the daguerreian era, cheap mats with fancy overall patterns were produced of very thin stamped brass foil.

The simplicity of the traditional daguerreotype package with its paper tape seal was deceptively effective. Recipes for making the paper tapes were given in daguerreotype manuals. These usually consisted of spreading an adhesive of gum arabic, gum tragacanth, isinglass, and a preservative such as benzoin on heavy tissue paper or writing paper. This prepared paper was then cut into strips when it was needed to seal a daguerreotype.[6] These seals acted both as a corrosion barrier and as protection for the daguerreotype against mechanical damage. While the appearance of the materials used for sealing daguerreotypes became more elaborate over the years, no real improvements were made to the actual methods of sealing daguerreotypes during the nineteenth century.

Packaging of contemporary daguerreotypes and repackaging of historic daguerreotypes has usually followed nineteenth-century tradition. Paper tapes can be used to seal the package. If at all possible, metal mats should be used or retained from a previous package. The brass mat may need to be smoothed if it is scratching the plate surface. The brass mat has no chemical effect on the daguerreotype and will not set up a galvanic couple with the plate that will initiate new corrosion. Corrosion at the mat edge is related to the geometry of the daguerreotype package itself and not to the brass used to make the mat. When buffered board is used as a mat material within the daguerreotype package, the buffering materials themselves eventually etch the plate. Alkaline buffered board or paper should not be used on daguerreotypes. A modern daguerreotype was submitted for cleaning because of a tarnish ring that had formed at the mat edge and was moving across the plate surface. Analysis of the corrosion ring showed that it was made up of calcium and magnesium compounds derived from the

Fig. 12.3. Lockets with daguerreotypes. Courtesy of M. Susan Barger.

calcium and magnesium carbonate buffers used in the board. Other instances of such papers being used under mats corroborate these findings. If a paper mat must be used, it should be made of neutral paper or board.

Cover Glasses and Reglazing

As it turns out, the most critical part of the daguerreian packaging system is the cover glass. The early daguerreotypists chose their sources of glazing primarily on the basis of the color and clarity of the glass.[7] The preferred glass was white French plate glass. Glass that was cloudy or blistered was to be avoided even at the price of choosing a glass that was off-color. The stability of the glasses against long-term chemical degradation was not a primary concern. The authors collected several dozen samples of cover glasses, spread them out on a table, and examined them under ultraviolet light. Most of these old glasses were fluorescent because of impurities in the raw materials from which they were made, and the samples glowed mostly with yellow luminescence of various hues and intensity. Here and there bright blue emission appeared. Later chemical analysis showed that the blue fluorescing glasses were lead glasses, whereas the yellow fluorescing glasses were variations of the basic soda-lime-silica composition generally used for plate glass. From the variation in chemical composition the authors concluded that early daguerreotypists used many different sources for the cover glass. All the nineteenth-century glasses tended to be unstable in the presence of any moisture that might condense within the daguerreotype package at any time in its history. As was discussed in much more detail in chapter 10, thin moisture films would leach alkali metals from the glass, corroding the glass and forming a variety of corrosion products. These flake off and fall onto the daguerreotype surface, where they mask the image and may cause further chemical damage.

Many varieties of plate glasses are now available. These are more complicated compositions than the old soda-lime-silica basic formula. Usually aluminum and other elements are added to tailor the glass for high resistance to moisture and corrosion by aqueous solutions so that fairly ordinary modern plate glasses have a vastly better corrosion resistance than old glasses. Likewise, quality control is much better, and manufacturers' specifications are more reliable. Contemporary daguerreotypes should be glazed with high-resistance glasses, and an argument can be made for reglazing historical daguerreotypes with new glass. Replacing the cover glass every twenty to thirty years prevents the buildup of corrosion products and allows one to take advantage of new glass technology as it may develop.

There have been general discussions about the use of specialty glasses for glazing daguerreotypes. These glasses include those with ultraviolet filtering or antiglare coatings and those with very low color. Since

daguerreotypes, unlike many other works of art, are not harmed by ultraviolet radiation and since they must be specially lighted to avoid glare when they are displayed, the use of coated glasses is an unnecessary expense. The slight greenish cast of many modern plate glasses will not be noticed by most viewers when daguerreotypes are displayed. In some situations a low-color glass may offer some aesthetic advantage in display and be chosen for that reason. Acrylic plastic glazing materials such as Lucite and Plexiglas have also been suggested for use on daguerreotypes. These materials offer no advantage over glass and may enhance corrosion because of their high gas permeability.

Reglazing historical daguerreotypes is a procedure of some controversy, and it turns on this point: Is the daguerreotype package part of the historical object that is to be preserved, or does it serve only to protect the real object—the daguerreotype plate itself? If the entire package is deemed to be of historic value, the only cleaning allowable is to open the package, brush the corrosion products from the cover glass, and reseal the package. The cover glass should not be washed, because washing will remove much of the hydrated surface layer of the glass, exposing the underlying pristine glass to renewed attack by moisture. On the other hand, if the cover glass can be sacrificed, much better protection for the daguerreotype itself can be provided by simply replacing the old glass with a carefully selected modern one.

Protective Coatings

Early experiments in protecting the daguerreotype image by coating it directly did not lead to an acceptable technology. Although some daguerreotypes were varnished or lacquered—for example, the Joseph Saxton daguerreotype—the consensus seems to be that there was too much loss of image quality, and the procedure was abandoned. Times have changed. Can modern technology be called upon to provide coatings that will protect daguerreotypes without the loss of image quality?

Protection of the daguerreotype image by coating the plate with a protective material sounds like an attractive procedure at first, but to date no coating procedure has appeared that can be given an unqualified recommendation. The difficulty has both a technical component and an ethical component. Technically, it has not proven possible to coat daguerreotype surfaces with a layer that provides protection to the image particles but at the same time does not disturb their optical properties. The refractive index of any coating material, regardless of its composition, will always be much greater than unity (one). Thus, the scattering contrast between the image particles and the surrounding medium will be reduced, and the image will appear with reduced apparent contrast, as Daguerre noted long ago. The ethical component arises because most coating procedures are irreversible. Once applied, most coatings cannot be removed without damage to the image structure beneath. Because image particles stand out in relief above the silver plate, they will be embedded in the coating material and thus easily plucked off when the coating is removed.

Despite the dismal history of previous efforts to produce a completely successful protective coating, there may be occasions when a carefully chosen protective coating would be desirable for the preservation of a daguerreotype. The prime function of such a coating should be to prevent corrosion, particularly that associated with the repeated cleaning and corrosion cycle. There is also the possibility that a coating might provide some enhancement of the optical properties and thus improve the apparent contrast of an overcleaned daguerreotype.

Table 12.1. Optical Properties of Coating Materials

Chemical Composition	Refractive Index (visible light)	Band Gap (eV)
Silica (SiO_2)	1.4	8 (300 K)
Titania (TiO_2)	2.75–2.5	3.12 (300 K)
Aluminum nitride (AlN)	2.2	6.3 (5 K)
Boron nitride (BN)	2.42	6–8 (300 K)
Lanthanum hexaboride (LaB_6)	———	metallic[a]
Parylene C[b]	1.64	unknown

[a] Plasma edge at 600 nanometers
[b] Polymonochloroparaxylylene

The authors' experiments with coatings included sputtered coatings of the insulating materials silica (SiO_2), titania (TiO_2), boron nitride (BN), and aluminum nitride (AlN), sputtered coatings of the metallic material lanthanum hexaboride (LaB_6), and vapor-deposited coatings of parylene C, a polymeric film.[8] These were chosen for their optical properties (see table 12.1), their chemical inertness, and their resistance to corrosion. Fifteen nineteenth-century daguerreotypes from the working collection (see chapter 8, note 3) were chosen to represent the commonly found variations in daguerreotypes caused by past handling, storage, and aging. Each was arbitrarily cut into three portions. One portion was left uncoated, one received a sputtered coating, and one was coated with parylene C.

The sputtered coatings were laid down in a radio frequency sputtering unit in the Materials Research Laboratory at The Pennsylvania State University (fig. 12.4).[9] Each daguerreotype was placed in the sputtering chamber without any special preparation. After the chamber was pumped down to a base pressure of 3×10^{-7} torr,

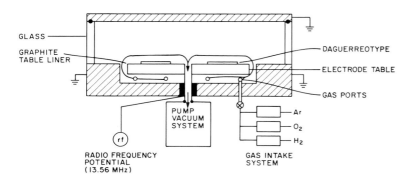

Fig. 12.4. Sketch showing main components of a sputtering apparatus. Not to scale.

various gases were injected to form the plasma. Table 12.1 gives the specific conditions for the various experiments. Substantial control over the composition, thickness, and physical structure of the coatings could be obtained by varying the plasma gas, the radio-frequency power, and the deposition time.

Parylene C is an unusual plastic film material used extensively in industry as a protective coating for electronic and biomedical components. Its principal advantages are nontoxicity, very low water and gas permeability, high strength, insolubility in all common solvents, transparency in the visible region, and ability to form adherent, pinhole-free conformal coatings. Parylene C films are highly crystalline and are produced by polymerization of a diradical species, di-monochloroparaxylylene, directly onto a surface. The diradical is obtained by vaporization of the solid dimer and pyrolysis of the vapor at high temperatures and low pressures. The highly reactive gaseous diradical will deposit directly from the vapor and will polymerize on any cool surface. Because deposition occurs at very low pressures, films formed by this process are uniform and continuous and follow the contours of the substrate. Parylene C films can be removed from surfaces with orthodichlorobenzene. The solvent has no effect on silver, the image microstructure, or the assorted corrosion products found on daguerreotype surfaces. The absorption spectrum of Parylene C is featureless from about 300 to 2000 nanometers.

Of the five sputtered coatings, two—silica and boron nitride—produced coatings of interest. Titania, aluminum nitride, and lanthanum hexaboride produced coatings that so modified the appearance of the daguerreotype that they were not deemed worthy of further investigation.

Lanthanum hexaboride is a very hard metallic material with a deep lavender metallic color. It was tested because in very thin layers it is transparent in visible light and because it provided an opportunity to examine metal-on-metal coatings. Lanthanum hexaboride produced a metallic, absorbing film that gave the daguerreotypes the appearance of detailed tintypes. In addition, there was poor adhesion between the lanthanum hexaboride and the underlying silver.

The failure of titania coatings was due mainly to nonstoichiometry. Stoichiometric titania is a colorless insulator with a high refractive index. Under the conditions in the sputtering chamber TiO_2 was reduced to TiO_{2-x} where x varied between 0.1 and 0.4. The reduced films were strongly colored and tended to overwhelm the image. Further, the coating structure was inconsistent and had poor adhesion to the daguerreotype surface.

The coatings of aluminum nitride were very consistent, and the film structure closely mimicked the underlying daguerreotype. Unfortunately, all these coatings were bright pink because of nonstoichiom-

Fig. 12.5. Total (diffuse plus specular) reflectance for silica coating, a parylene coating, and an uncoated portion of sample Dag 19-60. All reflectance measurements were made on a uniform piece of the image background.

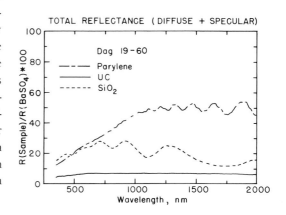

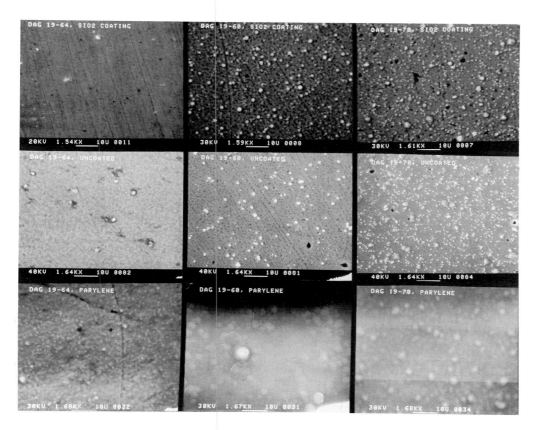

Fig. 12.6. Scanning electron micrographs of both coated and uncoated portions of Dag 19-64, Dag 19-60, and Dag 19-70. The top row of micrographs shows the microstructure of silica coatings. The bottom row shows parylene coatings. Uncoated portions of the coated daguerreotypes are shown in the middle row. Each vertical column shows the data for one daguerreotype. The scale bar in each micrograph is equal to 10 micrometers.

etry of the film. The aluminum nitride coating maintained the specularity characteristic of the daguerreotype image; in some cases the apparent contrast of the image was increased by the aluminum nitride coating.

More successful results were obtained with silica, which sputtered to form uniform coatings. These had a pinkish to greenish cast due to interference effects, but the coloration was not pronounced. The films retained the specular, metallic appearance of uncoated daguerreotypes. A typical set of reflectance spectra illustrating the difference between the uncoated and silica-coated portions of one of the sample daguerreotypes is shown in figure 12.5. Ideally, the visible spectrum of a coated daguerreotype should match that of an uncoated sample. Mismatches indicate that the daguerreotype's optical properties have been altered because of the coating. The degree of matching is directly related to the appearance of the coated daguerreotype. The short-dash line in figure 12.5 shows the typical reflectance spectrum of a silica film on a daguerreotype surface. The regular undulations in the spectrum are due to interference fringes. Generally, a silica film on

a daguerreotype surface is slightly more absorbing than an uncoated daguerreotype.

The silica coatings also have a shimmery appearance that comes about because of their film structure. Silica tends to form sheets of small spheres whose diameters are less than 1 micrometer. The micrographs in the top row of figure 12.6 show the general sphere structure of these coatings and the changes in film thickness. The silica film shown for sample Dag 19-64 does not show the spherical structure because of the deposition conditions.[10] The thickest film has the highest spherical silica particle density. This structure, typical of silica, is what gives rise to the fire in gem opals. Figure 12.7 shows individual silica spheres at two high magnifications. Even though these silica spheres are about the same size as daguerreotype image particles, they are transparent and nonmetallic; thus, their presence does not seriously interfere with the light scattering from the image. On the contrary, these films tend to slightly enhance the apparent contrast of the image. The spherelike structure does not provide a continuous coating for the daguerreotype. Spaces between the spheres are likely places for the

Fig. 12.7. Scanning electron micrographs of individual silica spheres that make up the microstructure of the silica coatings at two different magnifications. Scale is shown by the bars in each micrograph—1 micrometer in the upper photo and 0.5 micrometers in the lower photo.

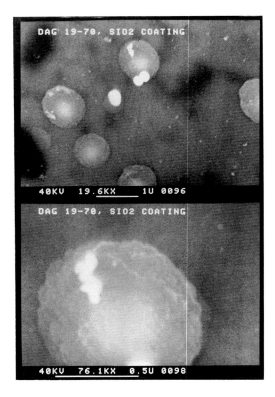

initiation of corrosion. This effect is demonstrated in accelerated aging tests.

Boron nitride forms a consistently dense, continuous film that closely follows the underlying daguerreotype surface, as seen in the first row of micrographs in figure 12.8. Thicker films tend to be the most dense. These coatings compare favorably with the uncoated portions of the sample daguerreotypes (middle row of fig. 12.8). Although boron nitride is usually formed with a soft, graphitelike hexagonal structure and is susceptible to moisture attack, proper control of the sputtering conditions produces a very hard, diamondlike phase that, not surprisingly, has many of the properties of diamond: optical transparency, resistance to chemical attack, and high hardness.[11]

Freshly deposited boron nitride coatings have a gold or rose-gold cast. As the coatings age, this coloration tends to dissipate, and the films become colorless. This change in appearance usually occurs in a matter of weeks, and the exact reason for this change

has not been determined. Thicker films may not become colorless for a few months. A typical spectrum for boron nitride coatings is shown by the short-dash line in figure 12.9. This spectrum closely matches that of the uncoated sample. All the boron nitride spectra follow the curve shape of the uncoated daguerreotypes, but some of these coatings were from 5 to 10 percent more absorbent than the uncoated daguerreotypes. The visual appearance of these films, disregarding the coloration, is very close to that of the uncoated samples. The more absorbing films cause some enhancement of the apparent contrast of the daguerreotype.

One portion of each of the fifteen daguerreotype samples used in this study was coated with Parylene C.[12] This coating is transparent and colorless, but many of the Parylene C–coated samples have noticeable interference fringes. However, their occurrence and severity may be controlled by altering the coating thickness. Unlike the other coating materials, Parylene C may be removed using orthodichlorobenzene.

The structure of the Parylene C films varied from sample to sample. Some coatings have many clumps of unreacted dimer material distributed over the film surface, as seen in the upper micrograph of figure 12.10. A few of the Parylene C films were totally structureless and appear to be draped over the image microstructure, as shown in the lower micrograph of figure 12.10. The micrographs show the extremes of Parylene C film structure. Other Parylene C films are shown in the bottom rows of micrographs in figures 12.6 and 12.8. Some of the micrographs appear indistinct because of interactions between the nonconducting polymer and the electron beam of the scanning electron microscope. These films were not gold-coated before scanning electron microscope examination because the gold would have interfered with any subsequent visual examination of the coatings.

Parylene C films are more absorbent than

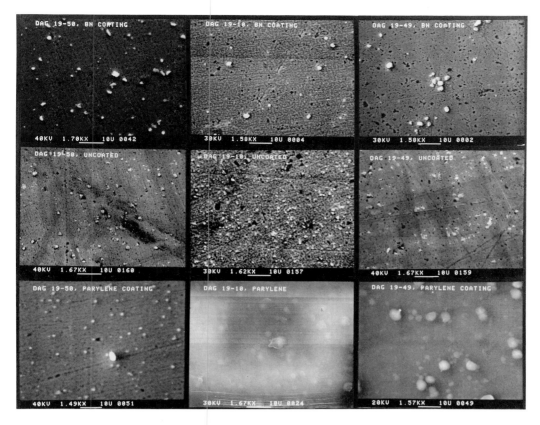

Fig. **12.8.** Scanning electron micrographs of both coated and uncoated portions of Dag 19-50, Dag 19-10, and Dag 19-49. The top row of micrographs shows the boron nitride coatings. Uncoated portions of the daguerreotype are shown in the middle row. The bottom row shows the parylene coatings. Each column shows the data for one daguerreotype. The scale bar in each micrograph is equal to 10 micrometers.

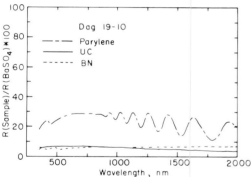

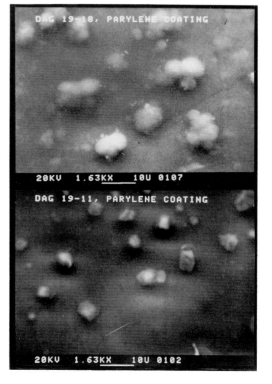

Fig. **12.9.** Total (diffuse plus specular) reflectance for a boron nitride coating, a parylene coating, and an uncoated portion of sample Dag 19-10. All reflectance measurements were made on a uniform piece of image background.

Fig. **12.10.** Scanning electron micrographs showing the extreme variations in parylene film structure. The upper micrograph (Dag 19-18) shows crystalline clumps embedded in the polymer coating. The lower micrograph (Dag 19-11) shows the conformal, structureless polymer film. The scale bar in each micrograph is equal to 10 micrometers.

the other coatings, primarily because they are 5 to 40 times as thick as the sputtered films. The long- and short-dash lines in the spectra of figures 12.5 and 12.9 show typical reflectance data for these films. The undulations in the curves are due to interference fringes. All the Parylene C films have similar spectra, although some coatings are less absorbing than others because of variation in film thickness among the samples. Most of the Parylene C films reduce the specularity of the daguerreotype and cause the image to appear dulled or matte. The image remains visible, but there is a loss of contrast and detail. In some cases the image tone was shifted from a warm to a cold cast. In the most severe cases the image contrast appears almost reversed (positive to negative). Thus, while the image is still visible, to a large extent the characteristic image quality associated with daguerreotypes is lost.

A preliminary test of the ability of boron nitride, silica, and Parylene C films to provide protection against corrosion of daguerreotypes was carried out using sulfur-containing gases. These tests were carried out in sulfur gas atmospheres because the kinetics of sulfur attack on silver are much faster than those of oxygen attack. Thus, in a relatively short time these tests gave a very good qualitative clue as to the protective characteristics of the coatings, particularly with regard to mechanical completeness or freedom from pinholes and flakes. These tests are not very effective for assessing the resistance of the films to corrosion initiated by diffusion of gases through the film. They also give very little information about chemical attack on the film itself. Since it is expected that daguerreotypes will always be sealed, these last two points are of minor interest compared with questions of the integrity of the films.

Samples of unimaged daguerreotype plates coated with boron nitride and silica were subjected to gaseous hydrogen sulfide (vapor over a saturated aqueous K_2S_x solution) and sulfur dioxide (vapor of a 0.5 M $NaHSO_3$ solution) in separate sealed beakers over a 72-hour period. A Parylene C–coated daguerreotype was exposed only to the hydrogen sulfide atmosphere. Uncoated areas on the samples served as controls. The concentrations of hydrogen sulfide and sulfur dioxide were not measured, but these exposure conditions are orders of magnitude more severe than would be expected during the normal storage of daguerreotypes. Within one hour the uncoated controls showed evidence of corrosion. At the end of the test most of the boron nitride–coated plates and the Parylene C–coated sample were unaffected, but the silica films did not prove protective.

The mechanical adhesion of the boron nitride and Parylene C coatings to the daguerreotype surface was tested using an ASTM test.[13] A razor blade was used to score the coating; a piece of transparent tape was placed over the scored area and burnished using the eraser of a pencil; finally, the tape was pulled off the surface. The scored area was inspected under a light microscope for removal or peeling of the coating. The tests for both the boron nitride and Parylene C films showed either no peeling or removal (classification 3B) or only trace peeling or removal along the scored line in the film (classification 4B). This is good evidence of these films' strong adhesion to the daguerreotype surface.

Overall the judicious use of protective coatings on certain daguerreotypes can be a useful approach to their long-term preservation. Coatings do offer a way to break the harmful cycle of repeated corrosion and cleaning that has been used to treat so many daguerreotypes. Of the two coatings that offered the best protective advantage, Parylene C modified the optical properties of the daguerreotypes and darkened the image; however, it can be removed by solvent treatment. Sputtered boron nitride coatings best preserve the optical characteristics but produce a rose-gold color that fades through time and can in no way be regarded as removable coatings.

Storage and Display Environment

In spite of the mechanical weaknesses of daguerreotypes, which requires them to be handled, displayed, and stored in protective cases, the daguerreotype image is remarkably robust to the long-term threats of aging and reaction within the storage environment. The only real enemy is moisture, which is a threat more to the cover glass than to the daguerreotype itself. Daguerreotypes in museum collections, gallery collections, and private collections should be stored under dry conditions— that is, in a comfortable humidity that is well below the dew point for any expected range of temperature fluctuations. Humidity is harmful only if it reaches values that permit the condensation of liquid water films within the daguerreotype package.

Temperature, as such, is not a problem within the normal range of room ambients and below. The components that make up the daguerreian system are not temperature-sensitive, and only at temperatures well above those that would be tolerated in a human habitat would diffusion of silver and perhaps traces of residual mercury begin to modify the geometry of the image particles.

Daguerreotypes do not fade. No bleaching reaction occurs, and display lighting conditions should not cause any damage. Cased daguerreotypes are automatically stored in the dark, but there is no reason why daguerreotypes should not be left in display situations even in the presence of strong ambient lighting. Of course, lighting has much to do with the viewing of the image. The daguerreotype image will not be seen well in bright direct lighting, but such lighting should not cause physical harm. Silver sulfide formation is enhanced in light, all other things being equal, so that if contaminants from the atmosphere are present, their degradation of the silver surface can be enhanced by photochemical reactions, and this consequence should be taken into account.

Lighting can be modified to optimize the effective contrast of the daguerreian image. Optimum lighting is extremely important for display,[14] but it can also be used to enhance the images of daguerreotypes that have been damaged to the point where the image has become indistinct under ordinary lighting.[15] The latter is applicable to historically or artistically important daguerreotypes whose image content can sometimes be preserved by making copies using conventional photographic methods.

For most photographs the image seen by an observer is the result of varying degrees of light absorption by a silver (or possibly a dye) microstructure. Optimal display conditions for these materials are not critical as long as the illumination is sufficient and the image surface has no glare. Preservation factors such as light fading do not affect display conditions for daguerreotypes. The arrangement of the observer, the daguerreotype, and the illumination source determines whether the viewer sees a positive or negative image as well as the contrast range of the image seen by the viewer. Thus, the viewing geometry as well as the state of the microstructure of the daguerreotype surface itself are important to the display.

Viewing and illumination geometry is also important for copying daguerreotypes by conventional photographic methods. The illustrations used in this book were obtained using the arrangement shown in figure 12.11. The daguerreotype is illuminated by two lamps angled so that no specular reflection reaches the camera. The camera itself is mounted on a copy stand and views the daguerreotype through a hole cut into a piece of stiff, flat black mat board, which should be twice or three times the width of the daguerreotype. Its purpose is to prevent secondary illumination in the vicinity of the camera from striking the daguerreotype at an angle that will permit it to be specularly reflected into the camera. The authors have experimented with other arrangements, including attempts to photograph the daguerreotypes in their negative image phase, but the daguerreotype image was not en-

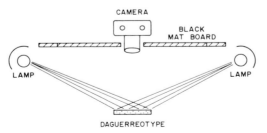

Fig. 12.11. Sketch showing arrangement for photographing daguerreotypes.

hanced more than in the arrangement described above.

Contrast range is the maximum variation of tones from black to white seen in a photograph. The contrast range of daguerreotypes changes with the viewing geometry. As long as the angle of view is not equal to the angle of incident illumination, the daguerreotype will appear as a positive. However, as the value of these two angles becomes equal, the daguerreotype will appear to fade or become less positive. Where the two angles are equal, the daguerreotype appears as a negative. The highlights of a daguerreotype always appear with the same relative brightness because their dominant imaging mode is diffuse reflectance from the image particles. Light scatter from a diffusely reflecting surface is not highly dependent on the angle of view, whereas the dominant imaging mode of the shadow areas is specular reflectance. The relative brightness of these areas changes dramatically with the angle of view.

Thus, in order to best display a daguerreotype, the illumination source should be placed so that the angle of illumination and the angle of view will not be equal for most viewers. Each daguerreotype should be spaced so that it is illuminated by only one source and not by overlapping illumination from neighboring daguerreotypes. Preventing the illumination from one daguerreotype from falling on its neighbors is even better and can be easily accomplished by using a shade or housing on the lamps. Displaying the daguerreotypes in a dark field is also helpful. Both these precautions will help minimize any indirect illumination of the plate surface from light reflected by surrounding objects. A clever way of accomplishing these objectives is the use of fiber optic lighting to place the illumination exactly where it is needed.[16] It is also a very nice blend of the oldest photographic technology with the most modern optical technology. However, small spotlights work very well, are much cheaper, have better color balance, and are more easily obtained than fiber optic systems. Fiber optics do reduce heat production, if that is a concern.

On a more sophisticated level, improvements in the efficiency of light scatter from the image particles will result in the perception of enhanced image contrast. Such improvements can be made by matching the dominant illumination wavelength to the image particle size distribution curve. This matching may be accomplished by choosing either a narrow-band illuminant that peaks near the scattering curve maximum or a broad-band illuminant that covers the whole scattering curve distribution. In either case certain aspects of the human visual system need to be considered in addition to the optical properties of the daguerreian system. If the direction of illumination is placed so that it is parallel to the polish lines on the plate, the image will also appear somewhat better to the viewer.

In general, the daguerreotype scattering curve covers the entire visible spectrum. If the daguerreotype is in pristine condition, the scattering curve maximum is in the green region of the spectrum (i.e., near 550 nanometers). However, if the daguerreian microstructure has been disturbed by mechanical or chemical means, the scattering curve maximum is usually shifted toward the red and near-infrared regions of the spectrum. The vast majority of nineteenth-century daguerreotypes have microstructural damage caused by tarnish removal treatments, evident in a shift in the scattering curve maximum. These data indicate that the best narrow-band illuminant colors would be green, yellow, or red.

While green or red monochrome illumination may be aesthetically unpleasing for display purposes, yellow lighting is a good choice for a variety of reasons.[17] Much of the lighting that is commonly used has a distinctly yellow cast, and people today are generally adapted to distinguishing colors in yellow light. In addition, the maximum sensitivity of the human visual system is in the yellow-green region of the spectrum. In most cases a low-temperature tungsten bulb used with a light-yellow filter (something like a Kodak safelight filter 00) would not only enhance the contrast of a daguerreotype on display but also provide a relatively low-cost and readily available lighting combination. Further, this type of illumination would not noticeably interfere with the viewer's perceptions of colored daguerreotypes. A few precautions need to be taken when a narrow-band illumination source is used for exhibition purposes.

When a contrasting pattern—for example, a daguerreotype illuminated using a narrow band source—is viewed in a dark field the viewer may experience persistent after-images that are the color complement of the pattern once the viewing conditions are changed. This is called the McCollough effect and has been well studied.[18] Thus, when daguerreotypes are seen in a display illuminated by a narrow-band source, viewers may experience varying degrees of color-reversed after-images when they leave the display area. This effect may be lessened in the following ways: (1) by separating individual daguerreotype displays with another visual activity (e.g., a "normally" lighted wall label); (2) by painting the display area the color complement of the illumination source output (e.g., if a yellow illuminant is used, then the walls or surrounding should be painted a dark blue or a blue-gray); and (3) by controlling the exit from the display area so that the viewer does not go directly into another display area.

The effect of illumination by monochromatic light is most striking in the examination of damaged daguerreotypes. The authors tried a number of sources throughout the visible spectrum including a sodium-vapor lamp (yellow, 589 nanometers), and

a mercury-vapor lamp (filtered to blue, 435.8 nanometers). We also examined a daguerreotype illuminated with a helium-neon laser (632.8 nanometers) using a lens to spread the laser beam. The idea was that the scattering of coherent radiation from the laser might cause some additional enhancement. However, the effect was similar to incoherent red light. Daguerreotypes viewed in blue light are very difficult to see because of the characteristic absorption of blue light by silver. The sodium-vapor lamp was quite effective and was the most extensively investigated. We made use of image enhancement in this context to obtain good copies of the daguerreotypes by conventional photography.

The contrast in a damaged daguerreotype also appears to increase when the daguerreotype is viewed at oblique angles from normal to the image plane. This increase can be easily observed by placing the daguerreotype in a rotating holder and illuminating it with one stationary light source. As the daguerreotype is turned about the axis of rotation, the viewer can see the apparent density of the image change. By selecting a monochromatic light source and an optimum viewing angle, the apparent contrast can be maximized. In addition, since any conventional photographic copy of a daguerreotype will increase the contrast of the image by eliminating its mirrorlike behavior, the overall effect of image enhancement can be quite substantial. If the damaged daguerreotype is photographed on conventional black-and-white film, an image-enhanced copy of the daguerreotype will be produced.

The photographs shown in figure 12.12 demonstrate the effectiveness of this procedure. Figure 12.12a shows a badly scratched daguerreotype viewed perpendicular to the image plane in white light. It should be noted that the image in this daguerreotype is barely visible, and even the conventional copy improves its appearance.

a.

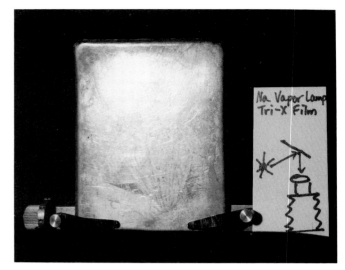

b.

c.

d.

Fig. 12.12. Image en-
hancement of a surface-
damaged daguerreotype.
Figure *a* shows a badly
scratched daguerreotype
photographed in white
light. Figures *b*, *c*, and *d*
show the same daguerreo-
type photographed using
sodium-vapor light and
different viewing geome-
tries. Figure *d* shows the
best image enhancement.

Figures 12.12b, 12.12c, and 12.12d show the same daguerreotype viewed at different angles in the monochromatic light from a sodium-vapor lamp. The positions of the camera, light source, and the daguerreotype are shown in the schematic drawing next to the daguerreotype in each photograph. Any kind of panchromatic black-and-white film may be used for this procedure. A fine-grained, moderate contrast developer—for example, D-76—is recommended for processing of the copy negative. The photographs obviously show that in figure 12.12d the appearance of the copy is visibly improved over the appearance of the daguerreotype copy in figure 12.12a.

A special effect appears when observing severely damaged daguerreotypes that also have a tarnish layer. The tarnish can act as an optical coating, a thin film that modifies the optical properties of a surface. Antireflection coatings on lenses are an example. Other coatings create interference effects or modify the reflection characteristics of the coated surface. Some tarnish layers obscure the specularity of the daguerreotype plate, making the viewing geometry less critical for seeing the image. For some daguerreotypes in this condition, the presence of a tarnish layer may make it possible to see any sort of image at all. If a light source of the proper wavelength is used, the tarnish layer will be transparent, and the image can be copied through the coating. Previous examination of the optical properties of silver sulfide and silver oxide tarnish have shown that if the tarnish layer is fairly thin, a monochromatic light source in the yellow or red regions of the spectrum might be used to look through the layer. However, if the tarnish is very thick, the "window" in the tarnish layer will be in the infrared. In that case an infrared source and infrared film could possibly be used to copy the image, although this idea has not been tested.

13

Daguerreotypy: A Model for the Interpretation of Early Technology and Works of Art

This subject may be of but little interest to the experienced and scientific artist: such persons have a theory of their own. It is however, a matter of very great importance that *beginners* should start right. We would not advise them to be tied down to the foregoing views; but we do most earnestly urge upon them the importance of their possessing themselves of intelligent ideas of the profession they have chosen. A man with no correct knowledge of the steam engine, who should attempt to construct and set in operation a locomotive, would soon run off the track. I never knew a skillful daguerrian artist who worked at random; but among the many unskillful operators whom I have met, I never knew one who understood the first principles of the business.

Levi L. Hill, *A Treatise on Daguerreotype*[1]

THIS WAS LEVI HILL'S ADVICE to those who would take up the daguerreotype process. The authors would like to commend his sentiments to those wishing to take up interdisciplinary studies aimed at the elucidation of early technological processes or the science behind art and archaeological objects, their manufacture, preservation, and decay. While this study has centered on the daguerreotype process, some measure of its lasting importance will be derived from the fact that it can be used as a paradigm for the fruitful application of science to problems in the history of science and technology and to the conservation and preservation of art and artifacts.

The study described here in its complete form for the first time has been extremely prolific and is the most extensive and successful scientific study ever undertaken on the daguerreotype process. At the outset the authors were primarily interested in determining the best way to care for daguerreotypes. However, to accomplish that task we needed to determine what a daguerreotype is, how it is formed, and how it ages. Experimental findings raised new questions, and the curiosity to find answers to these questions led to further experimental work. The daguerreotype itself was our guide to the direction of the experimental work. Our task was to learn to find and decipher the code carried in each daguerreotype that tells us how it was made and how it has changed over the years. In effect, we were looking for a Rosetta Stone. As the first part of this book attests, we were not the first scientists to look at this process; we used and built on the observations of those investigators who went before us. We believe, however, that we have been the most successful, largely because of our experimental point of view.

We are materials scientists. Our outlook is centered on materials as they exist as whole systems and not as they ought to exist as ideal systems. This means that we use the tools of the chemist, the physicist, the biologist, the geologist, and the engineer as they are needed to describe the structure and properties of materials. Thus, ours is an applied science. The so-called materials revolution has aroused new interest in our field and put it in the limelight as one of the burgeoning new professions. There is presently great activity—experimenters are discovering new materials and finding new methods for making materials that require less energy in forming, that are more efficient, or that are less polluting.

The same tools that are used to discover new materials and improve known materials can also be profitably applied to understanding the process of aging and the preservation of art and archaeological materials. These tools can also be employed to understand problems in the history of technology, especially with regard to elucidating materials processing from earlier times. The materials approach is a liberating method for these problems. The material or object is the focus of the work, and the work can be measured by the effectiveness in matching what is actually there. Solving problems by doing armchair chemistry or measuring older technologies against current technologies is easy. What ought to be is often easier to think about or model than what actually is.

The problems associated with the art and artifacts of history need to be approached on several fronts. Often materials of earlier technologies are shrouded in myths derived from secondary documents and hearsay rather than from data derived from the careful materials characterization of these materials. Thus, it is desirable to start with actual examples of the material in question and examine as many samples as needed to determine what the usual state of the material is, what its state is when first manufactured, and then what happens to it during the normal course of aging. It is also essential to actually make materials in the

Fig. 13.1. The moon, daguerreotyped by the Lunar Daguerreian Society, June 18, 1986, using the 24-inch reflector telescope at the United States Naval Observatory. Source: Collection of M. Susan Barger. Courtesy of M. Susan Barger, Jan Kenneth Herman, Kenneth E. Nelson, and Theodore Rafferty.

laboratory and become experienced with how they are processed. This work gives a depth of knowledge that cannot be gained in any other way. Contemporary written records need to be studied to know what those actually involved in the making of the materials in question observed. These earlier observations can help resolve quandaries brought about in laboratory production of the materials or can give clues as to what experimental paths are useful to explore.

The daguerreotype story is a peculiar one. There was no pressing need for the technology that photography presented. Why would anyone want to take photographs under awkward and time-consuming conditions when one could hire an artist who would prepare a skillfully arranged illustration and delete superfluous detail? It was only after Niépce and Daguerre made their discovery that people realized the implications of the new technology, and its use spread like wildfire.

This phenomenon has been repeated throughout history. The most amazing modern analogy is the computer. In the 1940s, when the computer was being designed and invented by Alan Turing, John von Neumann, and others, there was a real debate as to whether there was any practical point in spending the large sums of money that would be needed to develop the computational machine. At one time there was an estimate that six to ten computers would satisfy the computing needs of the entire United States. Once the technology was in place, its implications were apparent to everyone, and the computer revolution spread like wildfire. The daguerreotype demonstrated a new way of viewing the world, just as in modern times the computer has demonstrated a new means of managing information.

One must distinguish between two types of technological innovation. The first type is that driven by an imminent need. A little earlier in the industrial revolution than the invention of the daguerreotype, a need for a power source better than animal power for driving ships and pumping mines was perfectly obvious. Steam engine technology was indeed driven by technological need. The second type of technological innovation in foresight seems to be largely unnecessary. Once the invention is realized, the implications are suddenly boundless, as can be seen in tracing the development of photography and its implications through the nineteenth century and into our own time. Leaving aside all the other uses to which photography has been applied, modern science would be unrecognizable without it. Modern photography is not the technological descendent of daguerreotype, as has been shown in this book. But it is very much the intellectual descendent of daguerreotypy because the daguerreotype demonstrated the implications and pointed the way.

Notes

Chapter 1: Beginnings

1. Antoine Claudet, "The Progress and Present State of the Daguerreotype Art," *Journal of the Franklin Institute*, 40 (1845): 49.

2. Richard Rudisill, *Mirror Image: The Influence of the Daguerreotype on American Society* (Albuquerque: University of New Mexico Press, 1971), 198.

Chapter 2: Image Making before Photography

1. Anthony van Leeuwenhoek to Henry Oldenburg, first secretary of the Royal Society, March 26, 1675. Quoted in Clifford Dobell, *Anthony van Leeuwenhoek and His "Little Animals"* (New York: Dover Publications, 1960), 342.

2. *Transactions of the Society of Arts,* 34 (1817): xi.

3. Government organizations and premium societies, such as the Society for the Encouragement of Arts, Manufacture, and Commerce in England and its companion society in France, the Société d'Encouragement des Arts et Métiers, offered prizes and premiums for improvements and innovations in all areas of human endeavor—from industrial and scientific concerns to artistic works. The various premium societies published yearly lists of the prizes that would be offered in the coming year. Each listing for these prizes included a description of the innovation being sought and any constraints or rules placed on each particular prize. The premium system was very successful, particularly in the area of manufactures. Some people used the prize listings as incentives to begin work on specific problems or in specific areas. The fact that medals were offered provided an additional encouragement because these endorsements could be used to promote new products.

4. This was the less formal name of the Society for the Encouragement of Arts, Manufacture, and Commerce. It was also referred to as the Premium Society.

5. Thomas Allason, in *Transactions of the Society of Arts*, 145–47.

6. Ibid., 147.

7. The terms *silhouette* and *profile* refer to two different types of portraits. By convention, a silhouette is merely an outline of the sitter's profile

with no or minimal interior detailing of facial features. Silhouettes are often black cutouts or hollow-cuts. A profile includes drawn or painted facial and other details within the outline of the sitter's profile.

8. Rudolf Kingslake, *A History of the Photographic Lens* (Boston: Academic Press, 1989), 23–26.

9. Basil Hall, *Forty Etchings, from Sketches Made with the Camera Lucida in North America in 1827 and 1828* (Edinburgh: Cadell & Co., 1829), memorandum.

10. Ibid.

Chapter 3: Toward Points de Vue *and the Daguerreotype*

1. "Pictorial delineations by light: solar, lunar, stellar, and artificial called Photogenic and the art of Photography," *American Journal of Science* 37 (1839): 169.

2. Marie Boas Hall, *All Scientists Now: The Royal Society in the Nineteenth Century* (Cambridge: Cambridge University Press, 1984).

3. Phlogiston was thought to be a substance that exists in all combustible materials and that is disengaged or released during the process of combustion. It was an important chemical concept introduced at the beginning of the eighteenth century and maintained until the death of Joseph Priestly in 1802. Although Priestly discovered oxygen in 1774, he did not recognize that it was the substance responsible for combustion. Antoine Lavoisier made that connection and in so doing sounded the death knell for phlogiston.

4. François Arago, *Biographies of Distinguished Scientific Men*, trans. W. H. Smyth, Baden Powell, and Robert Grant, 2d series (Boston: Tricknor & Fields, 1859).

5. Line spectra do not give any information about the physics of light; rather, the lines are related to transitions between various energy levels in the atoms making up the compound analyzed. The resolution of what gives rise to the lines observed in line spectra had to wait until after the introduction of quantum mechanics and the work of Niels Bohr.

6. Talbot to John Herschel, March 27, 1833, Royal Society manuscript RS MS HS 17.270.

7. For an extensive discussion of early spectral analysis, see Frank A. J. L. James, "The Debate on the Nature of the Absorption of Light, 1830–1835: A Core-Set Analysis," *History of Science*, 21 (1983): 335–68; and M. A. Sutton, "Spectroscopy and the Chemists: A Neglected Opportunity?" *Ambix*, 23 (1976): 16–26.

8. There are conflicting reports as to when Wedgewood gave up his experimentation because of ill health. John Werge noted without documenting his source that Wedgewood had given up his activities as early as 1792 (*Evolution of Photography with a Chronological Record of Discoveries, Inventions, Etc.* [London: Piper and Carter, 1890], 9). Wedgewood's biographer, R. B. Litchfield, places the date later, in 1799 (*Tom Wedgewood, the First Photographer* [Covent Garden: Duckworth and Co., 1903]).

9. Thomas Wedgewood and Humphrey Davy, "An account of a method of copying painting on glass and of making profiles by the agency of light on nitrate of silver, invented by T. Wedgewood, Esq., with observations by H. Davy," *Journal of the Royal Institution (London)* 1 (1802): 170.

10. J. M. Eder, *History of Photography*, trans. Edward Epstean (New York: Dover Publications, 1972), 272–73.

11. Isidore Niépce, *Historique de la Découverte Improprement Nommé Daguerreotype, procédé d'une notice sur son véritable inventeur feu M. Joseph-Nicéphore Niépce de Chalon-sur-Saône* (Paris: Astier, Libraire, 1841), 21.

12. Nicéphore Niépce to F. Lemaitre, February 2, 1827, in *Nicéphore Niépce, lettres et documents choisis par Paul Jay* (Paris: Centre National de la Photographie, 1983), 96–97.

13. These items were shown several times during the nineteenth century but were thought lost since 1898. In 1952, through the efforts of Helmut and Alison Gernsheim, they were found in a trunk that had belonged to the Bauer family and had been in storage since 1917.

14. Initially, Daguerre also used strong salt solutions to stop the action of light on his plates.

15. Robert Lassam, *Fox Talbot, Photographer* (Tisbury, Wiltshire: The Compton Press, 1979), 16.

16. Mungo Ponton, "Notice of a cheap and Simple Method of Preparing Paper for Photographic Drawing, in which the Use of any Salt of Silver is Dispensed with," *Edinburgh New Philosophical Journal* 27 (1839): 169–71.

17. *Comptes Rendus* 8 (1839): 838.

18. Gail Buckland, *Fox Talbot and the Invention of Photography* (Boston: David R. Godine, 1980), 53–54. The same letter is quoted by Beaumont Newhall in *The History of Photography from 1839 to the Present* (New York: Museum of Modern Art, 1982) as "Excuse this ebullition!" (p. 23).

Chapter 4: The Technological Practice of Daguerreotypy

1. Quoted in Theodore F. Marburg, "Management Problems and Procedures of a Manufacturing Enterprise, 1802–1852; A Case Study of the Origin of the Scovill Manufacturing Company" (Ph.D. diss., Clark University, 1945), 246. The "Prof. Morse" referred to is Samuel F. B. Morse.

2. Henry Hunt Snelling, *The History and Practice of the Art of Photography* (New York: G. P. Putnam, 1849; reprint, Hastings on Hudson, N.Y.: Morgan and Morgan, 1970), 24.

3. See, for instance, Albert Bisbee, *The History and Practice of Daguerreotyping* (Dayton, Ohio: L. F. Claflin & Co., 1853; reprint, New York: Arno Press, 1973), 26–27.

4. For a discussion of the dating of this plate, see William F. Stapp et al., *Portraits from the Dawn of Photography* (Washington, D.C.: Smithsonian Institution Press, 1983), 47.

5. *American Journal of Photography* 6, no. 14 (1864): 335–36.

6. "Daguerreotype Miniatures," *National Gazette,* June 26, 1840, p. 2.

7. Jabez Hughes, *Photographic Journal* 10 (1863): 417–19.

8. John Frederick Goddard, "Valuable Improvements in Daguerreotype," *Literary Gazette,* no. 1,247 (1840): 803.

9. *Wiener Zeitung,* no. 19 (January 19, 1841): 139–40.

10. Josef Berres, *Wiener Zeitung,* no. 83 (March 24, 1841): 610.

11. *Comptes Rendus* 2 (1841): 1059–60.

12. F. A. P. Barnard is the astronomer for whom Barnard's Star is named and also the educator after whom Barnard College is named.

13. F. A. P. Barnard, *American Journal of Science,* 41 (1841): 353–54.

14. Dr. John W. Draper of the University of the City of New York worked actively on the daguerreotype process for many years. He made numerous valuable contributions to daguerreian theory and to photochemistry in general, but he also confused many historical issues by his fervor in establishing his priority for various discoveries. Many histories of photography cite Draper as having produced the first portrait from life made by the daguerreotype process—the daguerreotype of Dorothy Draper, his sister. This daguerreotype was taken in the spring of 1840, about the time Wolcott and Johnson were setting up their studio and months after many portraits had been made in Philadelphia and elsewhere. Draper sent his daguerreotype to Sir John Herschel with a letter about how he had made the first portrait from life. It may have been the first daguerreotype portrait seen in England but certainly not in the United States. This supposed priority was mentioned by Draper in all his writings. His stature as a scientist meant that many historians have referred to his writings on photography, and those not familiar with other events in photography at the time have repeated this false statement.

15. Editorial Department, *American Journal of Photography* 6 (1864): 335–36.

16. W. H. Goode, *American Journal of Science* 4 (1840): 137–44.

17. Hippolyte Fizeau, *Comptes Rendus* 12 (1841): 1189–90.

18. ———, "Photographie: Nouvelles découvertes de M. Daguerre," *Comptes Rendus* 12 (1841): 1228–29.

19. François Arago, *Comptes Rendus* 13 (1841): 23.

20. [François Arago and Louis Jacques Mandé Daguerre], "Photographie," *Comptes Rendus* 13 (1841): 26–27.

21. *Atheneaum,* no. 692 (1841): 95–96.

22. *American Philosophical Society Proceedings* 2 (1842): 144.

23. *American Philosophical Society Proceedings* 3 (1843): 180.

24. For a discussion of the background of the Cornelius-Goddard history, see Marian S. Carson, "The Eclipse and Rediscovery of Robert Cornelius," in Stapp et al, *Robert Cornelius: Portraits from the Dawn of Photography,* 13–24.

25. Fizeau, "Note sur un moyen de fixer les images photographiques," *Comptes Rendus* 11 (1840): 237–38.

26. United States Patent no. 9354 assigned to Charles L'Homdieu, October 26, 1852.

27. For a detailed description of French hand-coloring techniques and materials, see Alice Swan, "Coloriage des Épreuves: French Methods and Materials for Coloring Daguerreotypes," in Janet E. Buerger, *French Daguerreotypes* (Chicago: The University of Chicago Press, 1989), 150–63.

28. *Chemical Gazette* 1 (1843): 334–35.

29. United States Patent no. 2,826, granted October 22, 1842, to John Plumbe, Jr., assignee of Daniel Davis, Jr., Boston, Mass.

30. United States Patent no. 3,085, granted to Montgomery P. Simons, assignee of Warren Thompson, Philadelphia, Pa., May 12, 1843.

31. Edmond Becquerel, "Rémarques sur quelques points de la théories des radiations, en réponse à une Lettre de M. E. Becquerel, lué à la denière séance, et inserée au *Compte rendu*; par M. Biot," *Comptes Rendus* 9 (1839): 719–26.

32. ———, *Comptes Rendus* 11 (1840): 702–4.

33. ———, "De L'Image Photographique Colorée du Spectre Solaire," *Annales de Chimie et de Physique* 22 (1848): 451–59; Becquerel, "De l'Image Photochromatique du Spectre Solaire," *Annales de Chimie et de Physique* 25 (1849): 447–74; Becquerel, "Nouvelles Récherches sur les Impressions Colorées Produites lors de l'Action Chimique de la Lumière," *Annales de Chimie et de Physique* 42 (1854): 81–106.

34. Claude Felix Abel Niépce de Saint-Victor, "Heliochrome," *Photographic Art-Journal* 2 (July 1851): 32–37.

35. See Josef Maria Eder, *History of Photography* (New York: Dover, 1972), 664–66; and Beaumont Newhall, *History of Photography* (New York: The Museum of Modern Art, 1982), 267–68.

36. Levi L. Hill, *A Treatise on Heliochromy; or The Production of Pictures by means of Light, in Natural Colors, embracing a full, plain, and unreserved description of the process known as the Hillotype* (New York: Robinson & Caswell, 1856; reprint, State College, Pa.: The Carnation Press, 1972), 26.

37. Samuel D. Humphrey, "Hillotypes," *The Daguerreian Journal* 1 (1851): 209–11.

38. Senate Report no. 427, 32nd Congress, 2nd Session; U.S. Serial Set no. 671 (1853).

39. Joseph Boudreau, "Color Daguerreotypes: Hillotypes Recreated," in Eugene Ostroff, ed., *Pioneers of Photography: Their Achievements in Science and Technology* (Springfield, Va.: The Society for Imaging Science and Technology, 1987), 189–99.

40. For this reason daguerreotypes made by one daguerreotypist show up in collections from other artists's studios. For example, copies of Southworth and Hawes portraits of famous Americans are now part of the Mathew Brady Studio Set of daguerreotypes held by the Library of Congress.

41. See William F. Stapp, "Daguerreotypes onto Stone: The Life and Work of Francis D'Avignon," in Wendy Wick Reaves, ed. *American Portrait Prints: Proceedings of the Tenth Annual American Print Conference* (Charlottesville: University Press of Virginia, 1984), 194–231.

42. Alfred Donné, "Transformation en planches gravées des images formées par le procédé Daguerre," *Comptes 'Rendus* 9 (1839): 485–86.

43. *Atheneaum*, no. 656 (1840): 418–19.

44. William R. Grove, "On the Voltaic Process for Etching Daguerreotype Plates," *Philosophical Magazine* 20 (1842): 18–24.

45. See Noel Marie Paymal Lerebours, *A Treatise on Photography*, trans. J. Egerton (London: Longman, Brown, Green, and Longmans, 1843; reprint, New York: Arno Press, 1973), 117–25.

46. John W. Draper, "Note on the Tithonotype," *Philosophical Magazine* 23 (1843) 175–76.

47. *Philosophical Magazine* 37 (1841): 202.

48. Marcus A. Root, *The Camera and the Pencil; or The Heliographic Art* (Philadelphia: M. A. Root, 1864; reprint, Pawlet, Vt.: Helios, 1971), 358–59.

49. Louis Jacques Mandé Daguerre, *An Historical Account and Descriptive Account of the Various Processes of the Daguerreotype and the Diorama* (London: McLean, 1839; reprint, New York: Kraus Reprint Co., 1969), 1.

50. On December 30, 1839, J. M. Lamson Scovill wrote to his brother that American buyers of their plate in New York City wanted to know the quantity of silver on their plates as was indicated on French plates. Evidently the Scovill agent in New York wanted to know the weight of silver per copper so that the plates for sale in New York could be stamped with numbers like those used by the French. See Marburg, 249.

51. See Marburg, 281–2; 336–7.

52. Unpublished letter, Joseph Pennell to Al-

bert Southworth, January 25, 1848, International Museum of Photography at George Eastman House, Rochester, N.Y.

53. The word *doublé* in a French plate mark merely indicates that the silver is plated; it does not indicate the method of plating.

54. Charles Christofel, *Histoire de la d'orure et de l'argenture éléctrochemiques* (Paris: D'E. Duverger, 1851).

55. M. Lyons and W. Milward, "Bright Silver Deposition," British Patent 1847/11632.

56. For an example, see *American Journal of Pharmacy* 7 (1842): 102–4.

57. The word *kelp* is now used as a general term for seaweeds; however, it originally meant the ashes left from burning specific seaweeds found off the coasts of the British Isles and Norway.

58. Archibald and Nora L. Clow, "The Natural and Economic History of Kelp," *Annals of Science* 5 (1947): 297–316.

59. David Alter, "Bromine," *The Daguerreian Journal* 2 (1851): 345.

60. United States Patent no. 5,658, July 5, 1848, to David Alter and Edward Gillespie; United States Patent no. 62,464, February 26, 1867, to David Alter.

61. *The Daguerreian Journal* 2 (1851): 345.

62. "Bromine," *American Journal of Pharmacy* 5 (1875): 69–71. See also Edwin T. Freedley, *Philadelphia and Its Manufacturers; A Handbook of the Great Manufactories and Representative Mercantile Houses of Philadelphia in 1867* (Philadelphia: Edward Young and Co., 1867).

Chapter 5: Scientific Interest in the Daguerreotype during the Daguerreian Era

1. William Henry Fox Talbot, remarks in "Report of the Proceedings of the Ninth Meeting of the British Association for the Advancement of Science," *American Journal of Science* 37 (1839): 97–100.

2. François Arago, "Considerations relative to the Chemical Action of Light," *Scientific Memoirs* 3 (1843): 558–63, 558.

3. John William Draper, "Experiments on Solar Light," *Journal of the Franklin Institute* 19 (1837): 469–79; 20 (1837): 38–46; 114–25; 250–53.

4. See note 14, chapter 4.

5. Draper, "On the Process of Daguerreotype, and its application to taking Portraits from the Life," *Philosophical Magazine* 17 (1840): 217–25.

6. ———, "An account of Experiments made in the South of Virginia, on the light of the Sun," *Philosophical Magazine* 16 (1840): 81–84.

7. This came to be known as *chemical focus*. It is the result of using lenses made of nonachromatic glass so that light of differing wavelengths is dispersed and comes to a focus at different points along the central axis of the lens. Draper was not the first to point this out; see, for instance, John T. Towson, "On the Proper focus for the Daguerreotype," *Philosophical Magazine* 15 (1839): 381–85. For more modern discussions of the problems associated with the differences in chemical and visual focus, see Gunter D. Roth, *The Amateur Astronomer and His Telescope*, trans. Alex Helm (Princeton, N.J.: D. Van Nostrand Company, 1963), 20–21, or George Z. Dimitroff and James G. Baker, *Telescopes and Accessories*, (Philadelphia: The Blakiston Company: 1945), 88–91.

8. John Herschel, "On the Chemical Action of the Rays of the Solar Spectrum on Preparations of Silver and other Substances, both Metallic and Nonmetallic, and on some Photographic Processes," *Philosophical Transactions* 130 (1840): 1–60.

9. For an example, see Herschel, "New Researches on the Solar Spectrum and in Photography," *Philosophical Magazine* 16 (1840): 239; 331–33.

10. The Herschel effect is a bleaching of an image caused by a rearrangement of image silver. In this case the image is broken down by the adsorption of silver atoms onto ions and electrons in locations other than where the image silver was initially formed.

11. Jean-Baptiste Biot, "Sur le nouveau procédés pour étudier la radiation solaire tant direct que diffuse," *Comptes Rendus* 8 (1839): 259–72.

12. In spite of this work with chlorine and bromine, Becquerel did not establish a method for using these halogens to shorten camera exposure times in the routine processing of daguerreotypes.

13. Edmond Becquerel, "Sur les éffects électriques que se produitsent sous l'influence solaire," *Comptes Rendus* 9 (1839): 711–13.

14. ———, *Comptes Rendus* 11 (1840): 702–4.

15. The Becquerel effect is now thought to be the result of weak spectral sensitization by image silver in contact with silver halide in the rest of the light-sensitive layer; however, the exact mechanism is not fully understood.

16. Alfred Donné, "Sur ce qui se passe pendent les diverses parties de l'opération," *Comptes Rendus* 9 (1839): 376–78.

17. Golfier Bessyre, "Sur les phénomènes produits par daguerreotype," *Archives des Découvertes* (1839): 135–36.

18. Augustus Waller, "Memoire sur la Photochemie, 1ᵉ partie," *Comptes Rendus* 11 (1840): 568.

19. Donné, *Cours de Microscopie complèmentaire des Études Médicales, Anatomie Microscopique et Physiologie des Fluids de L'Économie. Atlas éxécute d'après natur au microscope-daguerreotype Par Al. Donné et Léon Foucault* (Paris: Chez J.-B. Bailliere, 1845).

20. ———, "Mikroskopische Daguerre-Bilder," *Annalen der Physik* 57 (1842): 176.

21. Robert Hunt, *A Popular Treatise on the Art of Photography, including the Daguerreotype and all of the new methods of producing pictures by the chemical agency of light* (Glasgow: Richard Griffin and Company, 1841).

22. Ibid., 63–64.

23. Draper, "On Some Analogies Between the Phenomena of the Chemical Rays and those of Radiant Heat," *Philosophical Magazine* 19 (1841): 195–210.

24. Theodore Freiherr von Grotthuss made the same observation as Draper in 1817. His report of this observation did not receive a great deal of attention, and Draper's independent formulation was the first time this law was widely promulgated. This law was called the Draper Law until the end of the nineteenth century, when Josef Maria Eder drew attention to the previous work of Grotthuss. The name was thus amended to reflect the work of both men.

25. Draper, "Analogies Between Chemical Rays and Radiant Heat," 195.

26. Friedreich Wilhelm Bessel, "On a very Curious Fact Connected with Photography, Discovered by M. Moser of Koenigsberg," *Philosophical Magazine* 21 (1842): 409–11.

27. "Eleventh Meeting of the British Association for the Advancement of Science. Section A.

Mathematical and Physical Science," *Athenaeum* 770 (1842): 687.

28. Ludwig Moser, "On Vision and the Action of Light on all Bodies," *Scientific Memoirs* 3 (1843): 422–61; Moser, "Some Remarks on Invisible Light," *Scientific Memoirs* 3 (1843): 461–64; Moser, "On the Power which Light Possesses of Becoming Latent," *Scientific Memoirs* 3 (1843): 465–72; Moser, "Sur les images que se formant sur la surface d'une glace ou de tout autre corps poli, et réproduisent les contours d'un corps placé très-près de cette surface, amis sans contact immédiate," *Annales de Chimie* 7 (1843): 237–39.

29. Draper, "On certain Spectral Appearances, and in the discovery of Latent Light," *Philosophical Magazine* 21 (1842): 348–50.

30. ———, "On a new Imponderable Substance, and on a Class of Chemical Rays analogous to the Rays of Dark Heat," *Philosophical Magazine* 21 (1842): 453–61.

31. Herschel, "On the Action of the Rays of the Solar Spectrum on the Daguerreotype Plate, *Philosophical Magazine* 22 (1843): 120–32.

32. Ibid., 121.

33. Hunt's latent light is probably what is now called fluorescence.

34. Hunt, "On Thermography, or the Art of Copying Engravings, or any printed Characters from Paper on Metal Plates; and on the recent Discoveries of Moser, relative to the formation of Images in the Dark," *Philosophical Magazine* 21 (1842): 462–68.

35. ———, "Some Experiments and Remarks on the Changes which Bodies are capable of undergoing in Darkness, and on the Agent producing the changes," *Philosophical Magazine* 17 (1843): 270–79.

36. Horatio Prater, "Experiments and Observations on Moser's Discovery," *Athenaeum* (1843): 485; 598.

37. Hippolyte Fizeau, "Sur les causes que concourent à la production des images de Moser," *Annales de Chimie* 7 (1843): 240–41.

38. Ibid., 359–60.

39. Hunt, "On the Spectral Images of M. Moser; A Reply to his Animadversions, etc.," *Philosophical Magazine* 23 (1843): 415–26.

40. Draper, "On Mr. Hunt's Book entitled 'Researches on Light,'" *Philosophical Magazine* 25 (1844): 49–51.

41. Ibid., 51.

42. Arago, "Phosphorescence du sulfate de bar-yte calcine, communication de M. Arago sur quelques expériences de M. Daguerre," *Comptes Rendus* 8 (1839): 243–46.

43. Becquerel, "Memoir on the Constitution of the Solar Spectrum, presented to the Academy of Sciences at the Meeting of the 13th of June 1842," *Scientific Memoirs* 3 (1843): 537–57.

44. Nathan Reingold et al., *The Papers of Joseph Henry* (Washington: D.C.): Smithsonian Institution Press, 1985), 5: 337–39. See also Joseph Henry, "On Phosphorogenic Emanation," *Proceedings of the American Philosophical Society* 3 (1843): 42.

45. Hunt, "Reply to Prof. Draper's Letter, in the preceding number of the *Philosophical Magazine* on a work entitled 'Researches on Light,'" *Philosophical Magazine* 25 (1844): 119–22.

46. Ibid., 122.

47. There appears to be no published biographical information about either of these men.

48. Charles Choiselat and Stanislas Ratel, "Sur une manière d'envisager les phenoménes du daguerreotype," *Comptes Rendus* 16 (1843): 1436–39; Choiselat and Ratel, "De l'action des substances accélératrices dans les opérations du daguerreotype," *Comptes Rendus* 17 (1843): 173–77.

49. Waller, "Memoire sur la Photochemie," *Comptes Rendus* 11 (1840): 568.

50. ———, "Observations on certain Molecular Actions of Crystalline Particles &c; and on the Cause of the Fixation of Mercurial Vapours in the Daguerreotype Process," *Philosophical Magazine* 18 (1846): 94–105.

51. Léon Foucault and Hippolyte Fizeau, "L'action des rayons rouges sur les plaques daguerriennes," *Comptes Rendus* 23 (1846): 679–82.

52. Becquerel, "Observations sur les expérience de MM. Foucault et Fizeau, relatives a l'action des rayons sur les plaques daguerriennes," *Comptes Rendus* 23 (1846): 800–804.

53. Antoine Claudet, "Researches on the Theory of the Principal Phenomena of Photography in the Daguerreotype Process," *Daguerreian Journal* 1 (1850): 33–38. This article also appears in *Photographic Art Journal* 2 (1851): 37–40.

Chapter 6: The Daguerreotype as a Scientific Tool

1. Ludwig Moser, "On Vision and the Action of Light on All Bodies," *Scientific Memoirs* 3 (1843): 423–24.

2. For a discussion of the relation of photography to seventeenth-century Dutch art, see Svetlana Alpers, *The Art of Describing: Dutch Art in the Seventeenth Century* (Chicago: The University of Chicago Press, 1983), especially chapter 2.

3. In recent years there has been a great deal of discussion on both sides of the question as to whether photographs are works of art or merely documents and which fall into each category. For assessments of the aesthetic and cultural implications of this question, see Susan Sontag, *On Photography* (New York: Farrar, Straus and Giroux, 1977); Joel Snyder, "Picturing Vision," *Critical Inquiry* 6 (1980): 499–526; Roger Scruton, "Photography and Representation," *Critical Inquiry* 7 (1981): 577–603; John Tagg, *The Burden of Representation: Essays on Photographies and Histories* (Amherst: University of Massachusetts Press, 1988).

4. *Excursions daguerriennes: vues et monuments les plus rémarquables du globe*, ed. N. P. Lerebours (Paris: Author, 1841–43). This series of 111 plates engraved after daguerreotypes by various artists was published in parts of four plates each. Two plates were engraved directly from daguerreotypes etched by Hippolyte Fizeau.

5. When Edward Anthony returned from the northeastern border survey, he changed his profession, first becoming a daguerreotypist and later, along with his brother, founding E. and H. T. Anthony Company, a major supplier of daguerreotype and photographic apparatus and equipment.

6. U.S. Serial Set 420 (27–3), H. Doc 31.

7. See Harold Francis Pfister, *Facing the Light: Historic American Portrait Daguerreotypes* (Washington, D.C.: Smithsonian Institution Press, 1978), 184–91.

8. Charles Preuss, *Exploring with Frémont: The Private Diaries of Charles Preuss, Cartographer for John C. Frémont on His First, Second, and Fourth Expeditions to the Far West*, trans. and ed. Erwin G. and Elisabeth K. Gudde (Norman: University of Oklahoma Press, 1958), 32.

9. Ibid., 35.

10. Solomon N. Carvahlo, *Incidents of Travel and Adventure in the Far West; With Col. Frémont's Last Expedition* (New York: Derby & Jackson, 1857), 32–33.

11. U.S. Serial Set 705, Senate Misc. Doc. 67 (33–1): 3.

12. Robert V. Hine, *In the Shadow of Frémont; Edward Kern and the Art of Exploration, 1845–1860* (Norman: University of Oklahoma Press, 1981); originally published as *Edward Kern and American Expansion* (New Haven: Yale University Press, 1962).

13. U.S. Serial Set 1068, H. Rpt. 208 (36–1):1.

14. John Lloyd Stephens, *Incidents of Travel in Yucatan* (Norman: University of Oklahoma Press, 1962) 1: 65–68.

15. Ibid., 1: 117.

16. Ibid., 1: 117.

17. For an indication of the ridicule heaped on Gall and Spurzheim's ideas in Scotland before Combe's conversion, see the anonymous review "The Doctrines of Gall and Spurzheim," *Edinburgh Review* 25 (1815): 227–68.

18. For more detailed discussion of phrenology in Edinburgh, see the series of essays "Phrenology in Early Nineteenth-Century Edinburgh: An Historiographical Discussion," *Annals of Science* 32 (1975): 195–256.

19. Marmaduke B. Sampson, *Rationale of Crime and Its Appropriate Treatment, being a Treatise on Criminal Jurisprudence considered in Relation to Cerebral Organization* (New York: D. Appleton and Co., 1846).

20. Charles Dickens, *American Notes for General Circulation* (London: Chapman and Hall, 1842).

21. Samuel G. Morton, *Crania Americana* (Philadelphia: J. Dobson, 1839), and *Crania Aegyptica* (Philadelphia: John Penington, 1844).

22. Edward Lurie, *Nature and the American Mind: Louis Agassiz and the Culture of Science* (New York: Science Publications, 1974).

23. For a complete discussion of Agassiz and Morton and their roles in the promotion of the ideas of polygeny, see Stephen Jay Gould, *The Mismeasure of Man* (New York: W. W. Norton & Co., 1981).

24. Elinor Reichlin, "Faces of Slavery," *American Heritage Magazine* 28 (June 1977): 4–11.

25. George Clark and Frederick H. Kasten, *History of Staining,* 3rd ed. (Baltimore: Williams and Wilkins, 1983).

26. François Arago, *Oeuvres Complètes*, t. 7: 458, 1858.

27. This was not a problem for reflector telescopes. In a reflector telescope all light comes to a focus at the same spot because the focusing and shaping of incoming light is accomplished using curved mirrors so that radiation of all wavelengths converges at the focal plane.

28. Minutes, New York Lyceum of Natural History, March 23, 1840, New York Academy of Sciences. See also Daniel Norman, "John William Draper's Contributions to Astronomy," *The Telescope* 5 (1938): 11–16; and Don Trombino, "Dr. John William Draper," *Journal of the British Astronomical Association* 90 (1980): 565–71.

29. The whereabouts of the lunar daguerreotype Draper presented to the Lyceum seem to be in doubt. It has long been held that the daguerreotype was destroyed during a devastating fire at the Lyceum in 1866; see Daniel Norman, "Draper's Contributions to Astronomy," and "The Development of Astronomical Photography," *Osiris* 5 (1938): 560–94. The Lyceum evolved into the New York Academy of Sciences, and its records are held by that institution. Trombino (see "Dr. John William Draper") has recently called this fact into question, claiming that an albumen copy print of a daguerreotype found in a Greenwich Village used-book store during the late 1970s can be none other than Draper's original lunar daguerreotype. His argument is laid out in his paper, but there have been no other investigations to substantiate his claims.

30. See chapter 5, note 7.

31. Gérard de Vaucouleurs, *La Photographie Astronomique du Daguerreotype au téléscope éléctronique* (Paris: Albin Michel, 1958), 18.

32. François Arago, *Astronomie Populaire*, ed. J. A. Barral (Paris: Gide et J. Baudry, Editeurs, 1854), 2: fig. 163.

33. See de Vaucouleurs. Biographical data on Majocchi is available in J. C. Poggendorff, *Biographisch-Literarisches Handwörterbuch zur Geschichte de exacten Wissenshaften*, Leipzig (Johann Ambrosius Barth, 1863), b. 2, p. 19. The citation for Majocchi's report of his work on the 1841 eclipse is contained in the Poggendorff entry.

34. Samuel D. Humphrey, in *The Daguerreian Journal* 1 (1850): 14.

35. See the detailed description of these events in Richard Rudisill, *Mirror Image: The Influence of the Daguerreotype on American Society* (Albuquerque, University of New Mexico Press, 1971), 83–91.

36. Charlene Stephens, "'The Most Reliable Time': William Bond, the New England Railroads, and Time Awareness in 19th-Century America," *Technology and Culture* 30 (1989): 1–24.

37. Bessie Zolan Jones and Lyle Gifford Boyd, *The Harvard College Observatory: The First Four Directorships, 1839–1919* (Cambridge: Belknap Press of Harvard University, 1971), 71.

38. George P. Bond to William Mitchell, July 6, 1857, quoted in Norman, "Development of Astronomical Photography," 571–72.

39. Jones and Boyd, 77.

40. Sir John Herschel et al., "Suggestions to Astronomers for the Observation of the Total Eclipse of the Sun on July 28, 1851," *Report of the British Association for the Advancement of Science* 20 (1850): 364–65. The entire report is found on pages 360–72.

41. George P. Bond, "Solar Eclipse of July 28, 1851," *The Astronomical Journal* 2 (1851): 49–51.

42. Busch, who became the director of the Königsberg Observatory after the death of Friedrich Bessel in 1846, is given credit by many sources for taking the eclipse daguerreotype; however, he himself reported that he observed the eclipse from Rixhoft in the Baltic and that Berkawski carried out his plans to daguerreotype the eclipse at Königsberg. See *Astronomische Nachrichten* 33 (1852): 230–34; and Norman, "The Development of Astronomical Photography," 564, n. 20.

43. Dorrit Hoffleit, *Some Firsts in Astronomical Photography* (Cambridge, Mass.: Harvard College Observatory, 1950), 18–20. Father Secchi was quite active in this area and reported several successes, including daguerreotyping a nebula; however, even at the time his reports were considered somewhat exaggerated. See also Norman, "The Development of Astronomical Photography," 570.

44. Stephen Alexander, "Suggestions Relative to the Observation of the Solar Eclipse of May 26, 1854," *The Astronomical Journal* 3 (1854): 169–72.

45. ———, "Observation of the Annular Eclipse of May 26, in the Suburbs of Ogdensburgh, N.Y.," *The Astronomical Journal* 4 (1854): 29–32.

46. W. H. C. Bartlett, "On the Solar Eclipse of 1854, May 25," *The Astronomical Journal* 4 (1854): 33–34.

47. Several reports from observatories during the 1850s mention problems with initiating photographic recording in their routine observation. Of these, the Whipple-Bond reports from Harvard College Observatory are the most well known. See also John Lankford, "The Impact of Photography on Astronomy," in Owen Gingerich, ed., *Astrophysics and Twentieth-Century Astronomy to 1950: Part A* (Cambridge: Cambridge University Press, 1984).

48. This was the first time that such an international effort was made for the scientific recording of an astronomical event. The 1851 solar eclipse did have many individuals recording the progress of the moon; however, the Transit of Venus photographic teams were coordinated by de la Rue with much advanced preparation.

49. Richard A. Proctor, *Transits of Venus: A Popular Account of Past and Coming Transits* (London: Longmans, Green, and Co., 1883), 226–27.

50. *Récueil de Mémoires, Rapports et Documents Rélatifs a l'Observation de Passage de Vénus sur le Soleil* (Paris: L'Institut de France, 1876).

51. Pierre Jules César Janssen (1824–1907) made many advances in physical astronomy, solar and stellar spectroscopy, and astronomical photography. Between 1876 and 1903 he compiled an atlas of solar photographs. He referred to photographic plates as "the retina of the scientist." His *revolver photographique* was the forerunner of cinematographic cameras.

52. Simon Schaffer, "Astronomers Mark Time: Discipline and the Personal Equation," *Science in Context* 2 (1988): 115–45.

53. Early in the 1830s Gauss had determined that the primary source of the earth's magnetic field was deep within the earth's core.

54. Sir Edward Sabine, "Report on Kew Magnetographs," *Report of the British Association for the Advancement of Science* 21 (1851): 360–64.

55. John P. Gassiot, "Report of Kew Committee of the BAAS for 1856–1857; a) Re-establishment of self-recording magnetic instruments at Kew submitted to the Comm. by General Sabine on Jul 22, 1856," *Report of the British As-*

sociation for the Advancement of Science 27 (1857): xxxi–xxxvi.

56. John William Draper, "Description of the Tithonometer, an Instrument for Measuring the Chemical Force of Indigotithonic Rays," *Philosophical Magazine* 23 (1843): 401–15.

57. Robert Wilhelm von Bunsen and Henry Enfield Roscoe, "Photo-chemical Researches.—Part II. Phenomena of Photo-chemical Induction," *Philosophical Transactions* 147 (1857): 381–402.

58. M. A. Sutton, "Spectroscopy and the Chemists: A Neglected Opportunity?" *Ambix* 23 (1976): 16.

59. William A. Hamor, "David Alter and the Discovery of Spectrochemical Analysis," *Isis* 22 (1934–35): 507–10.

60. David Alter, "On certain Physical Properties of Light, produced by the combustion of different Metals, in the Electric Spark Refracted by a Prism," *American Journal of Science* 68 (1854): 55–57; and David Alter, "On certain Physical Properties of the Light of the Electric Spark, within certain Gases as seen through a Prism," *American Journal of Science* 69 (1855): 213–14.

61. James C. Booth, *The Encyclopedia of Chemistry* (Philadelphia: Henry C. Baird, 1850).

62. Gérard de Vaucouleurs, *Astronomical Photography from the Daguerreotype to the Electron Camera*, trans. R. Wright (New York: The Macmillan Company, 1961). The quotation as it appears in *Oeuvres Complètes* is as follows: ". . . quand des observateurs appliquent un nouvel instrument à l'étude de la nature, ce qu'ils en ont espère est toujours peu de chose rélativements à la succession de découvertes dont l'instruments dévient l'origine. En ce genre, c'est sur l'imprévu qu'on doit particulièrement compter" (François Arago, *Oeuvres Complètes* [Paris: Gide, 1858], t. 4: 500).

Chapter 7: Scientific Interest in Daguerreotypy after the Daguerreian Era

1. "Daguerreotype," in *A Dictionary of Photography*, eds. Thomas Sutton and George Dawson (London: Sampson Low, Son and Marston, 1867), 87.

2. James Clerk Maxwell, *British Journal of Photography* 8 (1861): 270.

3. See chapter 4, notes 31, 32, and 33.

4. Of course, this phenomenon also applies to all types of waves found in the electromagnetic spectrum. The relationship and continuity of all types of wave phenomena—sound, light, heat, and so forth—was also one of the great physics problems that was addressed at the end of the nineteenth century and beginning of the twentieth century.

5. Hippolyte Fizeau, "Prix Bordin," *Comptes Rendus* 66 (1868): 932–36.

6. For a general description of the ins and outs of the story of interference and color photography, see Pierre Connes, "Silver Salts and Standing Waves: The History of Interference Colour Photography," *Journal of Optics* (Paris) 18 (1987): 147–66.

7. See, for instance, Robert W. Wood, *Physical Optics,* 3rd ed. (New York: The Macmillan Company, 1934), 176.

8. Wilhelm Zenker, *Lehrbuch der Photochromie* (Berlin, 1868; reprint, Braunschweig: Friedrich Vieweg und Sohn, 1900).

9. Carl Schultz-Sellack, "Ueber die Farbung der trüber Medien, und die sogenannte farbige Photographie," *Annalen der Physik und Chemie* 143 (1871): 449–51; Schultz-Sellack, "On the Chemical and Mechanical Changes Undergone by Silver Haloid Salts When Acted Upon by Light," *Photographic News* 15 (1871): 308–9, 321–22.

10. Mathew Carey Lea, "On red and purple chloride, bromide, and iodide of silver; on heliochromy and on the latent photographic image," *American Journal of Science* 33 (1887): 349.

11. John Strutt (Lord Rayleigh), "On the Maintenance of Vibrations by Forces of Double Frequency and on the Propagation of Waves Endowed with a Periodic Structure," *Philosophical Magazine* 24 (1887): 145–59. For a summary of Lord Rayleigh's work, see R. Bruce Lindsey, *Lord Rayleigh, the Man and His Works* (London: Oxford University Press, 1970).

12. Otto Wiener, "Stehende Lichtwellen und die Schwingungsrichtung polarisirten Lichtes," *Annalen der Physik und Chemie* 40 (1889): 203–43.

13. Gabriel Lippmann "La photographie des couleurs," *Comptes Rendus* 112 (1891): 274–75.

14. Edmond Becquerel, "Observations de M. Edm. Becquerel sur la Communication de M. Lippmann au sujet de la réproduction photographique des couleurs," *Comptes Rendus* 112 (1891): 275–77.

15. Two papers were published in *Eder's Jahrbuch für Photographie*—one in 1891, p. 294, and one in 1893, p. 114. These are cited in Connes, "Silver Salts and Standing Waves," notes 83 and 84.

16. Wiener, "Farbenphotographie durch Körperfarben und mechanische Farbenanpassung in der Natur," *Annalen der Physik und Chemie* 55 (1895): 225–49; also translated as "Color Photography by Means of Body Colors and Mechanical Color. Adaption in Nature," *Smithsonian Annual Report* (1895–96): 167–205.

17. ———, "Eine Beobachtung von Streifen beim Entwickeln belichteter Daguerre'scher Platten mit keilformiger Jodsilberschicht," *Annalen der Physik und Chemie* 68 (1899): 145–49.

18. Hermann Scholl, "Ueber Veranderungen von Jodsilber im Light und den Daguerre'schen Process," *Annalen der Physik und Chimie* 68 (1899): 149–82.

19. In particular, see figures for Wiener's 1889 paper.

20. The most heavily exposed regions in Scholl's samples are black because he used conventional photographic developers instead of mercury for image formation. An image produced in that way is more like a conventional photograph except that it is on silver plate.

21. During the 1880s Wilhelm Hallwach and Heinrich Hertz discovered independently that incident light can eject electrons from metals. *Hallwach's effect* is now an archaic term for the *photoelectric effect*.

22. J. C. Maxwell-Garnet, "Colours in Metal Glasses and in Metallic Films. I," *Philosophical Transactions* 203 (1904): 385–420; Maxwell-Garnet, "Colours in Metal Glasses and in Metallic Films. II," *Philosophical Transactions* 204 (1905): 237–86.

23. See Bengt Nagel, "The Discussion Concerning the Nobel Prize for Max Planck," in *Science, Technology, and Society in the Time of Alfred Nobel; Nobel Symposium 52* (Oxford: Pergamon Press, 1982), 352–76.

24. See Connes, "Silver Salts and Standing Waves," 163.

25. Ferdinand Hurter and Vero Charles Driffield, "Photochemical Investigations and a New Method of Determination of the Sensitiveness of Photographic Plates," *Journal of the Society of the Chemical Industry* 3 (1890): 455.

26. C. S. McCamy, "History of Sensitometry and Densitometry," in Eugene Ostroff, ed. *Pioneers of Photography: Their Achievements in Science and Technology* (Springfield, Va.: The Society for Imaging Science and Technology, 1987), 169–88.

27. The name "Lüppo-Cramer" was the only name used by Lüppo Henricus Cramer. He was an active, prolific author and experimenter in all areas of photographic science. Lüppo-Cramer, "Uber Kornumbau und Kornigkeit," *Photographische Korrespondenz* 75 (1939): 1–4.

28. An uncited reference to Bronginart in the translation of Eder's *History* used the 0.04 millimeter measurement. See Josef Maria Eder, *History of Photography*, trans. Edward Epstein (New York: Dover, 1945), 262.

29. Heinz Jaenicke, "Eur Korngrosse der Daguerreotypie," *Photographische Industrie* 38 (1940): 530.

30. J. L. Tupper, "Report on the Reflectance Characteristics of an Old Daguerreotype," Kodak Research Laboratories Report no. 9452 (UDC Classification: 772.11:771.534.5), March 4, 1952.

31. Irving Pobboravsky, "Study of Iodized Daguerreotype Plates," Report no. 142 (Rochester, N.Y.: Graphic Arts Research Center), 1971.

32. See chapter 5, note 23.

33. Pobboravsky, op. cit., p. 44.

34. G. Thomas, *History of Photography, India 1840–1980*, Pondicherry, India: Andhra Pradesh State Akademi of Photography, 1981.

35. James Waterhouse, "Further Notes on Electro-Chemical Reversals with Thio-Carbamide," *The British Journal of Photography* 38 (1891): 261–62; Waterhouse, "Electro-Chemical Reversals with Thio-Carbamide," *The British Journal of Photography* 38 (1891): 630–31.

36. ———, "Teaching of Daguerreotype," *British Journal of Photography* 66 (1899): 740–45, 756–57.

37. Julius Alois Reich, "Zur Theorie der photographischen Vörgange," *Chemiker-Zeitung* 51 (1927): 38.

38. See Julian H. Webb, "The Photographic Latent Image Considered from the Standpoint of

the Quantum Mechanics Model of Crystals," *Journal of the Optical Society of America* 26 (1936): 367–83.

39. For a review of the early history of electro-photography, see Edward K. Kaprelian, "Pioneering Electrostatic Printing in the United States," in *Pioneers of Photography: Their Achievements in Science and Technology,* ed. Eugene Ostroff (Springfield, Va.: The Society for Imaging Science and Technology, 1987), 129–34; and R. W. Gundlach, "The World's Second Photography," *Pioneers of Photography,* 135–41.

40. Georges Simon, "Development of Daguerre Plates by Cathode Pulverization," *Comptes Rendus* 186 (1928): 139.

41. Walter Limberger and Rudolf Wendt, "Photographic Method of Producing a Copy from an Endless Silver Surface," U.S. Patent no. 3,416,920, December 17, 1968.

42. R. Prohaska and A. Fisher, "X-ray Response of Daguerreotype Photographic Plates," *Applied Physics Letters* 40 (1982): 283–85.

43. Ivor Brodie and Malcom Thackray, "Photocharging of Thin Films of Silver Iodide and Its Relevance to the Daguerre Photographic Process," *Nature* 312 (1984): 744–46.

44. Brian Coe, "The Daguerreotype Image," *The Photographic Journal* 112 (1972): 119.

45. Leon Jacobson and W. E. Leyshon, "The Daguerreian Measles Mystery," *Graphic Antiquarian* (Spring 1974): 14–15.

46. Alice Swan, "Conservation Treatments for Photographs," *Image* 21 (1978): 24–31; Swan, "The Preservation of Daguerreotypes," in *AIC Preprints* (Washington, D.C.: The American Institute for Conservation of Historic and Artistic Works, 1981), 164–72.

47. Alice Swan, C. E. Fiori, and K. F. J. Heinrich, "Daguerreotypes: A Study of the Plates and the Process," *Scanning Electron Microscopy* 1: 411–23.

Chapter 8: The Daguerreotype Image Structure

1. Edward Weston, "Art Photography," in *Encyclopaedia Brittanica,* 14th ed. (New York: Encyclopaedia Brittanica, 1940), 17: 798.

2. Alice Swan, C. E. Fiori, and K. F. J. Heinrich, "Daguerreotypes: A Study of the Plates and the Process," *Scanning Electron Microscopy* 1 (1979): 411–23. These authors were primarily interested in chemical composition, and most of their attention was turned to the chemical details of the daguerreotype, particularly the corrosion products formed by daguerreotype deterioration. Their published work does not make the connection between the physical distribution of amalgam particles and the optical appearance of the daguerreotype plate.

3. The Materials Research Laboratory collection of nineteenth-century daguerreotypes was built up from diverse sources including both vintage (i.e., nineteenth-century) and new (i.e., laboratory-made) daguerreotypes. In the end the study collection had a total of 150 vintage daguerreotypes. In addition, some fifty more daguerreotypes were loaned by collectors and museums for study. Vintage daguerreotypes came from several sources: some were purchased for the study, some were donated to the study by interested parties outside the university community, and one set of twenty-one daguerreotypes consisted of "recycled" samples that had been used in a previous daguerreotype preservation study at the Canadian Conservation Institute, Ottawa, by Siegfried Rempel and Marilyn Laver. See Siegfried Rempel, "Recent Investigations on the Cleaning of Daguerreotypes," *AIC Preprints* (Washington, D.C.: The American Institute for Conservation of Historic and Artistic Works, 1980), 99–105.

Several measures were taken to assure that no daguerreotypes of esthetic, historical, or scientific value beyond their intrinsic worth as examples of nineteenth-century daguerreotypes were harmed during the study. Each daguerreotype was photographed, and an inventory was maintained of pertinent data concerning daguerreotypist, subject identification, plate mark, plate size, condition, and source. Each daguerreotype was assigned a number that began "Dag 19-" followed by an individual sample number. The daguerreotypes from the Rempel-Laver study were designated "Dag 19-S-" followed by the numbers used in their study. William F. Stapp, curator of photography at the National Portrait Gallery of the Smithsonian Institution, reviewed the inventories from time to time as a further check that no daguerreotypes of value were destroyed. A partial inventory of the sample daguerreotypes is given in Appendix A of M. Susan Barger, "The Daguerreotype: Image

Structure, Optical Properties, and a Scientific Interpretation of Daguerreotypy" (Ph. D. diss., The Pennsylvania State University, 1982).

In 1983 and 1984, during the trials on the electrocleaning method for daguerreotypes, the following collectors and institutions also provided items from their collections to use as test specimens: William B. Becker, Pamela Kistler, Lou Kontos, Irving Pobboravsky, Willman Spawn, The Huntingdon County (Pa.) Historical Society, and The Maryland Historical Society. These daguerreotypes were examined in the laboratory, cleaned, reexamined, and returned to their owners. An agreement was made that any daguerreotypes used in the cleaning study would be monitored and returned should any unforeseen problems arise in the future.

Another set of early daguerreotypes by Joseph Saxton, Robert Cornelius, and Walter Rogers Johnson was provided for examination at the request of William F. Stapp during preparations for the exhibition "Robert Cornelius: Portraits from the Dawn of Photography." The results of those analyses are described extensively in several publications: William F. Stapp with Marion S. Carson and M. Susan Barger, *Robert Cornelius: Portraits from the Dawn of Photography* (Washington, D.C.: Smithsonian Institution Press, 1983); M. Susan Barger and William F. Stapp, "The Evolution of the Daguerreotype Art of Robert Cornelius: A Scientific Retrospective by Scanning Electron Microscopy," in *Application of Science in Examination of Works of Art,* ed. Pamela A. England and Lambertus van Zelst (Boston: The Research Laboratory, Museum of Fine Arts, 1985), 164–73.

In addition to the nineteenth-century daguerreotypes used in the study, modern daguerreotypes were made in the laboratory for study purposes. The methods used for the manufacture of daguerreotype plates and the production of daguerreotype step tablets used in experiments is described in chapter 4 and Appendix B of Barger's dissertation. Modern daguerreotypes used in the study were designated with numbers that began "Dag 20-" with a suffix that indicated the original plate and section number of the daguerreotype plate.

4. At the time this research was done there were three scanning electron microscopes in the Materials Research Laboratory. Most of the micro-

graphs were taken on an ISI DS 130 high-resolution microscope.

5. The term *image particle*, which can include Becquerel-developed daguerreotypes (i.e., those developed without mercury), is used as a more general term than *amalgam particle*, used by Swan and her colleagues.

6. The statistical analysis of image particle properties was done with the JOELCO 50A scanning electron microscope interfaced with a PDP11/20 computer system. This equipment design allows the computer to control the electron beam of the microscope as the beam is swept across the sample surface. An energy-dispersive x-ray detector and a multichannel analyzer were used to monitor the signal coming from the sample surface. A particle was recognized when the x-ray signal coming from the sample surface was amplified above a certain preset background level. The computer then stopped the sweeping motion of the beam, and the beam was used to interrogate the particle. The particle size, shape, and average diameter were recorded as well as certain chemical information. After repeated evaluation of various areas of the sample surface, the computer gave the total particle count and other statistical data. The evaluation is independent of visual display or recognition. Typically, evaluations of different areas of one sample were repeated until two or three hundred individual particles had been counted on the sample surface. This number was required to give reliable statistical data. For a more complete discussion of the methodology, see E. W. White, K. Mayberry, and G. G. Johnson, "Computer Analysis of Multi-Channel SEM and X-ray images from Fine Particles," *Journal of Pattern Recognition* 4 (1972): 173–93.

7. A Rank Taylor Hobson Talysurf 10 profilometer was used to examine overall surface roughness and measure particle heights and spacing. It also provided a check for the computer-SEM analysis of particle properties.

8. M. Susan Barger, R. Messier, and William B. White, "A Physical Model for the Daguerreotype," *Photographic Science and Engineering* 26 (1982): 285–91. M. Susan Barger and William B. White, "Optical Characterization of the Daguerreotype," *Photographic Science and Engineering* 28 (1984): 172–74.

9. John William Draper, "On some analogies between the phenomena of the chemical rays

and those of radiant heat," *Philosophical Magazine*, 3rd series, 19 (1841): 195–210. Irving Pobboravsky, "Study of Iodized Daguerreotype Plates," Report no. 142 (Rochester N.Y.: Graphic Arts Research Center, 1971).

10. Baden Powell, "Observations on certain cases of elliptic polarization of light by reflexion, Part 1," *Philosophical Transactions* 133 (1843): 35–43.

11. Otto Wiener, "Farbenphotographie durch Korperfarben und mechanische Farbenanpassung in der Natur," *Annalen der Physik und Chemie* 55 (1895): 225–49.

12. J. L. Tupper, "Report on the Reflectance Characteristics of an Old Daguerreotype," Report no. 9452 (Rochester, N.Y.: Eastman Kodak Research Laboratories, 1952).

13. George D. Cody and Richard B. Stephens, "Optical Properties of a Microscopically Textured Surface," *AIP Conference Proceedings* 40 (1978): 225–39.

14. Metals contain free electrons. Impinging electromagnetic waves at low frequency interact with the electrons in the metal in such a way that much of the wave is reflected from the surface, thus giving metals their "metallic" reflectivity. There exists a frequency above which the electrons in the metal can no longer move in phase with the electromagnetic wave; the metal becomes more absorbing, and less of the wave is reflected. The threshold frequency where this phenomenon first appears is called the *plasma frequency*, and the corresponding wavelength is the *plasma edge*. White metals such as silver and aluminum have plasma edges in the ultraviolet, while for metals such as copper and gold the plasma edge is in the visible region of the spectrum and is responsible for their characteristic colors. For plasma edge data on silver, see D. C. Skillman, "Optical Constants of Silver," *Journal of the Optical Society of America* 61 (1971): 1264–67.

15. See LeRoy E. DeMarsh and Edward J. Giorgianni, "Color Science for Imaging Systems," *Physics Today* 42 (September 1989): 44–52.

16. Milton Kerker, *Scattering of Light and Other Electromagnetic Radiation* (New York: Academic Press, 1969).

17. G. W. W. Stevens, "The Blackness of Developed Silver," *Journal of Photographic Science* 13 (1965): 228–32. L. R. Solman, "The Cov-

ering Power of Photographic Silver Deposits. III. Comparison of Chemical and Physical Development," *Journal of Photographic Science* 18 (1970): 179–84.

18. Apparent density is formally defined as

$$D_{app} = log\ 1/R_d = log\ 1/(R - R_s)$$

where R is the total reflectance from the daguerreotype surface, R_d is the diffuse reflectance, and R_s is the specular reflectance. When a daguerreotype is viewed as a positive, the R_s term is very small and approaches zero. While this definition is essentially the same as the formal definition of density for conventional reflecting materials, there is negligible light absorption in the daguerreian system, and, therefore, the basis of the conventional definition, light absorption, does not apply. Thus, the term *apparent density* is used to distinguish between the appearance of density due to light reflection and scatter and true optical density due to light absorption.

19. Reflection measurements were made using a Beckman DK 2A reflectance spectrophotometer with $BaSO_4$ optical paint as the reflectance standard. A PbS detector was used in the wavelength range of 2000 to 550 nanometers and a photomultiplier tube detector from 550 to 350 nanometers.

20. Because the term *solarized* has been traditionally used to indicate the reversal of apparent density in the highlight areas of daguerreotypes, the authors have retained its use. As indicated in the text, however, this image reversal is due to changes in the microstructure and is not to be confused with solarization as strictly defined. See C. E. K. Mees and T. H. James, *The Theory of the Photographic Process*, 3d ed. (New York: Macmillan, 1966), 150.

21. Shadow particle agglomerates can grow to a size that can be distinguished by the human eye. When this happens, shadow areas can appear grainy when viewed at close distances.

22. Samuel D. Humphrey, *American Hand Book of the Daguerreotype*, 5th ed. (New York: Author; reprint, New York: Arno Press, 1973), 37.

23. Kenneth Nelson, personal communication to M. Susan Barger, 1979.

24. Richard B. Stephens and George D. Cody, "Optical Reflectance and Transmission of a

Textural Surface," *Thin Solid Films* 45 (1977): 19–29.

25. J. R. Zeidler, R. B. Kohles, and N. M. Bashara, "High Precision Alignment Procedure for an Ellipsometer," *Applied Optics* 13 (1974): 1115–20.

26. R. F. Miller, A. J. Taylor, and L. S. Julien, "The Optimum Angle of Incidence for Determining Optical Constants from Reflectance Measurements" *Journal of Physics D: Applied Physics* 3 (1970): 1957–61.

Chapter 9: Image Formation

1. James Waterhouse, "Teachings of Daguerreotype," *British Journal of Photography* 46 (1899): 740.

2. For a more complete discussion of latent image formation, see Grant Haist, *Modern Photographic Processing* (New York: John Wiley and Sons, 1979), 1: Chapter 3; B. H. Carroll, G. C. Higgins, and T. H. James, *Introduction to Photographic Theory: The Silver Halide Process* (New York: John Wiley and Sons, 1980), Chapter 7; J. W. Mitchell, "Chemical Sensitization and Latent Image Formation: A Historical Perspective," *Journal of Imaging Science* 33 (1989): 103–14.

3. H. J. Metz, "On the Mechanism of Photographic Development," *Journal of Photographic Science* 20 (1972): 111–19.

4. Waterhouse, 740–45, 756–57.

5. Irving Pobboravsky, "Study of Iodized Daguerreotype Plates," Report no. 142 (Rochester, N.Y.: Graphic Arts Research Center, 1971), 53.

6. Metz, 111–19.

7. The developer formulas used in these experiments are as follows:

Developer A Acid pyro developer

Pyrogallic acid	4 g
Citric acid	8 g
Water to make	1 liter

Add a few drops of 4 percent stock silver nitrate solution immediately before use.

Developer B Acid iron developer

Ferrous sulfate	10 g
Citric acid	20 g
Water to make	1 liter

Add a few drops of stock silver nitrate solution (above) immediately before use.

Developer C Mercury intensifier

Potassium bromide	22.5 g
Mercuric chloride	22.5 g
Water to make	1 liter

Dip the plate in this solution and then follow development in developer A.

Kodak Developer D-76

Metol	2 g
Sodium sulfate	100 g
Hydroquinone	5 g
Borax	2 g
Water to make	1 liter

Developer D Metz high-potential physical developer

Solution 1

Metol	20 g
Citric acid	100 g
Water to make	1 liter

Solution 2

10 percent solution of silver nitrate

Developer E Metz low-potential physical developer

Metol	10 g
Potassium thiocyanate	0.5 g
Sodium sulfite	80 g
Water to make	1 liter

The formulas for developers A and B were taken from W. de W. Abney, *A Treatise on Photography*, 10th ed. (London: Longmans, Green, and Co., 1907); developers C and D76, from E. J. Wall and F. J. Jordan, *Photographic Facts and Formulas* (Garden City, N.Y.: Amphoto, 1975); developers D and E, from Metz.

8. An amalgam is, in general, an alloy of mercury and some other metal. In its strictest sense an amalgam is an alloy of mercury and silver only. In this work the authors use the strict definition.

9. Alice Swan, C. E. Fiori, K. F. J. Heinrich, "Daguerreotypes: A Study of the Plates and the Process," *Scanning Electron Microscopy* 1 (1979): 411–23.

10. Historically, the hexagonal, closest packed ϵ-phase ($Ag_{11}Hg_9$) has also been called the β-phase.

11. K. D. Jorgensen, *National Bureau of Standards Special Publication No. 354* (1972): 33.

12. It is interesting to note that the French word *fixer* means to fix, fasten, stick, or settle. In the daguerreian era fixing the image always meant gilding, not removing the silver halide.

13. R. W. Henn and B. D. Mack, "A Gold Protective Treatment for Microfilm," *Photographic Science and Engineering* 9 (1963): 378–84.

14. Swan and her colleagues (op. cit.) measured

the thickness of a gold layer on a daguerreotype and found it to be on the order of 0.75 ± 0.3 micrometers. This measurement was made using x-ray line scans on a daguerreotype cross section. The accelerating voltage they used was 7.5 KeV, and at that voltage the dispersion of the electron beam in gold and silver would not permit accurate measurement of a gold layer of the thickness reported. Further, if there was a gold layer of the reported thickness on a daguerreotype, it would be opaque and would appear gold colored.

15. Various sources report that the light used in this system after the latent image is formed should be yellow, green, or red. The exact color of the light is relatively unimportant, as long as it is spectral light of longer wavelength than the initial exposure.

16. Pobboravsky (op. cit.) discussed a speed criterion for daguerreotypes based on more rigorous conditions than the ones used in this study. For the purposes of the present study it was considered sufficient to determine a daguerreotype's relative speed rather than its absolute speed. The goal here was to determine the relative increases or decreases in speed derived from changes in the sensitizing process for the daguerreotype.

17. By convention these mixed silver halide salts are named so that the major halide present is named last. Thus, the term *silver bromoiodide* implies that the crystal is predominantly silver iodide that contains some dissolved bromine.

18. It is possible to use other halogens for the second sensitizing step; however, bromine was the one used most commonly. For this reason this discussion centers primarily on bromine.

19. K. Shahi and J. B. Wagner, Jr., "Fast Ion Transport in Silver Halide Solid Solutions and Multiphase Systems," *Applied Physics Letters* 37 (1980): 757–59.

20. The photoconductivity of thin films of silver halide is the phenomenon investigated by both nineteenth- and twentieth-century workers who have looked for the connection between electricity and the daguerreian system. The ionic conductivity of silver halide is the quality that makes silver halide such an effective material for photography.

21. Increases in speed as high as sixty times have been reported in the literature. The discrepancies

between those reports and the authors' findings are probably due to differences in the individual practice of daguerreotypy and in how the speed is determined. For instance, the plates used in our speed calculations were not gilded; gilding would enhance an image and possibly increase the number of steps that could be easily counted.

22. A set of eighteen spectral sensitizing dyes was provided for the study by Dr. Paul B. Gilman of the Eastman Kodak Research Laboratories, Rochester, N.Y. The specific dyes were chosen because each one works by a slightly different mechanism; it is possible, then, to use this particular set of dyes to infer unknown sensitizing mechanisms. See W. West and Paul B. Gilman, "Spectral Sensitivity and the Mechanism of Spectral Sensitization," in T. H. James, *The Theory of the Photographic Process*, 4th ed. (New York: Macmillan Publishing Company, 1977), 251–90.

23. Pobboravsky, op. cit.

24. John M. Blocher, Jr., "Structure/ Property/ Process Relationships in Chemical Vapor Deposition (CVD)," *Journal of Vacuum Science and Technology* 11 (1974): 680–86. J. Schlichting, "Chemical Vapor Deposition of Silicon Carbide," *Powder Metallurgy International* 12 (1980): 141–47.

25. Some of the best descriptions of this method of silver extraction are found in older metallurgy texts such as M. Eissler, *The Metallurgy of Silver* (London: Crosby, Lockwood and Son, 1889) and J. Scoffern, W. Turan, and W. Clay, *The Useful Metals and Their Alloys* (London: Holston and Wright, 1869).

26. Harald Schafer, *Chemical Transport Reactions*, trans. Hand Frankfort (New York: Academic Press, 1964), 75–85.

27. Gerald W. Sears, "A Growth Mechanism for Mercury Whiskers," *Acta Metallurgica* 3 (1955): 361–66; Sears, "Growth of Mercury Platelets from the Vapor," *The Journal of Chemical Physics* 25 (1956): 637–42; Sears, "Nucleation and Growth of Mercury Crystals at Low Supersaturation," *The Journal of Chemical Physics* 33 (1960): 563–67.

28. When a sensitized plate is inspected in the scanning electron microscope, it is obviously altered by the electron beam in the same way as when it is exposed to light. In order to keep the alteration by the electron beam to a minimum,

the micrographs were taken as quickly as possible and the sample was moved frequently.

29. William F. Stapp, Marian S. Carson, and M. Susan Barger, *Robert Cornelius: Portraits from the Dawn of Photography* (Washington, D.C.: Smithsonian Institution Press, 1983).

Chapter 10: Image Deterioration

1. Abraham Bogardus, "Trials and Tribulations of a Photographer," *British Journal of Photography* 36 (1889): 184.

2. J. M. Bennett, J. L. Stanford, and E. J. Ashley, "Optical Constants of Silver Sulfide Tarnish Films," *Journal of the Optical Society of America* 60 (1970): 224–32. B. T. Reagor and J. D. Sinclair, "Tarnishing of Silver by Sulfur Vapor. Film Characteristics and Humidity Effects," *Journal of the Electrochemical Society* 128 (1981): 701–5.

3. All thermodynamic data used in assessing silver corrosion chemistry were taken from Donald D. Wagman, William H. Evans, Vivian B. Parker, Richard H. Schumm, Ira Halow, Sylvia M. Bailey, Kenneth L. Churney, and Ralph L. Nuttall, "The NBS Tables of Chemical Thermodynamic Properties," *Journal of Physical and Chemical Reference Data* 11 (1982): Supplement no. 2.

4. The studies of daguerreotype corrosion were undertaken by the authors in the Materials Research Laboratory and other laboratories of The Pennsylvania State University. The specific equipment used was (a) a Physical Electronics Industries model no. 15-25G Auger electron spectrometer with a double-pass cylindrical mirror analyzer; (b) a Kevex energy-dispersive x-ray spectrometer, model no. 7077, which was attached to an ISI model DS-130 scanning electron microscope also used for the microstructural analysis; (c) an IBM Instruments model 98A Fourier transform infrared spectrometer equipped with both specular and diffuse-reflectance attachments; (d) a Rigaku x-ray diffractometer, model DMAX-IIA with a copper K-alpha x-ray source, curved crystal graphite monochromator, and Tl-drifted NaI scintillation detector; (e) a Huber Gandolfi x-ray camera also attached to a copper K-alpha x-ray source (identification of unknown phases was accomplished by the Johnson-Vand [Version 20] search and

match program utilizing data sets 1 to 34 of the Joint Committee on Powder Diffraction Standards); (f) a Spectrametrics Spectrascan III DC plasma source multichannel atomic emission spectrometer used for solution analysis of metals; and (g) a Spex Ramalog Raman spectrometer with a Spectra Physics Model 164 argon ion laser used to measure Raman spectra. Both blue laser lines (488 nanometers) and green laser lines (514.5 nanometers) were used as an excitation source.

5. Howard Brenner, "Silver Dips," *Soap and Sanitary Chemicals* 29 (1953): 161–67. A thiourea cleaner devised specifically for daguerreotypes at the Missouri Historical Society was first published in 1956. See Charles van Ravensway, "An Improved Method for Restoration of Daguerreotypes," *Image* 5 (1956): 156–59. See also Siegfried Rempel, "Recent Investigations on the Cleaning of Daguerreotypes," *AIC Preprints* (Washington, D.C.: The American Institute for Conservation, 1980), 99–105. A modification of this formula was published by James L. Enyeart, "Reviving a Daguerreotype," *The Photographic Journal* 110 (1970): 338–44.

6. Brenner, op. cit.

7. Leon Jacobson and W. E. Leyshon, "The Daguerreian Measles Mystery," *Graphic Antiquarian* (Spring 1974): 14–15.

8. B. H. Loo, "Molecular Orientation of Thiourea Chemisorbed on Copper and Silver Surfaces," *Chemical Physics Letters* 89 (1982): 346–50.

9. J. Fedarb, "Destruction of Daguerreotypes," *Photographic Journal* 2 (1856): 315.

10. David E. Clark, Carlo G. Pantano, and Larry L. Hench, *Glass Corrosion* (New York: Books for Industry, 1979), 4.

11. H. E. Simpson, "Study of Surface Structure of Glass in Relation to Its Durability," *Journal of the American Ceramic Society* 41 (1958): 43–49. Simpson, "Study of Surface Durability of Container Glasses," *Journal of the American Ceramic Society* 42 (1959): 337–43. R. M. Tichane, "Initial Stages of the Weathering Process on a Soda-Lime Glass Surface," *Glass Technology* 7 (1966): 26–29. H. V. Walters and P. B. Adams, "Effects of Humidity on the Weathering of Glass," *Journal of Noncrystalline Solids* 19 (1975): 183–99.

12. A complete listing of all the products found on the cover glass surfaces, including all of the

x-ray powder diffraction data, may be found in M. Susan Barger, Deane K. Smith, and William B. White, "Characterization of Corrosion Products on Old Protective Glass, Especially Daguerreotype Cover Glasses," *Journal of Materials Science* 24 (1989): 1343–56. This paper also contains an extensive discussion of the "molds" found on daguerreotype cover glasses and other reactive glasses.

13. Alice Swan, C. E. Fiori, and K. F. J. Heinrich, "Daguerreotypes: A Study of the Plates and the Process," *Scanning Electron Microscopy/1979/I* (1979): 411–24.

14. Barger, Smith, and White, op. cit.

15. Ibid.

16. Ibid.

Chapter 11: Corrosion Removal

1. "How to Clean Daguerreotypes," *The Photographic News* 21 (1877): 338–39.

2. As an example of this type of study, see "Durability of Daguerreotype," *The Daguerreian Journal* 1 (1851): 113–14.

3. James T. Foard, "Permanence of Daguerreotypes," *Journal of the Photographic Society* 3 (1856): 91–92.

4. Daguerreotype cleaning formulas appeared frequently during the nineteenth century and this century. See, for example, "Copying and Cleaning Daguerreotypes," *The Photographic News* 4 (1860): 310–11; "Restoring Faded Daguerreotypes," *The British Journal of Photography* 22 (1875): 504; "How to Clean Daguerreotypes," *Practical Photographer* 1 (1877): 224–25; "On the Treatment of Faded Daguerreotypes," *The British Journal of Photography* 27 (1880): 386–87.

5. "On the Treatment of Faded Daguerreotypes," 386–87.

6. Montgomery P. Simons, "A Few Words on Cleaning the Daguerreotype," *Anthony's Photographic Bulletin* 6 (1875): 300.

7. "Photography in Court," *The British Journal of Photography* 22 (1875): 514.

8. See James L. Enyeart, "Reviving a Daguerreotype," *The Photographic Journal* 110 (1970): 338–44. In 1969 the Taft papers were given to the Kansas State Historical Society and with them the Draper daguerreotype. In the late 1960s Enyeart made an attempt at restoring the

daguerreotype in a thiourea cleaning solution and reported that the image returned at about 50 percent of the original intensity.

9. Howard Brenner, "Silver Dips," *Soap and Sanitary Chemicals*, 29 (1953): 161–67, 183.

10. Charles van Ravensway, "An Improved Method for the Restoration of Daguerreotypes," *Image* 5 (1956): 156–59.

11. Hirst Milhollen to Edgar Breitenbach, memorandum, June 14, 1957, Prints and Photographs Division, Library of Congress.

12. Edgar Breitenbach to assistant director, Reference Department, memorandum, June 21, 1957, Prints and Photographs Division, Library of Congress.

13. Edgar Breitenbach to Dorothy Meserve (Mrs. Phillip B.) Kunhardt, letter, October 13, 1958, Prints and Photographs Division, Library of Congress.

14. *Annual Report* (Washington, D.C.: Library of Congress, 1959), 38.

15. Leon Jacobson and W. E. Leyshon, "The Daguerreian Measles Mystery," *Graphic Antiquarian* (Spring 1974): 14–15.

16. See especially Alice Swan, "The Preservation of Daguerreotypes," in *AIC Preprints* (Washington, D.C.: American Institute for Conservation, 1981): 164–72; Siegfried Rempel, "Recent Investigations on the Cleaning of Daguerreotypes," in *AIC Preprints* (Washington, D.C.: American Institute for Conservation, 1980): 99–105; T. J. Collings and F. J. Young, "Improvements in Some Tests and Techniques in Photograph Conservation," *Studies in Conservation* 21 (1976), 79–84.

17. Brenner, op. cit.

18. Van Ravensway, op. cit.

19. The ability of one metal to displace another depending on their relative positions in the electromotive series is described in most introductory chemistry textbooks. The specific half-cell reactions involved in the reduction of silver by aluminum are:

$$Ag^+ + e^- \rightleftarrows Ag\text{-metal} \qquad E° = +0.7996 \text{ volts}$$
$$Al^{3+} + 3e^- \rightleftarrows Al\text{-metal} \qquad E° = -1.706 \text{ volts}$$

$E°$ is the standard electrode potential for the reaction. Data are from Allen J. Bard and Larry R. Faulkner, *Electrochemical Methods* (New York: John Wiley & Sons, 1980), 699.

20. Vincent D. Daniels, "Plasma Reduction of Silver Tarnish on Daguerreotypes," *Studies in Conservation* 26 (1981): 45–49; M. Susan Barger, S. V. Krishnaswamy, and R. Messier, "The Cleaning of Daguerreotypes: I. Physical Sputter Cleaning. A New Technique," in *AIC Preprints* (Washington, D.C.: The American Institute for Conservation of Historic and Artistic Works, 1982): 9–20; Barger, Krishnaswamy, and Messier, "The Cleaning of Daguerreotypes: Comparison of Cleaning Methods," *Journal of the American Institute for Conservation* 22 (1982): 13–24; Mogens S. Koch and Anker Sjøgren, "Behandlung von Daguerreotypien mit Wasserstoffplasma," *Maltechnik Restauro* 90 (October 1984): 58–64.

21. The experiments described here were performed in a Materials Research Corporation Model no. SES 8632 radio frequency sputtering system with an MKS flow ratio controller to control gas flow and pressure in the sputtering chamber. The system was pumped to a base pressure of 200 nanotorr before it was backfilled with the prescribed sputtering gas (argon or argon plus hydrogen) in the 1–100 millitorr range.

22. Barger, Krishnaswamy, and Messier, "The Cleaning of Daguerreotypes" *Journal of the American Institute for Conservation* 22 (1982): 13–24.

23. M. Susan Barger, A. P. Giri, William B. White, and Thomas M. Edmondson, "Cleaning Daguerreotypes," *Studies in Conservation* 31 (1986): 15–28.

24. William John McGregor Tegart, *The Electrolytic and Chemical Polishing of Metals* (New York: Pergamon Press, 1958).

25. The field trials were carried out in the conservation shop of Thomas M. Edmondson, Connecticut.

26. Barger, Messier, and White, "A Physical Model for the Daguerreotype," *Photographic Science and Engineering,* 26 (1982): 285–91.

Chapter 12: Daguerreotype Preservation and Display

1. Thomas Sutton and George Dawson, *A Dictionary of Photography* (London: Sampson, Low, Son and Marston, 1867), 87.

2. The Joseph Saxton daguerreotype was submitted for analysis as part of the preparation for the Smithsonian Institution's exhibition "Robert Cornelius: Portraits from the Dawn of Photography," held at the National Portrait Gallery in the winter of 1983–84. During those examinations it was discovered that this daguerreotype had been varnished. During the nineteenth century recommendations for coatings to protect daguerreotypes were common, but no modern references to this practice have been noted. The presence of a varnish on the Saxton daguerreotype had never been noted, and there is no record of when this daguerreotype was varnished or by whom. To identify the varnish, its infrared (IR) spectrum was obtained using a Fourier transform infrared spectrometer. It is possible to obtain IR reflectance spectra of thin films and coatings on daguerreotype surfaces in situ because silver is totally reflecting in the infrared. No special sample preparation is needed. By comparing the spectrum with various references, the varnish coating was tentatively identified as Manila copal, considerably aged and oxidized. This daguerreotype is owned by the Historical Society of Pennsylvania, Philadelphia.

3. For an extensive discussion of daguerreotype cases and case manufacture, see Floyd Rinhart and Marion Rinhart, *American Miniature Case Art* (New York: A. S. Barnes and Co., 1969).

4. See Reese V. Jenkins, *Images and Enterprise: Technology and the American Photographic Industry, 1839 to 1925* (Baltimore: The Johns Hopkins University Press, 1975).

5. Clifford Krainick and Michele Krainick, with Carl Walvoord, *Union Cases: A Collector's Guide to the Art of America's First Plastics* (Grantsburg, Wis.: Centennial Photo Service, 1988).

6. See Henry Hunt Snelling, *The History and Practice of the Art of Photography or the Production of Pictures Through the Agency of Light* (New York: G. P. Putnam, 1849, p. 77; reprint, Hastings-on-Hudson, N.Y.: Morgan and Morgan, 1970). See also Samuel D. Humphrey, *American Hand Book of the Daguerreotype,* 5th ed. (New York: S. D. Humphrey, 1858, pp. 56–57; reprint, New York: Arno Press, 1973).

7. Cover glasses were sold by companies dealing in daguerreotype goods. The different grades of glass were variously labeled "French

plate," "German plate (white), half white, $\frac{3}{4}$ white," "Green English glass" or "crystal glass." These came in best, second quality, and third quality. Listings are given in "A Comprehensive and Systematic Catalogue of Photographic Apparatus and Material; Manufactured, Imported, and Sold by E. Anthony, New York," in Snelling, *A Dictionary of the Photographic Art,* Part 2 (New York: H. H. Snelling, 1854, p. 54; reprint, New York: Arno Press, 1979).

8. M. Susan Barger, A. P. Giri, William B. White, William S. Ginell, and Frank Preusser. "Protective Surface Coatings for Daguerreotypes," *Journal of the American Institute of Conservation* 24 (1984): 40–52.

9. The process of physical sputtering refers to the removal of material from a solid surface (target, cathode) when struck by ions or atoms in a gas discharge. This involves momentum transfer between the incident positive ions from the gas discharge and the target held at a negative potential on the order of 500 volts. Sputtering is usually used for the deposition of thin films, but the same technique using the same apparatus was applied to sputter cleaning as described in chapter 11. In a normal coating sequence the target is prepared from the source material that is to provide the coating. Target-ejected particles traverse the distance between the cathode and the anode in a vacuum chamber and are subsequently deposited on the substrate (i.e., the anode) as a thin film. The method can also be used for removal of material, in which case the sample itself is the target (i.e., the cathode). For coating or sputter cleaning of dielectric materials (e.g., oxides, sulfides, nitrides, and so forth) radio frequency voltage is applied instead of dc voltage to eliminate the charging problem resulting from electron emission.

10. During the deposition of this film a temporary power failure caused an uncontrollable drop in voltage.

11. Russell Messier, personal communication to M. Susan Barger, 1983.

12. The Parylene C films were applied at the Getty Conservation Institute, Santa Monica, California, as part of a cooperative research program on materials for protective coatings for art materials. See note 8. To remove the Parylene C film, a daguerreotype is immersed in a hot (165 degrees Centigrade) bath of orthodichlorobenzene, thus converting the film to a gelatinous mass that can be easily removed with gentle swabbing. This process must be carried out in a fume hood suitable for handling chlorinated solvents. Care should be taken to avoid inhaling the fumes.

13. Method B as described in ASTM Standard D3359-83, "Standard Methods for Measuring Adhesion by Tape Test." This is a simple test used to establish adequate adhesion of coating materials to metal substrates.

14. M. Susan Barger, R. Messier, and William B. White, "Daguerreotype Display," *Picturescope* 31 (1983): 57–58.

15. ———, "How to Enhance Daguerreotype Images," *History News* 39 (1984): 26–28.

16. An account of the daguerreotype display using fiber optic illumination at the Canadian National Gallery appears in *The Ottawa Citizen,* Thursday, May 19, 1988. This type of lighting has also been used at the Chrysler Museum (Norfolk, Virginia), Musée Carnavalet (Paris), and elsewhere.

17. Some nineteenth-century writers claimed that the best way to view a daguerreotype is by candlelight. This claim is, in fact, true. Candlelight is a yellow, broad-band illumination source; thus, it is a very effective illuminant for daguerreotypes.

18. Celeste McCollough, "Color Adaptation of Edge-detectors in the Human Visual System," *Science* 149 (1965): 1115–16.

Chapter 13: Daguerreotypy: A Model for the Interpretation of Early Technology and Works of Art

1. Levi L. Hill, *A Treatise on Daguerreotype; The Whole Art Made Easy, and All the Recent Improvements Revealed* (Lexington, N.Y.: Holman & Gray, 1850), 13–14.

References

Chapter 1: Beginnings

Figuier, Louis. *Les Merveilles de la Science ou Description Populaire des Inventions Modernes.* Vol. 3. Paris: Furne, Jouvet et Cie., n.d. [1867].

Gernsheim, Helmut. *The Origins of Photography.* New York: Thames and Hudson, 1982.

Rinhart, Floyd, and Marion Rinhart. *The American Daguerreotype.* Athens: The University of Georgia Press, 1981.

Rudisill, Richard. *Mirror Image: The Influence of the Daguerreotype on American Society.* Albuquerque: University of New Mexico Press, 1971.

Wood, John, ed. *The Daguerreotype: A Sesquicentennial Celebration.* Iowa City: University of Iowa Press, 1989.

Chapter 2: Image Making before Photography

Alpers, Svetlana. *The Art of Describing: Dutch Art in the Seventeenth Century.* Chicago: The University of Chicago Press, 1983.

Galassi, Peter. *Before Photography: Painting and the Invention of Photography.* New York: The Museum of Modern Art, 1981.

Gernsheim, Helmut. *The Origins of Photography.* New York: Thames and Hudson, 1982.

Hammond, John H., and Jill Austin. *The Camera Lucida in Art and Science.* Bristol, England: Adam Hilger, 1987.

A New and complete dictionary of arts and sciences; comprehending all branches of useful knowledge . . . extracted from the best authors in all languages by a Society of gentlemen. Vol. 1. London: W. Owen, 1754

Transactions of the Society of Arts, 34 (1817).

Twyman, Michael. *Lithography 1800–1850: The Techniques of Drawing on Stone in England and France and Their Application on Works of Topography.* London: Oxford University Press, 1970.

Chapter 3: Toward Points de Vue and the Daguerreotype

Arago, François. *Oeuvres Completes de François Arago, Tables.* Ed. J. A. Barral. Paris: Gide, Editeur, 1859.

Buckland, Gail. *Fox Talbot and the Invention of Photography*. Boston: David R. Godine, 1980.

Eder, Josef Maria. *History of Photography*. Translated by Edward Epstean. New York: Dover, 1972.

Gernsheim, Helmut. *The Origins of Photography*. New York: Thames and Hudson, 1982.

Gernsheim, Helmut, and Alison Gernsheim. *The History of Photography from the Camera Obscura to the Beginning of the Modern Era*. New York: McGraw-Hill, 1969.

————. *L. M. J. Daguerre: The History of the Diorma and the Daguerreotype*. New York: Dover, 1968.

Jay, Paul, and Michel Frizot. *Nicéphore Niépce: Lettres et Documents choisis par Paul Jay*. Paris: Centre National de la Photographie, 1983.

Lassam, Robert. *Fox Talbot, Photographer*. Tisbury, Wiltshire: The Compton Press, 1979.

Chapter 4: The Technological Practice of Daguerreotypy

Eder, Josef Maria. *History of Photography*. Translated by Edward Epstean. New York: Dover, 1972.

Daguerre, Louis Jacques Mandé. *An Historical Account and Descriptive Account of the Various Processes of the Daguerreotype and the Diorama*. London: McLean, 1839. Reprint. New York: Kraus Reprint Co., 1969.

Figuier, Louis. *Les Merveilles de la Science ou Description Populaire des Inventions Modernes*. Vol. 3. Paris: Furne, Jouvet et Cie., n.d. [1867].

Fisher, George Thomas. *Photogenic Manipulation: Part 1. Containing the Theory and Plain Instructions in the Art of Photography or the Production of Pictures Through the Agency of Light*. Philadelphia: Carey and Hart, 1845.

Hill, Levi L. *A Treatise on Heliochromy; or The Production of Pictures by means of Light, in Natural Colors, embracing a full, plain, and unreserved description of the process known as the Hillotype*. New York: Robinson & Caswell, 1856. Reprint. State College, Pa.: The Carnation Press, 1972).

Hunt, Robert. *Encylcopaedia Metropolitana*. Vol. 16 in *A System of Universal Knowledge: Photography, A Manual of Photography*. 4th ed. London and Glasgow: Richard Griffin and Co., 1854.

Jenkins, Reese V. *Images and Enterprise: Technology and the American Photographic Industry, 1839–1925*. Baltimore: The Johns Hopkins University Press, 1975.

Root, Marcus Aurelius. *The Camera and the Pencil; or the Heliographic Art*. Philadelphia: M. A. Root, 1864. Reprint. Pawlet, Vt: Helios, 1971).

Rinhart, Floyd, and Marion Rinhart. *The American Daguerreotype*. Athens: The University of Georgia Press, 1981.

Snelling, Henry Hunt. *A Dictionary of the Photographic Art, forming a Complete Encyclopaedia of All the Terms, Receipts, Processes, Aparatus and Materials in the Art; Together with a List of Articles of Every Description Employed in its Practice*. New York: Author, 1854. Reprint. New York: Arno Press, 1979.

Stapp, William F., Marion S. Carson, and M. Susan Barger. *Robert Cornelius: Portraits from the Dawn of Photography*. Washington, D.C.: Smithsonian Institution Press, 1983.

Taft, Robert. *Photography and the American Scene: A Social History, 1839–1889*. New York: Dover, 1964.

Chapter 5: Scientific Interest in the Daguerreotype during the Daguerreian Era

Becquerel, Edmond. "Memoir on the Constitution of the Solar Spectrum, presented to the Academy of Sciences at the Meeting of the 13th of June, 1842." *Scientific Memoirs* 3 (1843): 537–57.

Donné, Alfred, and Léon Foucault. *Cours de Microscopie Complementaire des Études Médicales, Anatomie Microscopique et Physiologie des Fluids de l'Économie. Atlas éxécute d'après au microscope-daguerreotype*. Paris: Chez J.-B. Bailliere, 1845.

Draper, John William. "On a new Imponderable Substance, and on a Class of Chemical Rays analogous to the Rays of Dark Heat." *Philosophical Magazine* 21 (1842): 453–61.

————. *Scientific Memoirs Being Experimental Contributions to a Knowledge of Radiant Energy*. New York: Harper & Brothers, 1878.

Figuier, Louis. *Les Merveilles de la Science ou Description Populaire des Inventions Modernes*. Vol. 3. Paris: Furne, Jouvet et Cie., n.d. [1867].

Fleming, Donald H. *John William Draper and the Religion of Science*. New York: Octagon Books, 1972.

Herschel, J. F. W. "On the Action of the Rays of the Solar Spectrum on the Daguerreotype Plate." *Philosophical Magazine* 22 (1843): 120–32.

Hunt, Robert. *Photography, Including Daguerreotype, and All the New Methods of Producing Pictures by the Chemical Agency of Light*. Glasgow: Richard Griffen and Company, 1841. Reprint. Athens: Ohio University Press, 1973.

Waller, Augustus. "Observations on Certain Molecular Actions of Crystalline Particles, &c; and on the Cause of the Fixation of Mercury Vapours in the Daguerreotype Process." *Philosophical Magazine* 18 (1846): 94–105.

Chapter 6: The Daguerreotype as a Scientific Tool

Alpers, Svetlana. *The Art of Describing: Dutch Art in the Seventeenth Century*. Chicago: The University of Chicago Press, 1983.

Alter, David. "On certain Physical Properties of Light, produced by the combustion of different Metals in the Electric Spark refracted by a Prism." *American Journal of Science* 69 (1854): 55–57.

Arago, François. *Astronomie Populaire*. Ed. J. A. Barral. Vol. 2. Paris: Gide et J. Baudry, Editeurs, 1854.

Booth, James C. *The Encyclopedia of Chemistry, Practical and Theoretical: Embracing Its Application to the Arts, Metallurgy, Mineralogy, Geology, Medicine, and Pharmacy*. Philadelphia: Henry C. Baird, 1850.

Boring, Edwin G. *A History of Experimental Psychology*. New York: Appleton-Century-Crofts, 1950.

Hawks, Francis L. *Narrative of the Expedition of an American Squadron to the China Seas and Japan Performed in the Years 1852, 1853, and 1854 under the command of Commodore M. C. Perry, United States Navy by the order of the Government of the United States*. New York: D. Appleton & Co., 1856.

Jones, Bessie Zolan, and Lyle Gifford Boyd. *The Harvard College Observatory: The First Four Directorships, 1839–1919*. Cambridge: Belknap Press of Harvard University Press, 1971.

Naef, Weston J. *Era of Exploration: The Rise of Landscape Photography in the American West, 1860–1885*, Buffalo: Albright-Knox Art Gallery, 1975.

Norman, Daniel. "The Development of Astronomical Photography." *Osiris* 5 (1938): 560–94.

Pierce, Sally. *Whipple and Black: Commercial Photographers in Boston*. Boston: The Boston Athenaeum, 1987.

Récuil de Mémoires, *Rapports et Documents Rélatifs a l'Observation de Passage de Venus sur le Soleil*. Paris: L'Institute de France, 1876.

Rudisill, Richard. *Mirror Image: The Influence of the Daguerreotype on American Society*. Albuquerque: University of New Mexico Press, 1971.

Sampson, Marmaduke B. *Rationale of Crime and Its Appropriate Treatment, being a Treatise on Criminal Jurisprudence considered in Relation to Cerebral Organization*. New York: D. Appleton and Co., 1846.

Snelling, Henry Hunt. *A Dictionary of the Photographic Art Forming a Complete Encyclopaedia of all the Terms, Receipts, Processes, Apparatus and Materials of the Art, Together with a List of Articles of Every Description Employed in its Practice*. New York: Author, 1854. Reprint. New York: Arno Press, 1979.

Stephens, Carlene E. "Partners in Time: William Bond & Son of Boston and the Harvard College Observatory." *Harvard Library Bulletin* 35 (1987): 351–84.

Stephens, John Lloyd. *Incidents of Travel in Yucatan*. 1843. Reprint. Norman: University of Oklahoma Press, 1962.

Chapter 7: Scientific Interest in Daguerreotypy after the Daguerreian Era

Coe, Brian. "The Daguerreotype Image." *The Photographic Journal* 112 (1972): 119.

Connes, Pierre. "Silver Salts and Standing Waves: The History of Interference Colour Photography," *Journal of Optics* (Paris) 18 (1987): 147–66.

Lichtenecker, Karl. "Otto Wiener." *Physikalische Zeitschrift* 28 (1928): 73–78.

Pobboravsky, Irving. "Study of Iodized Daguerreotype Plates." Report no. 142 (Rochester, N.Y.: Graphic Arts Research Center), 1971.

Scholl, Hermann. "Ueber Veranderungen von Jodsilber im Licht und den Daguerre'schen Process." *Annalen der Physik* 68 (1899): 149–82.

Sobel, Michael I. *Light*. Chicago: The University of Chicago Press, 1987.

Swan, Alice, C. E. Fiori, and K. F. J. Heinrich, "Daguerreotypes: A Study of Plates and the Process." *Scanning Electron Microscopy* 1: 411–23.

Wood, Robert W. *Physical Optics*. 3rd ed. New York: The Macmillan Company, 1934.

Chapter 8: The Daguerreotype Image Structure

Stapp, William F., Marian S. Carson, and M. Susan Barger. *Robert Cornelius: Portraits from the Dawn of Photography*. Washington, D.C.: Smithsonian Institution Press, 1983.

Chapter 9: Image Formation

Carroll, B. H., G. C. Higgins, and T. H. James. *Introduction to Photographic Theory: The Silver Halide Process*. New York: John Wiley and Sons, 1980.

Haist, Grant. *Modern Photographic Processing*. Volumes 1 and 2. New York: John Wiley and Sons, 1979.

James, T. H., ed. *The Theory of the Photographic Process*. New York: Macmillan Publishing Company, 1977.

Stapp, William F., Marian S. Carson, and M. Susan Barger. *Robert Cornelius: Portraits from the Dawn of Photography*. Washington, D.C.: Smithsonian Institution Press, 1983.

Vossen, John L., and Werner Kern, eds. *Thin Film Processes*. New York: Academic Press, 1978.

Name Index

Page references to illustrations (figures) are indicated by an *f* following the page number.

Agassiz, Louis, 80, 80*f*
Alexander, Stephen, 90
Allason, Thomas, 5, 5*f*
Alter, David, 53, 53*f*, 96–97, 96*f*
Anthony, Edward, 74, 76
Anthony, Henry T., 76
Arago, Dominique François Jean, xviii*f*, 1, 2, 11, 13–14, 23–25, 24*f*, 27–28, 35–36, 38–39, 46, 54–56, 58–59, 72, 84–85, 86*f*, 97, 99, 201
Archer, Frederick Scott, 91
Austin, Jill, 7*f*

Ballard, Antoine, 31
Barger, M. Susan, 42*f*, 101*f*, 188, 203, 217
Barlett, W. H. C., 91
Barnard, F. A. P., 35
Bartholow, Roberts, 53
Bauer, Francis, 21, 23–24
Bayard, Hippolyte, 26
Bayer, 90
Beard, Richard, 34, 39, 183
Becquerel, Antoine, 66
Becquerel, Edmond, 39, 59, 59*f*, 66–68, 66*f*, 70–71, 73, 85, 95–96, 99–101, 103, 106, 110–111, 122–123, 148, 149*f*
Beddoes, Thomas, 16
Berkawski, 90
Berres, Josef, 34, 43
Berzelius, J. J., 58
Bessel, Friedreich Wilhelm, 62
Bessyre, Golfier, 60
Bigelow, Henry J., 82, 82*f*
Biot, Jean Baptiste, 24, 26, 59, 112
Black, James Wallace, 88
Blocher, John M., 153
Bogardus, Abraham, 160
Boggs, Robert M., 90
Bond, George Phillips, 88–90, 89*f*
Bond, William Cranch, 87, 90*f*
Booth, James C., 97, 97*f*
Boudreau, Joseph, 42
Bouton, Charles Marie, 21
Boye, Martin Hans, 32
Brady, Mathew, 32*f*, 43, 75, 79, 79*f*, 185
Breitenbach, Edgar, 185
Brenner, Howard, 171, 186
Brewster, David, 13, 15, 24, 62
Bridgeman, Laura, 79, 79*f*
Brongniart, 107
Brown, Eliphalet, Jr., 76, 76*f*
Bunsen, Robert Wilhelm Von, 97
Busch, August Ludwig, 90

Cabot, Samuel, 77, 77*f*
Campbell, John L., 90

Carlson, Chester, 114
Carvahlo, Solomon Nunes, 75
Catherwood, Frederick, 76–77, 77*f*
Cauchy, Augustin Louis, 26
Chapée, Guillaume Edouard, 50
Chevalier, Charles, 19, 21, 29–30
Chevalier, Vincent, 19, 21
Choiselat, Marie Charles Isidore, 68–69
Chrétien, Gilles Louis, 6
Claudet, Antoine, 1, 34–35, 37, 43, 71, 94, 94*f*
Cody, George D., 129
Coe, Brian, 115, 115*f*
Combe, Andrew, 78–79
Combe, George, 78–79
Cornelius, Robert, 32–37, 33*f*, 52, 97, 97*f*, 111, 136, 157, 158, 158*f*, 159*f*, 172*f*
Courtois, Bernhard, 31

D'Avignon, Francis, 42
Daguerre, Louis Jacques Mandé, xviii*f*, 1, 2, 11, 19–30, 20*f*, 24*f*, 32, 35–37, 44, 48, 55–56, 58–59, 66, 84–85, 111, 134, 136, 136*f*, 139, 146, 159, 201, 218
Daniels, V. D., 188
Darwin, Charles, 80
David, Jacques Louis, 21
Davis, Daniel, Jr., 39
Davy, D. D. T., 41
Davy, Humphrey, 16, 31, 66
Dawson, George, 201
De Prangey, Girault, 74
De la Rue, Warren, 89, 92, 94–95
Degotti, Ignace Eugene Marie, 20
Donné, Alfred, 43, 60, 60*f*, 73, 83
Draper, Dorothy, 58, 184
Draper, John William, 35, 43, 46, 56–59, 57*f*, 61–66, 63–64*f*, 68, 70–71, 73, 85, 95–96, 99, 109–112, 122, 184–185
Driffield, Vero Charles, 107
Du Ponceau, Pierre Étienne, 158, 158*f*, 159*f*

Easterly, Thomas, 75, 83*f*
Eder, Josef Maria, 107
Edmondson, Tom, 199
Einstein, Albert, 114
Elkington, George R., 50–51
Elkington, Henry, 50–51

Farnham, Eliza, 79, 79*f*
Figuier, Louis, 40*f*
Fiori, C. E., 116, 116*f*, 118
Fisher, A., 115
Fisher, George Thomas, 31*f*, 47*f*
Fizeau, Hippolyte, 35–39, 36*f*, 43, 46, 65,

70–71, 73, 85, 86*f*, 93, 99–101, 112, 136*f*, 145
Foard, James T., 183
Foucault, Léon, 60, 60*f*, 70–71, 85–86*f*, 99
Fowler, Lorenzo N., 78–79
Fowler, Orson S., 78–79
Fraunhofer, Joseph, 96
Frémont, John C., 75
Fresnel, Augustin, 13, 100

Gall, Franz Joseph, 78
Gaudin, Marc, 36
Gauss, Karl Friedrich, 93
Gay-Lussac, Joseph Louis, 25, 31
Gillespie, Edward, 53
Gillespie, James, 53
Goddard, John Frederick, 34–37, 136*f*
Goddard, Paul Beck, 32–37, 34*f*, 52, 81*f*, 111, 136*f*, 158, 172*f*
Goode, W. H., 35
Gouraud, François Fauvel, 32–33
Grove, William R., 43
Gurney, Jeremiah, 41

Hall, Basil, 8
Hallwach, Wilhelm, 113–114
Hammond, John H., 7*f*
Hare, Robert, 33–35, 52, 57
Harrison, C. C., 44*f*
Harrison, William Henry, 74
Hawes, Josiah Johnson, 82
Hawks, Francis L., 76*f*
Hayden, David, 49
Heinrich, Kurt F. J., 116, 116*f*, 118
Henry, Joseph, 66–68, 111
Herman, Jan Kenneth, 217
Herschel, Caroline, 59
Herschel, Friedrich Wilhelm, 14
Herschel, John, 8*f*, 14–15, 26, 58–59, 61, 63–64, 64*f*, 70, 89, 94–96, 184–185
Hertz, Heinrich, 102, 112–113
Hill, Levi L., 39–42, 216
Holmes, Isreal, 49
Howe, Samuel G., 79, 79*f*
Hruschka, A., 102
Humphrey, Samuel D., 53, 86–88, 87*f*
Hunt, Robert, 44*f*, 60–61, 64–66, 68, 71, 95
Hurter, Ferdinand, 107
Huygens, Christiaan, 12–13

Isenberg, Matthew R., 39*f*, 43*f*, 45–47*f*, 84*f*
Ives, Frederick E., 103

Jacobson, Leon, 115, 186
Jaenicke, Heinz, 107–108

Janssen, Jules, 92–93*f*
Jeffreys, Thomas, 6*f*
Johnson, John, 30, 34–35, 45
Johnson, Walter Rogers, 32
Jones, William B., 88

Katz, Paul, 79*f*, 83*f*, 199*f*
Keller, Helen, 79
Kendall, Ezra Otis, 32
Kern, Edward, 75–76
Kirchoff, Gustav, 97
Koch, Mogens S., 188
Kratochwila brothers, 34–35
Krishnaswamy, S. V., 188
Krone, Hermann, 90
Kurnhardt, Dorothy Meserve, 185

L'Homdieu, Charles, 38
Laborde, Abbe Edmond, 45
Lagenheim, Frederick, 90, 91*f*
Lagenheim, William, 90, 91*f*
Lamarck, Jean Baptiste, 81
Languiche, Charles, 17*f*
Lavoisier, Antoine, 14
LeBlanc, Nicholas, 52
Lea, Mathew Carey, 73*f*, 101–103, 101*f*, 106, 112
Leclere, George Louis, 81
Leidy, Joseph, 81*f*
Lemaitre, A. F., 19–20, 22
Lenard, P. E. A., 113
Lerebours, Nicholas-Marie Paymal, 35, 42, 70, 73, 85–86
Leyshon, W. E., 115, 186
Lichtenecker, Karl, 102*f*
Limberger, Walter, 114
Lincoln, Abraham, 185
Lincoln, Mary Todd, 185
Linnaeus, Carolus, 81
Lippmann, Gabriel, 103, 106–107
Loomis, Elias, 90
Lumière brothers, 103
Lüppo-Cramer, 107, 114

Majocchi, Giovanni Alessandro, 86
Malus, Etienne Louis, 13
Maxwell, James Clark, 99, 112
Maxwell-Garnett, J. C., 106
McNeely, Isaac, 39*f*
Messier, Russell, 188
Metz, H. J., 137*f*, 138, 140
Mitchell, J. K., 57
Mitchell, William, 88
Moffat, John, 25*f*
Morse, Samuel, F. B., 16, 26, 28, 35, 41,

56, 58, 74
Morton, Samuel, 80
Morton, William T. G., 82, 82*f*
Moser, Ludwig, 62–65, 70–73, 113

Nash, Graham, 8*f*
Natterer brothers, 34–35
Nelson, Kenneth E., xiv*f*, 83*f*, 217
Newhall, Beaumont, 108
Newton, Isaac, 12
Niépce de Saint-Victor, Claude F. A.,
   39–40, 40*f*, 99
Niépce, Claude, 17–22
Niépce, Isidore, xviii*f*, 1, 20, 22–23, 25–26,
   29, 31, 55, 59, 84, 134
Niépce, Joseph Nicéphore, 2, 11, 17–24,
   17*f*, 19*f*, 27, 39, 43, 218

Oldenburg, Henry, 4
Olds, E. H., 90

Pemberton, 49
Pennell, Joseph, 49
Perry, Matthew Calbraith, 76
Petzval, Joseph Max, 29–30, 34
Planck, Max Karl Ernst, 106, 114
Plumbe, John, Jr., 39, 42
Pobboravsky, Irving, 108*f*, 109–111, 109*f*,
   122, 138, 151
Poe, Edgar Allan, 81*f*
Poisson, Simeon, 14
Ponton, Mungo, 26
Porta, Johann Baptista, 7
Powell, Baden, 122
Prater, Horatio, 65
Preuss, Charles, 75
Prévost, Pierre, 20–21
Prévost, Victor, 91
Priestly, Joseph, 14
Prohaska, R., 115

Rafferty, Theodore, 217
Ransom, Harry, xviii*f*, 19*f*, 25*f*
Ratel, Stanislaw, 68–69
Rayleigh. *See* Strutt, John William
Reich, Julius, 114
Renwick, James, 74
Ringgold, Cadwalader, 76
Ritter, J. W., 15
Root, Marcus Aurelius, 41, 90
Roscoe, Henry Enfield, 95
Roulz, Henri Catherin Camille, 50–51
Roy, Rustum, xiv*f*

Sabine, Edward, 94–95

Sachse, Julius F., 34*f*
Sampson, Marmaduke B., 79, 79*f*
Saxton, Joseph, 32–33, 32–33*f*, 37, 157, 201,
   204
Schaeffer, William L., ii*f*
Scheele, Karl Wilhelm, 14, 31
Scholl, Hermann, 104–106, 109–113
Schultz-Sellack, Carl, 101, 103, 106
Schulze, Johann Heinrich, 13–14, 16
Scovill, J. M. L., 28, 49
Scovill, W. H., 28, 49
Sears, Gerald W., 154
Secchi, Pietro Angelo, 90
Senefelder, Alois, 9
Shaw, Sarah White, 79, 79*f*
Simons, Montgomery P., 39, 184
Sinclair, T., 76*f*
Sjogren, Anker, 188
Smith, Louise Georgina, 20
Snelling, Henry Hunt, 7*f*, 44–47*f*, 94–95*f*
Southworth, Albert Sands, 49, 82
Sparks, Jared, 87, 88
Spurzheim, Johann Christoph, 78
Stanley, John Mix, 76
Stapp, William F., 157
Stephens, John Lloyd, 76–77, 77*f*
Stephens, Richard B., 129
Stevens, Isaac I., 76
Strutt, John William, 102, 106
Sutton, Thomas, 99, 201
Swaim, James, 33
Swan, Alice, 116, 116*f*, 118, 120, 144, 186

Taft, Robert, 185
Talbot, William Henry Fox, 11, 15, 24–27,
   25*f*, 55–56, 58, 62, 96
Taylor, B. F., 80*f*
Thackray, Malcom, 115
Thiesson, E., 20*f*
Thompson, J. J., 113
Thompson, Warren, 39, 44
Troost, Gerard, 33
Tupper, J. L., 108, 123
Turing, Alan, 217

Ure, Andrew, 52

Van Leeuwenhoek, Anthony, 4
Van Loan, Samuel, 45
Van Ravensway, Charles, 185
Vance, Robert H., 75
Voigtländer, Peter Friedrich, 29–30, 31*f*, 34
Von Bunsen, Robert Wilhelm, 95
Von Ettingshausen, A. F., 29
Von Fraunhofer, Joseph, 15

Von Friedrichsthal, Emmanuel, 77
Von Neumann, John, 217

Waller, Augustus, 60, 69, 69*f*, 70
Warren, John Collins, 82, 82*f*
Waterhouse, James, 111–113, 135, 138
Watkins, Carlton E., 75
Watt, James, 26
Weaver, K. S., 108
Wedgewood, Josiah, 16
Wedgewood, Thomas, 16
Wendt, Rudolf, 114
Weston, Edward, 117
Whipple, John Adams, 41, 82*f*, 87–90, 87*f*,
   89*f*, 90*f*
Wiener, Otto, 102–106, 102*f*, 109–112, 123
Wolcott, Alexander S., 30, 34–35, 58
Wollaston, Wiliam Hyde, 5, 7, 7*f*, 13, 15,
   22, 29
Wright, John, 50
Wyatt, Thomas, 81*f*

Young, Thomas, 12–14, 99

Zealy, J. T., 80, 80*f*
Zenker, Wilhelm, 100–101, 103, 106

Subject Index

Page references to illustrations (figures) are indicated by an *f* following the page number.

Abrasion, 161
Absorbance, 124
Absorption. *See* Spectra, absorption
Academie des Sciences, 23, 24*f*, 35, 37
Accelerated plates, 37
Acceleration effect, 59
Acceleration steps, 45
Accelerators, 69
Acid bath, 59
Acid etch, 19
Adelaide Gallery, 34
Adjustable rail, 30
Aether, 12, 61, 68, 100
Agate, 62
Agglomerates, 128
Alcohol lamp, 47*f*
Alkaline bath, 59
Alkaline developer, 112
Allotropic silver, 106
Amalgam, 60–61, 70, 118, 120, 144, 156, 233
Ambrotypes, 133
Ammonium carbonate, 70
Ammonium hydroxide, 14
Annealing, 48, 161
Anthropological applications, 76–81
Apparatus diagram, 3*f*, 5*f*, 7*f*, 31*f*, 44*f*, 47*f*, 92*f*, 94–95*f*, 97*f*, 130*f*, 194*f*, 206*f*, 211*f*, 214*f*, 215
Apparent density, 126–128, 232
Applications, 27
Archaeological applications, 74, 76–81
Arno River flood, 115
Arthropods, 174, 179, 179*f*, 181, 181*f*
Asphaltum, 19
Astronomical (firsts), 85–86, 88
Astronomical applications, 84–97
Athenaeum, 36
Auger electron spectroscopy, 118, 165

Band gap, 205
Band structure, 96, 168
Barilla, 52
Barite, 66, 66*f*
Barium sulfate, 66
Battery, 39, 44–45, 50
Becquerel development, 40, 60, 95, 122–123, 128, 139, 149*f*
Becquerel effect, 40, 62, 95, 110, 113
Becquerel images, 103, 110
Becquerel plates, 99, 101, 103, 123
Biconvex lens, 7
Bicycle, 17
Biological study, 77
Birth defect, 83, 83*f*
Bitumen of Judea, 19, 22, 31
Black silver iodide, 31
Black varnish material, 20

Blackness, 124–125, 164

Blistered, 177, 203

Blue color, 17, 61–62, 100, 127

Blue front daguerreotype, 100, 129–130

Blue glazing, 30

Blue light sensitivity, 58, 84, 110

Blue spectral region, 95, 105, 124, 128, 164

Blue vitriol solution. *See* Copper sulfate solution

Boat, 17

Bologna Stone, 66

Bone, 14

Book lice. *See* Arthropods

Botanical specimens, 73*f*

Brass tape, 38

Bromides, medicinal use, 52

Bromine, 16, 31–37, 45, 51–54, 62, 69–70, 111, 112

Bromine vapor, sensitization, 59, 112, 139–140, 151*f*

Buffered mat board, 173, 173*f*, 182, 202–203, 202*f*, 203*f*

Buffing paddle, 44*f*, 53

Buffing wheel, 44*f*

Calcite, 13

Caloric rays, 15

Calorific rays, 56–57, 66

Camera, 6–8, 13, 22, 27–28, 46, 47*f*, 53–54, 58, 85

Camera box, 47*f*

Camera design, 29, 35

Camera lens, 46

Camera lucida, 5, 7–8, 7–8*f*, 13, 24, 74–75, 77, 77*f*

Camera obscura, 5–6, 6*f*, 7–8, 7*f*, 16–18, 21–22, 24, 29, 40, 56, 65, 74

Camera, Giroux, 29, 30

Camera, Voigtländer, 29, 31*f*, 34

Camera, Wolcott, 30, 58

Camera, photographic, 47*f*

Camera, reflector lens, 58. *See also* Wolcott camera

Camera, rotating, 93, 93*f*

Camera, without lens, 30

Camphor vapors, 57

Candida. *See* Fungus

Case, daguerreotype, 37–38, 53, 160, 163, 163*f*, 165, 168–170, 182, 202–203, 202–203*f*, 211

Catalytic effect, 65

Cathan reflector, 43*f*

Cathode pulverization, 114

Cathode ray tube, 114

Chalked, 48

Chalon-sur-Saone, 18, 19, 19*f*

Characteristic curve, 128*f*

Chemical analysis, 14, 15, 165, 204

Chemical development, 62, 65, 137, 137*f*

Chemical focus, 84

Chemical rays, 15, 56–57, 64, 66, 70

Chemical reaction, 69, 164–169

Chemical spectrum, 66*f*, 67

Chemical vapor deposition, 153, 160

Chemicals needed, 48, 51–53

Chemicomechanical action, 65

Chippenham, 24

Chloride of gold. *See* Gold chloride

Chlorine, 16, 31–32, 34–37, 40, 59, 62, 69, 95, 99, 166

Christofel et Cie, 50–51

Chromate salts, 16

Chromatic aberration, 84

Cladding, 33, 48, 50. *See also* Cold-roll clad

Cleaning, daguerreotype, 1, 115, 132–134, 160–161, 168–172, 181, 183–200

Coad battery, 45*f*

Coating box, 46*f*

Coatings, 40, 44–45, 138, 204–210

Cold-roll clad, 33, 48, 50, 161, 195

Collodion, 91–92, 94, 98, 102, 107, 112, 137

Color photography, 103, 123

Color, characterization, 12–14, 39–40, 46, 57, 96, 103, 106, 109, 128

Color, pigments, 38, 186

Color, problems, 19, 84, 103

Color, silver iodide, 2, 109–110, 147

Color, tone, 2, 28, 44, 61, 128–130

Color, whiteness, blackness, 123–127

Colored halide layer. *See* Fixation

Colored image. *See* Image, colored

Colored interference photographs, 106

Colored reproductions, 38–42, 103

Colorific rays, 57

Colorimetric, 109

Coloring box, 39, 39*f*

Comptes Rendus, 23, 26, 35

Computer modeling, 136

Conservation, 115–116

Continuing rays, 40, 60, 70

Copper, 22, 39, 173–174, 173*f*

Copper electroplate, 43

Copper layer, 116*f*, 195

Copper plate engraving, 9, 42

Copper plates. *See* Plates, copper

Copper sulfate solution, 30, 40

Copy daguerreotypes, 42

Corpuscular theory, 12, 15

Corrosion, 118, 133, 160, 162, 162*f*, 163, 163*f*, 166–167, 179, 183–200, 208

Corrosion products, 116, 165–170, 172, 175, 180*f*, 181–182, 186, 192, 203–204

Cover glass, 38, 132, 160, 163, 163*f*, 171*f*, 173–179, 175–179*f*, 181–182, 202–204, 202*f*

Craniometrics, 78

Crystal defect. *See* Sensitivity, 136

Crystallization lines, 69*f*, 70

Cyanide cleaning, 46, 169–170, 172, 183–187

Cyanide plating baths, 50–51

Daguerreian Journal, 41

Daguerreotype processing steps, 1, 28, 44–46

Dark box, 47*f*

Dark lines. *See* Fraunhofer lines

Darkened room. *See* Camera obscura

Darkroom, ii, 7

Debris, 169–170, 174–175, 177, 177*f*, 179*f*

Densitometry, 107, 127

Dessin fume, 20

Deterioration, 160–164, 176, 181

Developer, chemical, 137–138

Developer, dry-plate mercury, 136–138, 138*f*

Developer, Metz, 138, 139–140*f*

Developers, photographic, 112, 136–138

Development mechanism, 112

Dictionary of the Photographic Art, 44*f*

Diffuse reflectance spectroscopy, 122, 123*f*, 124, 125*f*, 126–127, 127*f*, 128, 164, 164*f*

Diorama, 21

Display/storage environment, 201–215

Dissolution products, 177

Double-beam interference, 101

Drawing instruments, 5–6, 5*f*

Dye, 14, 17

E. and H. T. Anthony Company, 54

Eclipse, 86, 90, 90*f*, 91–92, 111

Economics, 23–27, 33

Edge corrosion. *See* Corrosion

Edinburgh Review, 15

Edinburgh Society of Arts, 26

Electric current flow, 39–40, 59

Electricity, 36, 39–40, 45, 53, 56, 58–59, 68, 104, 114

Electricity, image formation, 58, 111–113

Electrochemistry, 43, 50, 112

Electrocleaning, 192–200

Electrolyte solution, 192, 196

Electrolytic action, 112

Electrolytic cleaning, no potential, 187–188

Electrolytic fixing, 46, 140

Electrolytic gilding. *See* Gilding process

Electromagnetic spectrum, 96, 102, 112

Electronic effects, 113, 140

Electrophotography, 114

Electroplating, 43–44, 45f, 49–51, 161
Elemental data, 166–167
Emission spectra. *See* Spectra, emission
Emission theory, 12, 62
Encyclopedia Metropolitana, 44f
Energy-dispersive x-ray, EDX, 120, 121f, 151f, 177–178, 177f, 190, 195
Engraving, 19, 22, 40, 42–43, 74, 76, 79, 82, 97
Engravings, 77f, 79f, 92f, 97f
Equilibrium constant, 164
Etching, 43, 65, 170, 171f, 172, 180f, 187f, 197
Excursions daguerriennes, 42
Exposure, 28, 40, 62, 84, 127–128
Exposure time, 2, 22, 24, 27–29, 46, 59–60, 68, 86–87, 95, 95f, 111, 164
Exposure time reduction, 29–37, 69, 85, 111, 136
Exposure, acceleration, 35–37
Exposure, standing waves, 101, 103–104
Exposure, tracking mechanism, 84

Fastest iodized plate, 110
Fathers of photography, 16–27
Film. *See* Thin films
First star (Vega) daguerreotype, 88
Fixation, 38, 46, 60, 69–70, 95, 146, 233
Flame spectra. *See* Spectra, flame
Fluorescent materials, 101–102
Focus, chemical/visual, 84
Fourier transform. *See* Infrared spectroscopy
Frames. *See* Case, daguerreotype
Fraunhofer lines, 15, 61, 66f, 67, 96
Free energy of formation, 164
French plating, 49
Fresnel vector, 100
Fresnel's theory, 100, 102
Fuming. *See* Halogen fuming
Fungus, 174, 178

Gallery of Illustrious Americans, 43
Galvanic battery (cell), 39, 45f, 187, 196
Galvanizing, 44
Galvanometer, 112–114
Gandolfi x-ray diffraction, 173
Gas lamp, 94f
Gelatin coating, 109
Gelatin photography, 81f, 94, 101, 103, 109, 112–113
Geological applications, 4, 74–76
George Eastman House, 25f
German camera, 31f
Gernsheim Collection, 24f, 25f
Gilded daguerreotype, 43, 46, 119f, 120f, 121–122, 122f

Gilded step tablet, 118, 119f
Gilding process, 35, 38–39, 43, 46, 118, 128, 139, 145–146, 170
Gilding stand, 44f
Giroux camera, 30
Glass, chemical analysis, 13, 174, 175, 175f
Glass, corrosion, 174, 176, 179f, 180f
Glass, filter, 64
Glass, images on, 18, 57
Glass, making, 52
Glass, plates, 19, 174
Glazing, 174, 203–204
Gold chloride, 14, 16, 38–39, 46, 51
Gold plating, 50, 59
Goniophotometer, 108, 130–131, 130–131f
Grain boundary, 169–170, 171f, 197
Grain size and spacing control, 107–108, 125, 161
Grain structure, 106–108, 166
Graininess, 107–108, 125
Great Equatorial Telescope, 87, 89
Grotthuss-Draper Law, 61, 71, 96
Ground glass, 47f
Gum arabic, 9, 37–39, 43, 186, 203
Gum guaiacum, 19, 67, 67f
Gurney-Mott mechanism, 136

Hallwach effect, 106, 113–114
Halogen coating box, 46f
Halogen fuming, 49, 92f
Halogen treatment, 114. *See also* Bromine, Chlorine, Iodine
Halogen use (first), 31, 34, 36–37
Halogen vapor, 31, 44, 46f, 69
Halogens, 31–32, 46f, 51, 111
Heat, 14, 20, 38, 48, 62
Heliochrome, 40
Heliography, 19–23, 19f, 27, 31
Heliostat, 63f, 85
Herschel effect, 59, 70
Hillotype process, 41
Hillotypes, 40
Humanities Research Center, xviii, 17f, 19f, 24f, 25f
Hydrogen sulfide, 162–163, 163f, 164f, 165, 192
Hypo. *See* Sodium thiosulfate

Icelandic spar, 13
Illustrations, book (first use), 42, 60, 96
Image, area, 121
Image, bleaching, 70
Image, colored, 2, 29, 38, 41–42, 99, 186
Image, damage, 2, 37
Image, formation, 7f, 13, 16, 55–56, 62, 69,

114, 135–159
Image, formation, electricity, 111
Image, formation, role of light, 24, 30, 56–61, 99
Image, fragility, 2, 37–38, 144
Image, latent. *See* Latent image
Image, microstructure, 172, 197
Image, particle density, 119, 128–130, 134, 161, 168, 168f
Image, particle size and distribution, 107, 121–123, 123f, 124, 129–130, 134, 138
Image, particles, 119–121, 121f, 122, 138, 143f, 144
Image, permanent, 22, 27
Image, protection, 2, 37–38
Image, quality, 161
Image, stability, 29, 118
Image, structure, 107, 118–122, 125f, 126, 126f
Imponderables, 56
Indian rubber, 37
Indigo, 17, 58
Industry, 46–54
Infrared spectral region, 14, 95–96, 110, 164, 164f, 167, 168f
Infrared spectroscopy, 106, 165, 167–168, 168f
Ink, 14, 16, 18
Institute, L' (Paris), 1, 28
Interference, 13, 100–101, 105
Interference bands, 104–105f, 110
Interference phenomena, 100–103, 110, 162
Interference photography, 106
Iodides, medicinal use, 52
Iodine treatment, 34, 59, 151f
Iodine vapor, 1, 22, 28, 31, 34, 44, 109, 112, 139–140
Iodine, 16, 31–32, 35–36, 45, 51–52, 61–62, 65, 70. *See also* Halogen treatment
Iodized plate, 40, 65, 67f
Iron-pyrogallol developer, 112
Isinglass, 43, 186, 203

Journal of Royal Institute, 16

Kelp, 52
Kodak Research Laboratory, 107–108

Lacock Abbey, 24
Laser, 130, 168
Latent image, 22, 64–65, 111–115, 135–137
Latent light, 62, 64–65
Le Blanc method, 52
Leather, 14, 16
Lens aperture, 29

Lens cap, 47f
Lens design, 29, 35
Lens systems, 29, 30, 47f
Lens tube, 47f
Lens types, 29, 30
Lens, Petzval portrait, 29–30, 34
Lens, Wollaston, 7, 7f, 13, 29
Lens, achromatic, 15, 30, 58, 64, 71
Lens, biconvex, 7
Lens, double, 30
Lens, glass, 84
Lens, landscape, 29
Lens, large diameter, 34
Lens, meniscus-shaped, 7
Lens, reflector. *See* Wolcott camera
Lenses, 13, 30, 46, 53–54, 58, 85
Light, 12, 14–16, 23, 56, 99
Light ray, components, 13
Light waves, 12–13, 15, 61–62, 65, 70, 102, 105, 114
Light, absorption, 15, 57, 110
Light, action, 23–24, 27, 58, 60, 70
Light, behavior, 20, 56
Light, blue, 62
Light, chemical effects, 13, 14, 57–58, 60–61, 135, 164
Light, colored, 62, 70
Light, corpuscular theory, 15
Light, emission, 66
Light, exposure, 28, 46, 60, 68, 105, 105f, 107, 113–114
Light, images, 24
Light, latent, 62
Light, mechanical effects, 58–59, 61–62, 137
Light, nature of, 56, 68, 71, 95
Light, negative rays, 70
Light, quanta, 114
Light, recording medium, 23
Light, reflectance, 101, 108
Light, refraction, 84, 88
Light, scattering, 100, 110, 168
Light, sensitivity, 14, 31, 35, 95f, 103–104, 111–113
Light, theory, 12, 62
Light-sensitive film, 16, 25, 29, 31, 46, 103, 113
Light-sensitive ink, 16
Light-sensitive materials, 14–16, 19, 27, 30–31, 35, 59, 67, 100, 101
Light-sensitive media, 57, 136
Lighting arrangement, 21, 30, 34–35, 211, 211f
Limestone, 9, 18
Limitations, 84–85
Line spectra, 15, 66f, 67
Lippmann process, 103, 106–107, 112

Literary Gazette, 15, 23, 34
Lithograph, 20, 42, 64, 74, 76f, 86, 86f
Lithography, 8–10, 18–20, 22, 27, 42, 114
Lithography, regulations, 18
Luminescent materials, 68, 71, 203–204
Luminiferous aether, 12
Luminous radiation, 56, 96
Luminous spectrum, 67, 67f
Lunar daguerreotype, 55, 58, 84–87, 87f, 88–89, 89f, 217f

Magnetograph, 93–94f
Mat. *See* Buffered mat board
Maxwell-Garnett scattering theory, 106, 112
Measles, 171–172
Mechanical development. *See* Physical development
Mechanical fragility, 38, 132, 160
Medical (firsts), 82, 82f, 83
Medical applications, 60, 77, 81–84
Mercurizing bath, 47f
Mercurizing pots, 46
Mercury, 56, 58, 60–61, 68–70, 103, 110, 113–114, 118, 119f, 120, 120f, 122, 136–137, 184
Mercury arc-lamp, 130
Mercury baths, 44f, 47f, 53, 60–61, 92f
Mercury chloride, 14, 16
Mercury concentration, 121, 170
Mercury developer. *See* Developer, dry-plate mercury
Mercury iodide, 60, 68
Mercury oxalate, 16
Mercury oxide, 14
Mercury protoiodide, 68
Mercury vapor, 2, 22, 57, 60–62, 65, 69–71, 115
Mercury vapor, hot, 2, 28, 46, 140, 143
Mercury, cold, 128–129, 139
Mercury, image formation, 58, 68–71, 136
Metal salt, 38–39
Meteorological applications, 93
Metrologists, 74
Micrograph, 60, 60f, 83, 119, 121f, 122
Microscope, 13, 29, 83–84, 118
Microstructural control, 128
Microstructural spacing, 124–126, 125f
Microstructure, 107, 118, 119f, 122, 129, 190, 192, 207
Mirrors, 30, 34, 94f, 102–103
Mites. *See* Arthropods
Moisture effects, 57, 129, 164–165, 174–175, 178, 204, 210
Molds. *See* Fungus
Molecular change, 66
Monochromatic light, 15

Moon. *See* Lunar daguerreotype
Moser-bild, 62, 102
Multiple sensitization, 45, 151f, 155, 158, 158f
Multiple-exposure daguerreotype, 87f

Navigational applications, 93
Near-ultraviolet region, 124
Negative charged materials, 113
Negative electrode, 112
Negative process, 92
Negatives, 76
Nitric acid, 1, 43
Nitrous oxide, 16
Nobel Prize, 106, 114
Notre Dame cathedral, 23, 24f
Number of daguerreotypes made, 2

Oil of lavender, 31
Optical density, 125–126
Optical microscopy, 56, 60, 60f, 107, 175
Optical properties, 100, 102, 108–110, 118, 122–130, 164, 188, 205
Optics, 13, 61
Overexposure daguerreotypes, 71, 100
Overgilding, 129

pH, 171, 174, 176, 181, 193, 193f
Package, 173–182, 202–203, 202f. *See also* Case, daguerreotype
Panoramas, 20, 21
Paper, 14, 16, 61, 64, 94
Paper negatives and prints, 76
Paper tape, 38
Paris plate, 49
Particle diameter ratio, 121–122
Particle size and spacing, 124–126, 125f, 128, 144
Particle-light interaction, 118
Parylene C, 206, 208–210
Patents, 25–26, 30, 38–39, 41, 49–53, 56, 114
Peeling, 160–161, 195–196
Periodic structures, 102
Perspectograph, 5, 5f
Petzval lens. *See* Lens, Petzval portrait
Pewter, 19–20, 19f, 20, 22
Phase diagram, 144, 144f
Philosophical Magazine, 58, 62–63
Phlogiston, 14, 220
Phosphorescence, 62, 66–68
Phosphrescent materials, 67
Phosphorogenic rays, 66–67, 66f
Phosphorus, 21
Photochemical effects, 63, 66
Photochemical induction, 95
Photochemistry, 14, 56, 61
Photodecomposition, 168

Photoelectric effect, 106, 113–114
Photogenic drawings, 56, 73f
Photograph, 2, 46, 90, 213, 214f, 218
Photograph, gelatin copy, 81f
Photographic (firsts), 1, 19, 19f, 22, 25, 30, 34, 42, 58, 66, 74, 85, 102, 118, 184–185
Photographic firsts, 7, 18–19, 19f, 22, 25, 30–32, 34, 36–37, 42, 60, 66, 74, 83, 85, 97, 118, 184–185
Photographic images, 106–107
Photographic manuals, 30
Photographic material, 51, 59, 102, 107
Photographic phenomenon, 64
Photographic process, 42, 53, 61, 73f, 93, 103, 107, 111, 114
Photographic sensitivity, 110–111, 149
Photographic studios, 30
Photographometer, 95, 95f
Photography, 10–12, 16–17, 26, 29, 32, 35, 55, 58, 63, 66, 71–72, 75, 88, 125
Photography, astronomical, 85, 87, 89, 91–92
Photography, collodion, 102
Photography, color, 99–106
Photoheliograph, 94–95
Photomechanical printing, 99, 112
Photometry, 14
Phrenology, 6, 78–79, 79f
Physical damage, 160–164
Physical development, 62, 137, 137f
Physiognotrace, 6
Pigments, 38–39
Planck's constant, 114
Planished, 48
Planographic process, 9
Plasma edge, 124, 129, 232
Plate box, 44f
Plate damage, 180f
Plate holder, 44f, 47f, 83, 194f
Plate manufacture, 48
Plate mark, 49, 50
Plate preparation, 66
Plate sensitivity, 34, 61
Plate specifications, 2, 92f
Plate spectrum, 67f
Plated metal, 28
Platelets, 141–144
Plates, 39–40, 48–51
Plates, chlorinated, 40
Plates, copper, 1, 9–10, 19, 22
Plates, pewter, 19, 19f
Plates, polished, 1
Plates, silver-plated, 1, 22
Plates, zinc, 19
Plating vessel, 50
Plating, cold-roll clad, 33, 48–50, 161, 195
Plating, fusion, 48

Plating, silver process, 33
Platinum plates, 59
Platinum salts, 16
Pobboravsky work, 108–111
Points de vue, 11, 19
Polarization, 13
Polarized ray, 57
Polished plates, 1, 31, 44–45, 53, 126
Polygeny, 80–81
Portrait, 17f, 20f, 23f, 25f, 32f, 36f, 40f, 57f, 59f, 79–85f, 87f, 101f, 102f, 109f, 172f
Portrait photography, 6, 8, 32, 34–35, 42–43, 58, 77, 80–82, 85, 118, 185
Portrait studios, 34
Portraiture, 6, 35, 37–38, 54, 73, 77, 111
Positive-negative image, 18, 22, 26, 31, 59, 95, 108, 112
Positive-negative image reversal, 19, 30, 43, 112, 122, 127, 130, 134
Potassium cyanide solution, 50, 169
Potassium dichromate, 67, 73f
Potassium dichromate paper, 66, 67f
Pourbaix diagram, 193, 193f
Premium societies, 5, 5f, 17, 18, 21, 27, 30
Printing plates, 43
Printing plates (first), 43
Printmaking, 8, 9, 19, 76
Prints, paper, 76
Prism, 7f, 14–15, 29, 43, 43f, 64, 66, 71, 96, 99, 130
Prismatic effects, 84
Prix Bordin, 100–101
Prize system, 100, 102, 106, 114. *See also* Premium societies
Process, 22–23, 25, 27–28, 31–32, 36, 44, 58, 69, 136, 136f
Process, color, 40
Process, details, 1, 2, 28–29
Process, dichromate salts, 26
Process, hardening, 144–145
Process, silver halide, 135–136
Process, steps, 1, 28, 44–46, 136, 136f
Profiles, 6, 219
Profilometer data, 122–123
Protective coatings, 2, 204–210
Prussian blue pigment, 16
Pumice, slurry, 1
Pure silver, 43
Pyreolophore, 17, 18, 20, 21

Quantum mechanics, 114
Quicks, 45

Radiant electricity, 58
Radiant energy, 56, 64

Radio frequency, 188–189
Raman spectroscopy, 168, 168f, 172
Rays, 15, 57–58, 60, 62–66, 70, 95, 114
Reactivity test, 14
Recording media, 75
Red color, 84
Red light, 59–60, 70
Red stars, 84
Reduction-oxidation reaction, 136–137
Reflectance characteristics, 129–134, 131f, 132f, 133f
Rflectance densitometer, 108, 127
Reflectance measurements, 102, 108–109, 109f, 110, 130–133, 133f
Reflectance spectra, 110–111, 127f, 129, 129f, 187, 206f, 207, 209f
Reflected light, 101
Refraction, 13
Refractive index, 125, 132, 205
Refrangible rays, 15
Region, infrared, 14
Region, spectral, 14
Region, ultraviolet, 15
Reglazing. *See* Glazing
Resins, 18–19
Reversing prism, 43f
Rhomb, 13
Roll-clading, 49–50. *See also* Plating, cold-roll clad
Rolling mill, 48
Royal Botanical Gardens, 21
Royal Society, 21, 24, 34, 59

Salsola soda, 52
Salt, 2, 23–24, 26, 28, 31, 46, 70, 140, 170
Salts, dichromate, 26
Scale plates, 50
Scanning electron micrograph, 119f, 138–143f, 145f, 147–149f, 155–156f, 158–159f, 171–172f, 173f, 176–181f, 190–191f, 195f, 197f, 207f, 209f
Scanning electron microscope, 116, 118–119, 120, 122, 132, 138–140, 166, 170, 172, 178, 186, 197
Scanning electron microscopy, 165, 176, 190, 195
Scattering curves, 124, 125f, 126, 128
Scattering peak height and shift, 129
Scattering peak locations, 124–126, 125f
Scattering, light, 100
Schultz-Sellack theory, 101
Scovill Manufacturing, 54
Scratched daguerreotype, 133, 133f, 160–161
Seaweed, 31, 52

Secondary electron micrograph, 116f
Secondary ion mass spectroscopy, 118
Sensitive paper, 24, 72
Sensitivity, 84, 91, 95f, 104–105, 109–110, 113, 132, 136
Sensitization, multiple, 34–35, 37–38, 45–46
Sensitized plate, 1, 2, 46, 60, 70, 95, 105, 139, 142–143f
Sensitizing bath, 40
Sensitizing boxes, 44, 44f, 46f, 53, 129
Sensitizing steps, 45, 166
Sensitometry, 107, 127–128
Shadow particle agglomerates, 119
Sheffield plating, 48–50
Shield, blue filter, 30
Silhouettes, 6, 16, 219
Silver amalgam, 60–61, 68, 118
Silver bromide, 31, 35, 57, 68–69, 114, 136
Silver bromide photographic paper, 67
Silver bromoiodide, 111, 150, 234
Silver chloride, 50, 69
Silver chloride photographic papers, 14, 16, 18, 21, 24, 35, 57, 68
Silver cyanide, 132, 161, 166, 172, 186
Silver electroplating industry, 50
Silver halide, 30–31, 38, 58, 60, 98–99, 101, 103, 109–113, 135–137, 140, 169–170
Silver iodide, 2, 22, 28, 30–31, 35, 38, 58, 60, 68–69, 111, 113, 170
Silver iodide behavior, 60–61, 104, 110, 113–114
Silver iodide film, 103, 109–110, 115
Silver iodide image, 22
Silver iodide layer, 60, 104, 104f, 105–106, 110
Silver iodide thickness, 61
Silver iodide, interference bands, 105f
Silver iodide, spectrum, 64
Silver layer, 116f
Silver metal, 31, 43, 49, 107, 112–113, 138, 160, 164–165, 170, 188
Silver nitrate, 13, 14, 50
Silver nitrate paper, 14, 16–17, 24, 67f
Silver oxalate, 16
Silver oxide, 14, 132, 162, 164, 166, 169–170, 186, 193
Silver phosphate, 172
Silver plate, 22, 31, 42–43, 48–49, 59, 162
Silver plate, iodized, 65
Silver plate, light exposed, 113
Silver plate, support material, 64
Silver reflectance, 126
Silver salts, 31, 39, 49
Silver solution, 49, 186
Silver subchlorides, 101

Silver subhalides, 102, 106
Silver subiodide, 68, 101
Silver sulfide, 132, 162, 164–167, 164f, 168–170, 186–189
Silver-mercury amalgams, 118
Silver-mercury ratio, 118
Silver-plated copper, 28, 33, 48
Silver-plated copper plates, 1, 22
Silver-thiourea complexes, 132, 166, 171–172
Smee battery, 45f
Soap making, 52
Société d'Encouragement des Arts, 18, 27, 30
Society of Arts, 21
Soda, 52
Sodium chloride solution, 40
Sodium phosphate, 70
Sodium thiosulfate, 2, 16, 28, 38, 44f, 46, 51, 58, 61, 140, 170
Solar eclipse. *See* Eclipse
Solar light, 68
Solar microscope, 24, 84
Solar radiation, 57–58, 66–67
Solar spectrum, 15, 40, 55, 59, 63, 66, 85, 95–96
Solarization, 127–129
Solarized, 71, 85, 127–129, 232
Solution, copper sulfate, 30, 40
Solution, cyanide, 50
Solution, gold chloride, 38
Solution, salt, 2
Solution, silver, 49
Solution, silver chloride, 50
Solution, silver nitrate, 50
Solution, silver salt, 49
Solution, sodium chloride, 40
Solution, sodium thiosulfate, 2, 38
Solutions, cyanide, 46
Specific heat, 63f
Spectra, absorption, 63f, 96, 167, 168f
Spectra, emission, 96–97, 96f
Spectra, flame, 15, 96
Spectra, line, 15, 96
Spectra, reflectance, 109f
Spectral analysis, 14, 15
Spectral colors, 96–97, 99
Spectral radiation, 39
Spectral region, 14, 91
Spectral sensitivity, 14
Spectral studies, 24, 59, 66, 109
Spectrochemical analysis, 97
Spectrometer, 56, 66, 71
Spectrophotometer, 109, 111
Spectroscopic applications, 93, 120, 165
Spectroscopy, 14, 168, 190
Spectrum, 24, 96, 164, 164f
Spectrum sensitivity, 105

Spectrum, dark lines. *See* Fraunhofer lines
Spectrum, solar. *See* Solar spectrum
Specular reflectance, 124, 126, 127f, 130–134, 164, 164f, 186, 196
Sputter cleaning, 188–192, 190f
Sputter-induced photon spectroscopy, 118
Sputtering, 114, 238n.9
Standing waves, 101–106
Star, daguerreotype, 84, 88
Star, diameter measurement, 101
Star, distance measurement, 88
Stencil, 13, 39
Step tablet, 118, 119f, 120, 126–127, 127f, 139–140, 142, 173, 173f, 189–190
Storage. *See* Display/storage environment
Sulfuric acid solution, 45f
Sun daguerreotype, 85, 86f, 87
Sunspots, 86, 86f, 94
Super highlight, 129
Surface changes, 60
Surface debris, 166
Surface films, 132
Surface particle array, 118
Surveyors, 74
Systematics, 4, 81

Talbot exhibit, 24
Talbot, public disclosure, 24
Talbot's process, 62
Tarnish, 118, 133, 160–165, 168–170, 172, 183, 186, 188–190, 191f, 195–197, 215
Telescope, 5, 13, 29, 84, 86–87, 87f, 90
Telescope, focusing mechanism, 88, 92f
Telescope, reflector, 84, 87
Theodolite, 15
Theory, Young's, 12
Theory, corpuscular, 12, 15
Theory, emission, 12, 62
Theory, particle, 12
Theory, undulatory, 12, 13, 62
Theory, wave, 12–13, 15
Thermal effect, 62
Thermodynamics, 164–165, 170–171
Thin film, optical properties, 102, 109, 132, 162, 165, 169
Thin films, 103, 109, 169, 172
Thiocarbamide solution, 112
Thiourea cleaners, 115, 171–172, 172f, 185–186
Tintypes, 133
Tithonicity, 63–66, 68, 95
Tithonograph, 63f
Tithonometer, 64, 95
Tithonotype, 43
Transit of Venus, 92, 92f, 93, 93f, 111
Transmittance, 109
Transparent coatings, 40

Travel book illustrations, 42
Travel views, 74
Trinity College, 24
Tupper's experiment, 108

U.S. (firsts), 30, 32, 44, 74, 97, 184–185
Ultraviolet spectral region, 15, 95–96, 113–114, 124
Underdevelopment, 129
Undulatory theory, 12, 13, 62
Ungilded daguerreotype, 122–123, 200

Vapor images, 31, 57–58, 61–66, 69, 102
Varley telescope, 5
Varnish damage, 2, 168–169
Varnish process, 43, 201, 204
Varnishes, 2, 18, 37–38, 201
Viewing geometry, 108, 125, 127–128
Visible spectral region, 22, 67*f*, 110, 112, 164*f*, 169
Visual focus, 84
Voigtländer camera, 29, 31*f*, 34
Volatilization, 142
Voltaic electricity, 66
Voyages Pittoresques, 20, 42

Water baths, 40
Water vapor. *See* Moisture effects
Water-soluble films, 37, 165
Wave theory. *See* Theory, wave
Wavelength, 114, 124–126, 129–130, 215
Wax, 37
Weathering, 175, 177
Weeping glass, 176
Whipple-Bond alliance, 86–89
Whiteness, blackness, 123–127
Wiener-Scholl experiments, 106
Woad plant, 17
Wolcott camera, 30, 58
Wollaston lens. *See* Lens, Wollaston
Wood blocks, 9, 10

X-ray lithography, 114
X-ray spectrometer, 116
X-ray. *See* Energy-dispersive x-ray

Yeast. *See* Fungus
Yellow light, 30, 40
Yellow silver iodide layer, 110
Young's theory, 12–13

Zinc plates. *See* Plates, zinc
Zinc protective coating, 44
Zinc rod, 46
Zinc sheet, 45*f*